T0135462

Non-resonant Solutions in Hyperbolic-Parabolic Systems with Periodic Forcing

Aday Celik

Logos Verlag Berlin

λογος

Bibliographic information published by the Deutsche Nationalbibliothek

The Deutsche Nationalbibliothek lists this publication in the Deutsche
Nationalbibliografie; detailed bibliographic data are available
on the Internet at http://dnb.d-nb.de .

ISBN 978-3-8325-5172-8

Logos Verlag Berlin GmbH
Georg-Knorr-Str. 4, Geb. 10,
12681 Berlin

Tel.:   +49 (0)30 / 42 85 10 90
Fax: +49 (0)30 / 42 85 10 92
http://www.logos-verlag.de

Non-resonant Solutions in Hyperbolic-Parabolic Systems with Periodic Forcing

Dem Fachbereich Mathematik
der Technischen Universität Darmstadt
zur Erlangung des Grades eines
Doktors der Naturwissenschaften
(Dr. rer. nat.)
genehmigte

Dissertation

von

Aday Celik, M. Sc.

aus Nusaybin (Türkei)

Referent:	Prof. Dr. Reinhard Farwig
Korreferent:	Prof. Dr. Mads Kyed
Tag der Einreichung:	06. Februar 2020
Tag der mündlichen Prüfung:	26. Februar 2020

Darmstadt 2020
D 17

Acknowledgment

At this point I would like to express my gratitude to those people who supported me in writing this thesis.

First of all, I would like to thank my highly esteemed doctoral advisor, Professor Dr. Mads Kyed, for his extensive scientific support. I would like to thank him for his valuable professional advice, which has contributed significantly to the success of this work. I would also like to thank Professor Dr. Reinhard Farwig for his comprehensive and valuable support, that was especially helpful when Professor Dr. Kyed transfered to another University.

Furthermore, I would like to thank my colleagues from the Analysis Working Group, especially my highly appreciated colleague and "office-mate" Thomas Eiter. The numerous technical and non-technical discussions were both enlightening and of great help during my writing phase. I would also like to thank the other proofreaders Dr. Björn Augner, Sebastian Bechtel, Jens-Henning Möller, Junior Professor Dr. Amru Hussein, and Andreas Schmidt for the various helpful tips as well as for the inspiring and entertaining lunch and coffee breaks.

Finally, I would like to thank my family, especially my parents and my siblings including their spouses. Thanks to their unconditional and loving support, they have had my back continuously - not just during the time of my doctoral dissertation, but during my whole life so far.

Abstract

In this thesis we consider problems from nonlinear acoustics and fluid-structure interaction in a time-periodic framework.

We begin by studying two models from nonlinear acoustics, namely the Kuznetsov equation and the Blackstock-Crighton equation. Existence of time-periodic solutions to these systems are established for time-periodic data sufficiently restricted in size. We conclude that the dissipative effects in the Blackstock-Crighton equation and the Kuznetsov equation are sufficient to avoid resonance. The Blackstock-Crighton model is considered in a bounded domain with both non-homogeneous Dirichlet and Neumann boundary values, whereas the Kuznetsov equation is further studied in the whole space and in the half space. Existence of a solution is obtained via a fixed-point argument based on appropriate *a priori* estimates for the linearized equations. In order to deduce the L^q estimates, we decompose these systems into a stationary problem and a purely oscillatory problem, and consider the different Fourier modes separately. Via Fourier multiplier theory we obtain a strong time-periodic solution in an L^q framework. The investigation of these systems is carried out in Chapter 3.

In Chapter 4 the interaction of a viscous fluid with an elastic structure is studied. We consider a periodic cell structure filled with a viscous fluid, which interacts with the lower deformable boundary of the cell. The motion of the fluid is governed by the Navier-Stokes equitations and the deformable lower boundary is governed by the plate equation. Existence of a time-periodic solution to the linearized coupled system is deduced. Suitable L^q estimates are established for the linearized problem via Fourier multiplier theory and a localization argument. Finally existence of a solution to the nonlinear problem follow via a fixed-point argument.

Zusammenfassung

In dieser Arbeit beschäftigen wir uns mit Problemen aus dem Bereich der nichtlinearen Akustik sowie Fluid-Struktur-Kopplungs-Problemen.

Aus dem Forschungsgebiet der nichtlinearen Akustik beschäftigen wir uns mit der Blackstock-Crighton-Gleichung und der Kuznetsov-Gleichung. Die Existenz einer zeitperiodischen Lösung dieser Modelle unter einfluß zeitperiodischer äußerer Kräfte wird gezeigt. Die Blackstock-Crighton-Gleichung wird in einem beschränkten Gebiet untersucht, wohingegen die Kuznetsov-Gleichung im Ganzraum, im Halbraum, sowie in einem beschränkten Gebiet betrachtet wird. Beide Modelle werden sowohl mit inhomogenen Dirichlet- als auch Neumann-Randbedingungen untersucht. Existenz von Lösungen für die beiden Systeme wird über ein Fixpunktargument hergeleitet. Hierzu benötigen wir L^q-Abschätzungen für die Lösung des zugehörigen linearen Systems. Die Abschätzungen werden mit Hilfe von Fourier-Multiplikatoren bewiesen. Die Untersuchung dieser beiden Modelle findet in Kapitel 3 statt.

In Kapitel 4 dieser Arbeit untersuchen wir die Interaktion eines viskosen Fluids mit einer elastischen Struktur. Hierzu nehmen wir an, dass sich die Flüssigkeit in einer periodischen Zelle befindet, welche einen deformierbaren Boden hat. Die Strömung wird durch die Navier-Stokes-Gleichungen beschrieben, und der deformierbare Boden ist eine dünne elastische Platte. Es wird gezeigt, dass eine zeitperiodische Lösung existiert, welche die Interaktion beschreibt, wobei die elastische Platte durch äußere periodische Kräfte angeregt wird. Unter Verwendung von Fourier-Multiplikatoren sowie einem Lokalisierungsargument erhalten wir L^q-Abschätzungen, welche wiederum benutzt werden, um ein Fixpunktargument durchzuführen. Mit Hilfe des Fixpunktarguments zeigen wir, dass eine zeitperiodische Lösung zum nichtlinearen freien Randwertproblem existiert.

Contents

1 Introduction   **1**
  1.1   A Historical Background of Nonlinear Acoustics   3
  1.2   A Historical Background of Fluid-Structure Interaction . .   4
  1.3   Nonlinear Acoustics with Periodic Forcing   6
  1.4   Fluid-Structure Interaction with Periodic Forcing   8

2 Preliminaries   **9**
  2.1   General Notation   10
  2.2   Topology and differentiable structure   12
  2.3   Fourier Transform and Multiplier Theory   14
  2.4   Sobolev and Bessel Potential Spaces   15
      2.4.1   Functions on Spatial Domains   16
      2.4.2   Functions on Time-Space Domains   17
      2.4.3   Spaces of Solenoidal Functions   19
  2.5   Interpolation   19
  2.6   Embedding and Trace properties of Time-Periodic Sobolev
      spaces   24
      2.6.1   Embedding Properties   24
      2.6.2   Trace Space Operators   33
  2.7   Mathematical Tools from Fluid Mechanics   35
      2.7.1   Poincaré's Inequality in Layer Domains   35
      2.7.2   Solenoidal Vector Fields and the Divergence Problem   37

3 Nonlinear Acoustics   **39**
  3.1   Models   41
  3.2   The Damped Wave Equation   44
      3.2.1   The Steady State Damped Wave Equation   46
      3.2.2   The Purely Oscillatory Damped Wave Equation in
            the Whole Space   49
      3.2.3   The Purely Oscillatory Damped Wave Equation in
            the Half Space   54
      3.2.4   The Purely Oscillatory Damped Wave Equation in
            a Bounded Domain   57

	3.2.5	The Time-Periodic Damped Wave Equation	. . . .	66
3.3		The Kuznetsov Equation		69
3.4		The Blackstock-Crighton Equation		74
	3.4.1	The Linearized Blackstock-Crighton Equation	. . .	77
	3.4.2	The Nonlinear Problem		82

4   Fluid-Structure Interaction   **87**

4.1		Viscous Fluid Flow on an Elastic Plate		90
4.2		Reformulation in a Reference Configuration		93
4.3		Uniqueness		96
4.4		The Time-Periodic Stokes Equations		100
	4.4.1	The Stokes System in the Half Space		102
	4.4.2	The Stokes System in the Periodic Half Space	. . .	115
4.5		The Coupled Resolvent Problem		119
4.6		A priori Estimates		129
	4.6.1	The Periodic Half Space		129
	4.6.2	Estimates of the Pressure Field		135
	4.6.3	The Periodic Layer		139
4.7		The Stationary Linear System		145
	4.7.1	The Stationary Stokes System		145
	4.7.2	The Bi-Laplacian		147
	4.7.3	The Stationary Linearized Fluid-Structure System .		148
4.8		The Linear Fluid-Structure Problem		150
	4.8.1	The Linearized Fluid-Structure Problem with Homogeneous Divergence		151
	4.8.2	The Fully Inhomogeneous Fluid-Structure System .		152
4.9		The Nonlinear Problem		155

Appendix   **165**

A.1		Some Bounded Functions		165
	A.1.1	Functions in the Context of Nonlinear Acoustics . .		165
	A.1.2	Functions in the Context of Fluid-Structure Interaction		167
A.2		Fourier Multiplier		167
	A.2.1	Higher Order Partial Derivatives of Fractions		168
	A.2.2	Fourier Multiplier in the Whole Space		169
	A.2.3	Fourier Multiplier on the Torus Group		173

1 Introduction

Resonance in fluid-structure interaction, as well as in the study of wave propagation, is a well-known phenomenon occurring in nature. Resonance can be observed when the frequency of an applied time-periodic force is in harmonic proportion to a natural frequency. The dynamic parameters such as displacement, velocity and energy of the system will then oscillate with increasing amplitude.

Subject of this thesis is to study the occurrence (or rather the absence) of resonance in different problems from nonlinear acoustics and fluid-structure interaction. Resonance occurs naturally in undamped hyperbolic systems, but damping mechanisms can prevent this. In the following, we study two types of damped system. First, we study the hyperbolic equations governing the propagation of an acoustic wave in a viscous medium, which introduces a damping effect. Second, we study a fluid-structure-interaction problem, where the hyperbolic equation governing the motion of the structure is damped via the interaction with a viscous fluid. In those cases, resonance can be avoided if the energy from the external forces accumulated over a period is dissipated via the damping mechanism. The existence of a time-periodic solution would be a manifestation hereof.

In nonlinear acoustics, the propagation of sound waves through a viscous medium is studied. An acoustic wave propagates through a medium as a local variation of pressure. Nonlinear effects occur when the waves exhibit high amplitudes. The *Blackstock-Crighton* equation and the *Kuznetsov equation* are typically used to model this type of nonlinear wave propagation. In [7], Blackstock first introduced the model

$$(a\Delta - \partial_t)\left(\partial_t^2 u - c^2\Delta u - b\partial_t\Delta u\right) - \partial_t^2\left(\frac{1}{c^2}\frac{B}{2A}(\partial_t u)^2 + |\nabla u|^2\right) = f,$$

(1.0.1)

which later was also derived by Crighton in [18]. This model is used to describe the motion of a wave when viscous, heat-conducting fluids are considered. However, if temperature constraints are neglected, the Blackstock-Crighton equation is reduced to a nonlinear damped wave equation

$$\partial_t^2 u - \Delta u - \frac{b}{c^2}\partial_t\Delta u - \partial_t\Big(\frac{1}{\rho_0 c^4}\frac{B}{2A}(\partial_t u)^2 + |\nabla u|^2\Big) = f, \qquad (1.0.2)$$

called the Kuznetsov equation. This wave equation was first proposed by KUZNETSOV in [57] and is a widely used model to describe the propagation of sound in fluids. In both the Blackstock-Crighton and Kuznetsov equation, the damping term is due to a Kelvin-Voigt damping $\partial_t\Delta u$, where u denotes the acoustic potential. The constant a is the heat conductivity of the fluid, c the speed of sound, and ρ_0 the mass density. The diffusivity of sound b is a measure of energy dissipation due to viscosity and heat conduction in the fluid. Finally, B/A denotes the so-called (acoustic) parameter of nonlinearity, which is the quotient of the second and first coefficient in the Taylor expansion of the pressure-density relationship, see [6]. Chapter 3 is devoted to the investigation of (1.0.1) and (1.0.2) under periodic forcing. More specific, given a force that is periodic in time with period $\mathcal{T}$,

$$f\colon \mathbb{R} \times \Omega \to \mathbb{R}, \qquad \forall (t,x) \in \mathbb{R} \times \Omega\colon \quad f(t + \mathcal{T}, x) = f(t,x),$$

existence and L^q estimates of a $\mathcal{T}$-time-periodic solution u are studied. In case of the Kuznetsov equation, the whole space, half space and bounded domains are considered, whereas (1.0.1) is studied on a bounded domain. Inhomogeneous Boundary conditions of Dirichlet and Neumann type are examined.

The study of the interaction of a deformable structure with a viscous fluid is fundamental to many applications, for instance in the field of aeroelasticity, biomechanics or hydroelasticity, see for example [54, 78, 38]. In this thesis we carry out such a study for fluid-structure systems that are driven by a time-periodic force. Observe that the fluid domain, which is denoted by $\Omega_\eta(t) \subset \mathbb{R}^3$, changes in time, where η describes the evolution of the moving domain. The equations governing the motion of the fluid flow are given by the *Navier-Stokes* system

$$\begin{cases} \partial_t u - \mu\Delta u + (u \cdot \nabla)u + \nabla p = f, \\ \qquad\qquad\qquad\qquad\quad \operatorname{div} u = 0, \end{cases} \qquad (1.0.3)$$

where u denotes the fluid velocity, p the associated pressure field and $\mu > 0$ a constant. As a model for the deformable structure, we consider a thin elastic plate, whose motion is governed by the *plate equation*

$$\partial_t^2 \eta + \Delta'^2 \eta - \nu\Delta'\partial_t\eta = F - \mathrm{T}_\eta, \qquad (1.0.4)$$

where η is the displacement of the fluid-solid interface in transversal direction. The term T_η is the normal fluid stress tensor induced by the fluid on the elastic structure. Here, $\nu > 0$ is a constant. In Chapter 4, existence, regularity and uniqueness of a solution to (1.0.3)–(1.0.4) are established for suitable boundary values.

1.1 A Historical Background of Nonlinear Acoustics

Acoustics is the science of sound and is derived from the Greek word *akouein*, to hear. The study of sound goes back to the ancient world. Even then it was well known that sound propagates as a wave. However, the first experimental evidence of this phenomenon was in the seventeenth century. Most of the early acoustical investigations were closely tied to musical acoustics. It started with studying a vibrating string, over to a vibrating membrane, to the more complicated case of a vibrating plate. One of the first (theoretical) results in this field is due to D. BERNOULLI, EULER and D'ALEMBERT in the mid eighteenth century. BERNOULLI introduced a partial differential equation for the vibrating string and gave a solution thereto which was interpreted by d'ALEMBERT as a wave traveling in both directions along the string, see for example [68] and the references [5] and [21]. The one-dimensional model of the wave equation derived by D'ALEMBERT played a fundamental role in fluid mechanics and elasticity, see [73, Section 2.1 and Section 2.2] for a more detailed description of the connection between fluid mechanics and acoustics. After the works of BERNOULLI and d'ALEMBERT, EULER has derived an equation for nonlinear plane acoustic waves in air, which described the behaviour of gas at constant temperature, see [29]. However, the correct law of describing the propagation of plane progressive waves was found hundred years later by EARNSHAW. Mathematically, the works [63] and [74] of LAGRANGE and POISSON had an immense influence on solving nonlinear plane wave equations. Moreover, in the study of shock formation, STOKES already realized in 1848 that it is crucial to include viscosity in the description thereof. The main contribution here came from RANKINE and HUGONIOT, who first formulated the conservation laws (conservation equations for mass, momentum and energy) describing the connection of a flow field behind a shock with the flow field ahead of it. However, the first successful attempts to formulate a comprehensive model were made by RAYLEIGH and TAYLOR. All this together with the contributions of FAY (see for example [32]), and many other scientists in this field, leads to Burgers' equation (see [12]), which is still a classical model used to describe the propagation of plane waves. For more details on the historical evolution of acoustics before the middle of the twentieth century we refer to [68, 80].

After LIGHTHILL published his article [67] in 1956, the interest in the study of acoustic waves grows. In [67] LIGHTHILL describes propagation of shock waves, flood waves in rivers and traffic flow on highways, see [23, Chapter 1]. However, since there are nonlinear acoustical phenomena that can not be described adequately with Burgers' equation, more general wave equations are required. In [7] BLACKSTOCK introduced a special model of nonlinear wave equations of higher order, which models the propagation of sound in thermoviscous fluids. To be more precise, BLACKSTOCK followed the approach of LIGHTHILL, which is based on the assumption that effects of nonlinearity and dissipation are small. Keeping linear and quadratic nonlinear terms that do not involve viscosity or heat conduction terms, BLACKSTOCK derived (1.0.1) from the full equations of motion for a thermoviscous flow, but without the damping term $\partial_t \Delta u$. This system further appears in CRIGHTON's work [18]. If temperature constraints are neglected, the model leads to the Kuznetsov equation (1.0.2), which was first proposed by KUZNETSOV in [57]. Moreover, Kuznetsov's equation is a generalization of d'Alembert's wave equation with new terms due to nonlinearity and dissipation.

The Blackstock-Crighton and Kuznetsov equtions have been subject to increasing research over the last years, see for example [52, 53, 70] where just recently well-posedness of the corresponding initial-value problem for the Kuznetsov equation (1.0.2) was established. Optimal regularity results for (1.0.1) and (1.0.2) were given in [11]. Moreover, the corresponding initial-value problem (1.0.1) was subject in [10, 9]. For more details on the mathematical investigation of the Kuznetsov and Blackstock-Crighton equation, we refer to [51], where a brief overview is given, and the references therein.

1.2 A Historical Background of Fluid-Structure Interaction

Broadly speaking, fluid-structure interaction denotes the coupling between the laws that describe the dynamics of a fluid and structural mechanics.

More specifically, it is the interactions between a deformable structure and a surrounding or internal fluid flow. At the fluid-structure interface, stress is exerted on the solid object by the fluid and leads to deformations hereof. Fluid-structure interaction is a widespread phenomenon in nature, for example in blood flow in human arteries, see [54, 78].

The modern investigation of fluid dynamics started in the middle nineteenth century when NAVIER first proposed a system of equations describing the motion of an incompressible viscous Newtonian flow. Independent of NAVIER's work, STOKES published (for the first time in a scientific article) in [85] the same model, which is nowadays still the most widely used system to model the motion of a liquid. The Navier-Stokes equations are given by

$$\begin{cases} \partial_t u - \Delta u + (u \cdot \nabla)u + \nabla p = f & \text{in } (0,T) \times \Omega, \\ \qquad\qquad\quad \text{div}\, u = 0 & \text{in } (0,T) \times \Omega, \end{cases} \tag{1.2.1}$$

where the unknown $u \colon (0,T) \times \Omega \to \mathbb{R}^3$ and $p \colon (0,T) \times \Omega \to \mathbb{R}$ are the Eulerian velocity field and the pressure field of the liquid, respectively, and $f \colon (0,T) \times \Omega \to \mathbb{R}^3$ the external force. During this studies, STOKES further observed that the effective mass of a rigid body moving in a fluid increases. This phenomenon was first observed by BESSEL in 1828, when considering the motion of a pendulum in a fluid, and meant that the surrounding fluid increased the effective mass of the system. The observation of Stokes and his scientific contribution are known as the founding of (fluid mechanics and) fluid-structure interaction.

The first breakthrough in the mathematical analysis of the Navier-Stokes equations is due to LERAY in 1930. He showed existence of a weak solution to the Navier-Stokes equations, see [65, 66]. Later, HOPF [49] further developed this concept. In the context of time-periodic forcing, SERRIN [82] originally suggested to study time-periodic solutions to (1.2.1). However, the first complete results on existence of time-periodic solutions are due to PRODI [75], YUDOVICH [90] and PROUSE [76], who showed existence of weak time-periodic solutions. Over the past years, an increasing number of authors have investigated (1.2.1) in a time-periodic framework, see for example [35, 40, 41, 55, 60, 37, 59, 72]. When the domain $\Omega = \Omega(t)$ which occupies the fluid varies with time t, one of the first investigations is due to SATHER [81] and the work of FUJITA and SAUER [34].

In recent years the number of investigations into fluid-structure interaction have increased, see for example [39, 26, 25, 16, 19]. There are different types of fluid-structure-interaction problems. For example, one can consider a deformable body moving in a viscous fluid. These kind of free boundary problems were studied in [25]. Another case of fluid-structure interaction is to consider the fluid in a domain where the boundary or one part of it is an elastic structure like an elastic plate. This was subject to [16, 19]. In [16] existence of weak solutions to the initial-value problem corresponding to the coupled system (1.2.1) and

$$\partial_t^2 \eta + \Delta'^2 \eta - \nu\Delta'\partial_t \eta = f - \mathrm{T}_\eta \qquad \text{in } (0,T) \times \Gamma, \qquad (1.2.2)$$

in a three-dimensional cavity is obtained. Here, $\eta\colon (0,T) \times \Gamma \to \mathbb{R}$ is the transversal displacement of the fluid-solid interface $\Gamma \subset \partial\Omega$, $\nu > 0$ a constant and T_η the normal component of the stress induced by the fluid on the elastic plate. A solution and some further L^2 estimates were established by presenting a weak formulation to the nonlinear problem and utilizing Galerkin approximation. DA VEIGA considered a similar model of the elastic plate interacting with a viscous fluid and showed existence of a strong solution in an L^2-framework by a fixed-point procedure. Recently, DENK and SAAL [24] studied a similar model in the half space $\mathbb{R}_+^n$ where the boundary is given by a damped Kirchhoff plate model. The authors showed existence and uniqueness of a strong solution in an L^q-setting.

1.3 Nonlinear Acoustics with Periodic Forcing

To date, only the initial-value problem of the Blackstock-Crighton equation and Kuznetsov equation have been studied. In the following we carry out an investigation of the time-periodic version of these two problems. We start by considering the whole space case of the linearization of the Kuznetsov equation (1.0.2) given by

$$\partial_t^2 u - \Delta u + \partial_t\Delta u = f, \qquad (1.3.1)$$

and establish *a priori* L^q estimates. Instead of relying on a Poincaré map, we obtain the estimates directly via a representation formula for the solution. We hereby circumvent completely the theory for the corresponding initial-value problem, and develop a more direct approach. Moreover, the representation formula we establish seems interesting in the context of resonance, since it exposes the way how different modes of the solution are damped in relation to the modes of the forcing term. To this end, we replace the time axis with the torus group $\mathbb{T} = \mathbb{R}/\mathcal{T}\mathbb{Z}$ and reformulate (1.3.1) as a partial differential equation on $\mathbb{T}$. In this setting, it is possible to utilize the Fourier transform $\mathscr{F}_G$ in a framework of tempered distributions $\mathscr{S}'(G)$, which yields the representation formula

$$u = \mathscr{F}_G^{-1}\left[\frac{1}{|\xi|^2 - k^2 + ik|\xi|^2}\mathscr{F}_G[f]\right] \qquad (1.3.2)$$

for the solution. Here, $G := \mathbb{T} \times \mathbb{R}^n$, with $n \geq 2$. In order to obtain the desired L^q estimates, we decompose this formula into an "undamped" (the *steady state part* u_{s}) and "damped" (the *purely oscillatory part* u_{tp}) modes, *i.e.*,

$$u_{\mathrm{s}} + u_{\mathrm{tp}} = \mathscr{F}_G^{-1}\left[\frac{1}{|\xi|^2}\mathscr{F}_G[f]\right] + \mathscr{F}_G^{-1}\left[\frac{(1 - \delta_{\frac{2\pi}{T}\mathbb{Z}}(k))}{|\xi|^2 - k^2 + ik|\xi|^2}\mathscr{F}_G[f]\right].$$

Since one mode is damped and the other is not, we cannot expect u_{tp} and u_{s} to have the same regularity. Actually, by decomposing the solution into u_{tp} and u_{s}, we see that the Fourier multiplier in the representation of u_{tp} leads to better L^q estimates than can be expected for u. Based on the results deduced for the whole space problem, we investigate (1.3.1) in the half space and on a bounded domain. In the half space case, the linearized damped wave equation (1.3.1) is studied by a reflection principle. In the case of a bounded domain, a localization argument yields the desired L^q estimates. Finally, existence of a time-periodic solution to the nonlinear Kuznetsov equation is established by a fixed-point argument.

The linearization of the Blackstock-Crighton equation (1.0.1) is given by

$$(a\Delta - \partial_t)\left(\partial_t^2 v - c^2\Delta v - b\partial_t\Delta v\right) = f, \qquad (1.3.3)$$

and is considered in a bounded domain with both Dirichlet and Neumann boundary conditions. As for the Kuznetsov equation, we will decompose the linearized Blackstock-Crighton equation into a steady-state part and a purely oscillatory part. Instead of using Fourier multiplier theory to establish the desired *a priori* L^q estimates, we decompose (1.3.3) into a coupled system consisting of two equations of lower order, namely the time-periodic heat equation and the time-periodic wave equation. Based on known results for the heat and wave equations, we deduce existence of time-periodic solutions to (1.3.3). Similarly to the Kuznetsov equation, we obtain existence of a time-periodic solution to the nonlinear problem by utilizing a fixed-point argument.

1.4 Fluid-Structure Interaction with Periodic Forcing

Chapter 4 is devoted to the investigation of the Navier-Stokes equations (1.2.1) interacting with a thin elastic plate located at one part of the boundary, when a time-periodic external forcing is considered. The elastic structure is governed by the plate equation (1.2.2) in $\mathbb{T} \times \mathbb{T}_0^2$. Here, $\mathbb{T}$ denotes the torus given in the previous subsection and $\mathbb{T}_0^2 = (\mathbb{R}/L\mathbb{Z})^2$, with $L > 0$. The fluid problem and the solid problem are coupled by means of dynamic and kinematic interface conditions. For more details on the coupling we refer to Section 4.2. The coupled system (1.2.1) and (1.2.2) shall be studied in a layer domain $\Omega_\eta(t) = \mathbb{T}_0^2 \times (-\eta, 1)$, and the plate is located at the bottom of the domain. In the following we show existence of a solution to the coupled system. Instead of relying on a weak formulation of the free boundary problem, which yields solutions in an L^2 framework, we are interested in strong time-periodic solutions that obey an L^q estimate. To this end, we utilize a fixed-point argument, based on *a priori* estimates deduced for the corresponding linearized system. However, since the boundary of the fluid domain depends on the unknown η, a fixed-point argument cannot be utilized without any further modifications. For this reason, we first employ a coordinate transformation to reformulate the coupled system (1.2.1) and (1.2.2) on a *reference configuration*, where the boundary does not deform anymore. In this setting, we first consider the linearized system

$$
\begin{cases}
\partial_t^2 \eta + \Delta'^2 \eta - \nu \Delta' \partial_t \eta = F - e_3 \cdot \mathrm{T}(u,p)e_{3|x_3=0} & \text{in } \mathbb{T} \times \mathbb{T}_0^2, \\
\partial_t u - \mu \Delta u + \nabla p = f & \text{in } \mathbb{T} \times \Omega, \\
\operatorname{div} u = g & \text{in } \mathbb{T} \times \Omega, \qquad (1.4.1) \\
u(t, x', 0) = -\partial_t \eta(t, x')e_3 & \text{on } \mathbb{T} \times \mathbb{T}_0^2, \\
u(t, x', 1) = 0 & \text{on } \mathbb{T} \times \mathbb{T}_0^2.
\end{cases}
$$

Existence of a solution to (1.4.1) is established via the concept of weak solutions to the corresponding resolvent problem. The *a priori* L^q estimates follow by a Fourier multiplier argument in combination with a localization argument similar to the case of nonlinear acoustics. Based on these results, a solution to the nonlinear problem is obtained via a fixed-point argument, and due to the transformation ϕ, a solution to the coupled system on the time-dependent domain.

2 Preliminaries

This chapter is dedicated to introduce the basic notation and concepts we will use during this doctoral thesis. It is convenient to formulate $\mathcal{T}$-time-periodic problems in a setting of function spaces where the torus $\mathbb{T} := \mathbb{R}/\mathcal{T}\mathbb{Z}$ is used as a time axis. Indeed, via lifting with the quotient map $\pi \colon \mathbb{R} \to \mathbb{T}$, $\mathcal{T}$-time-periodic functions are canonically identified as functions defined on $\mathbb{T}$ and vice versa. Based on this observation we formulate the equations occurring in this thesis as systems defined on the torus $\mathbb{T}$ and decompose them into a so-called *steady-state* and *purely oscillatory* problem, which will be investigated separately. The necessity of this strategy will be obvious when investigating the regularity of a time-periodic solution. We will observe that the two parts of the solution solve various problems, and we will see that they do not have the same regularity properties. This concept of decomposing the time-periodic problems was first introduced in [58] and later generalized and further developed by KYED and some co-authors, see for example [62].

Equipped with the quotient topology, the time-space domain $G :=$ $\mathbb{T} \times \mathbb{R}^n$ is a locally compact abelian group and therefore has a Fourier transform $\mathscr{F}_G$ associated to it, see Subsection 2.3. Moreover, in Subsection 2.2 we introduce the so-called *Schwartz-Bruhat space* $\mathscr{S}(G)$ and its dual space $\mathscr{S}'(G)$ of tempered distributions. In this framework we can formulate the partial differential equations on the group G and employ the Fourier transform $\mathscr{F}_G$, which will be introduced in the second part of this chapter, to obtain a solution. Moreover, we need two more tools from harmonic analysis to examine the resulting solution. We need a multiplier theorem to deduce the desired L^q estimates. But since classical multiplier theorems are defined in a whole space setting $\mathbb{R}^n$ we further have to introduce the Transference principle. The Transference principle is a useful tool that allows us to investigate the resulting multipliers in a whole space setting where we can make use of the known multiplier theory, and conclude the claims in the group setting from this. In 1965, DE LEEUW was the first one introducing this principle (*cf.* [22]), which later was generalized by EDWARDS and GAUDRY, see [27].

Furthermore, the function spaces on the torus $\mathbb{T}$ are introduced in Section 2.4. Since our approach is based on concept of Fourier multipliers we will propose all the Sobolev spaces via Bessel potential spaces, and show that these function spaces coincide with the general Sobolev spaces. However, the corresponding trace spaces, namely the Sobolev-Slobodeckiĭ spaces, are defined via real interpolation. Furthermore, we introduce the function spaces of solenoidal functions in Subsection 2.4.3.

In the framework of fluid-structure interaction Fourier multipliers on a boundary surface occur and we shall estimate the solution with the boundary data in the corresponding Sobolev-Slobodeckiĭ norm. In order to adapt the multiplier theory to the Sobolev-Slobodeckiĭ spaces, interpolation theory for function spaces of time-periodic functions has to be introduced. This will be done in Section 2.5. There, we will extend the known interpolation results to the concept of time-periodic functions.

In Section 2.6 the embedding and trace properties of time-periodic Sobolev spaces utilized in this doctoral thesis are introduced. These embedding properties are later employed to deduce the necessary L^q estimates to carry out a fixed-point argument, when investigating the nonlinear problems occurring herein.

Finally, in the last section of this chapter we introduce two useful mathematical tools from mathematical fluid mechanics. More precise, the Poincaré inequality on layer-like domains and the Bogovskiĭ operator are proposed. Note, Poincaré's inequality is not only used in the framework of fluid mechanics. But first we start by introducing some general notation utilized in this thesis.

2.1 General Notation

Throughout this thesis $\Omega \subset \mathbb{R}^n$ is always a domain with a sufficiently smooth boundary or the whole space $\mathbb{R}^n$. As long as not further mentioned, the space dimension $n \in \mathbb{N}$ is always greater than 2, where $\mathbb{N} := \{1, 2, \ldots\}$ denotes the set of all positive integers. If the set further contains zero, we write $\mathbb{N}_0 := \mathbb{N} \cup \{0\}$. The domain Ω has a boundary of class $C^{k,1}$ or a $C^{k,1}$-smooth boundary, if $\partial\Omega$ can locally be expressed as the graph of a function $\omega \in C^k$ in the respective local coordinates, where the k-th order derivative of ω is Lipschitz continuous. That is, for any $x \in \partial\Omega$ there is a neighborhood $U \subset \mathbb{R}^n$ of x such that $\partial\Omega \cap \overline{U} \subset \mathrm{graph}(\omega)$. The outer unit normal vector on $\partial\Omega$ is denoted by ν, as far as not mentioned differently. Points in $\mathbb{T} \times \Omega$ are generally expressed by (t, x), with t being referred to as time, and x as the spatial variable. The time period $\mathcal{T} > 0$ remains fixed. In what follows we let $\mathbb{T}$ be the torus $\mathbb{R}/\mathcal{T}\mathbb{Z}$. For n-dimensional vectors $x \in \mathbb{R}^n$ and vector valued functions $f = (f_1, \dots, f_n)$ we will denote by $x' := (x_1, \dots, x_{n-1})$ and $f' := (f_1, \dots, f_{n-1})$ the first $n-1$ components of x and f, respectively.

As customary, the dot between two vector-valued elements is referred to as the scalar product. For two matrices this product is designated by a double dot. That is

$$x \cdot y := \sum_{j=1}^n x_j y_j \qquad \text{and} \qquad A : B := \sum_{i,j=1}^n A_{i,j} B_{i,j}$$

for all $x, y \in \mathbb{C}^n$ and matrices $A, B \in \mathbb{C}^{n \times n}$.

Here and in the following let ∂_{x_i} and ∂_t be the partial derivative $\frac{\partial}{\partial x_i}$ and the time derivative $\frac{\partial}{\partial t}$, respectively. Similarly to the notation of the n-dimensional vectors and vector-valued functions, we find it convenient to define the gradients in $\mathbb{R}^n$ and $\mathbb{R}^{n-1}$ separately as $\nabla := (\partial_{x_1}, \dots, \partial_{x_n})$ and $\nabla' := (\partial_{x_1}, \dots, \partial_{x_{n-1}})$, and the Laplace operators as

$$\Delta := \sum_{i=1}^n \partial_{x_i}^2 \qquad \text{and} \qquad \Delta' := \sum_{i=1}^{n-1} \partial_{x_i}^2,$$

respectively. For functions f defined on an $(n-1)$-dimensional domain we denote by ∇f the vector $(\nabla' f, 0)^\top$, as long as no confusions may arise. The operator Δ^2 is referred to as the Bi-Laplacian and is defined as

$$\Delta^2 := \Delta\Delta \qquad \text{and} \qquad \Delta'^2 := \Delta'\Delta', \tag{2.1.1}$$

where Δ'^2 is the corresponding operator for functions defined on an $(n-1)$-dimensional domain. Moreover, in the context of fluid mechanics, the symmetrical gradient D and the fluid stress tensor T appear in the equation of motion. These two tensors are defined as

$$\mathrm{D}(u) := \frac{1}{2}\mu(\nabla u + (\nabla u)^\top) \qquad \mathrm{T}(u, p) := 2\mathrm{D}(u) - p\mathrm{I}, \tag{2.1.2}$$

as long as the derivatives occurring on the right-hand side are well-defined. Here, u is the velocity field, p the corresponding pressure field and $\mu > 0$ the constant coefficient of viscosity. Letting $\alpha = (\alpha_1, \dots, \alpha_n) \in \mathbb{N}_0^n$ be a multi-index, we set $|\alpha| := \alpha_1 + \dots + \alpha_n$ and

$$\partial_x^\alpha u := \partial_{x_1}^{\alpha_1} \cdots \partial_{x_n}^{\alpha_n} u,$$

for any $x \in \mathbb{R}^n$ and $u \in C^{|\alpha|}(\Omega)$.

For $\mathcal{T}$-time-periodic functions f defined on time-space domains, we let

$$f_{\mathrm{s}} := \mathcal{P}f(t,x) := \frac{1}{\mathcal{T}} \int\limits_0^{\mathcal{T}} f(s,x) \, \mathrm{d}s, \quad f_{\mathrm{tp}} := \mathcal{P}_\perp f(t,x) := f(t,x) - \mathcal{P}f(t,x)$$

(2.1.3)

whenever the integral is well defined. Since f_{s} is independent of time t, we shall implicitly treat f_{s} as a function in the spatial variable x only and refer to it as the *steady-state* or *stationary* part of f. The remaining function f_{tp} is referred to as the *purely oscillatory* part of f.

For any $q \in (1, \infty)$ and $m \in \mathbb{N}$ we define the Lebesgue spaces $\mathrm{L}^q(\Omega)^m$ on a domain $\Omega \subset \mathbb{R}^n$ as usual. That is, for $q \in (1, \infty)$ we let

$$\|u\|_{\mathrm{L}^q(\Omega)} := \left(\int\limits_\Omega |u|^q \, \mathrm{d}x \right)^{\frac{1}{q}},$$

denote the L^q norm, and identify $\mathrm{L}^q(\Omega)^m$ as the set

$$\mathrm{L}^q(\Omega)^m := \{ u \colon \Omega \to \mathbb{C}^n \mid \|u\|_{\mathrm{L}^q(\Omega)} < \infty \}.$$

Note that $\|\cdot\|_{\mathrm{L}^q(\Omega)}$ is a norm on $\mathrm{L}^q(\Omega)^m$. The superscript m indicates the dimension of an element of the Lebesgue space and will be omitted in the case of scalar valued function, *i.e.*, when $m = 1$. The case $q = 2$ is a special case, since $\mathrm{L}^2(\Omega)$ is a Hilbert space. Moreover, $\langle \cdot, \cdot \rangle_\Omega$ denotes the scalar product on $\mathrm{L}^2(\Omega)$ and is defined for any vector-valued functions f and g as usual by

$$\langle f, g \rangle_\Omega = \int\limits_\Omega f \cdot g \, \mathrm{d}x,$$

where the Ω in the index will be omitted, as long as no confusion arises. Analogously, the function spaces L^q_{loc} are defined as usual, that is, L^q_{loc} is defined as the set of all functions u, such that $u \in \mathrm{L}^q(\Omega')$ for any bounded domain Ω' with $\overline{\Omega'} \subset \Omega$.

2.2 Topology and differentiable structure

We utilize $G = \mathbb{T} \times \mathbb{R}^n$ as a time-space domain. Equipped with the quotient topology induced by the quotient mapping

$$\pi : \mathbb{R} \times \mathbb{R}^n \to G, \ \pi(t,x) := ([t], x),$$

G becomes a locally compact abelian (LCA) group. We can identify G with the domain $[0, \mathcal{T}) \times \mathbb{R}^n$ via the restriction $\pi\big|_{[0,\mathcal{T}) \times \mathbb{R}^n}$. The Haar measure dg on G is the product of the Lebesgue measure on $\mathbb{R}^n$ and the Lebesgue measure on $[0, \mathcal{T})$. We normalize dg so that

$$\int_G u(g)\, dg = \frac{1}{\mathcal{T}} \int_0^{\mathcal{T}} \int_{\mathbb{R}^n} u(t, x)\, dx\, dt.$$

There is a bijective correspondence between points $(k, \xi) \in \widehat{G} := \frac{2\pi}{\mathcal{T}}\mathbb{Z} \times \mathbb{R}^n$ and characters $\chi : G \to \mathbb{C}$, $\chi(t, x) := e^{ix \cdot \xi + ikt}$ on G. Consequently, we can identify the dual group of G with $\widehat{G}$. The compact-open topology on $\widehat{G}$ reduces to the product of the Euclidean topology on $\mathbb{R}^n$ and the discrete topology on $\frac{2\pi}{\mathcal{T}}\mathbb{Z}$. The Haar measure on $\widehat{G}$ is therefore the product of the counting measure on $\frac{2\pi}{\mathcal{T}}\mathbb{Z}$ and the Lebesgue measure on $\mathbb{R}^n$.

The spaces of smooth functions on G and $\widehat{G}$ are defined as

$$C^\infty(G) := \{u : G \to \mathbb{R} \mid \exists U \in C^\infty(\mathbb{R} \times \mathbb{R}^n) : U = u \circ \pi\} \qquad (2.2.1)$$

and

$$C^\infty(\widehat{G}) := \left\{ u \in C(\widehat{G}) \;\middle|\; \forall k \in \frac{2\pi}{\mathcal{T}}\mathbb{Z} : u(k, \cdot) \in C^\infty(\mathbb{R}^n) \right\},$$

respectively. Derivatives of a function $u \in C^\infty(G)$ are defined by

$$\partial_t^\beta \partial_x^\alpha u := \left[\partial_t^\beta \partial_x^\alpha (u \circ \pi) \right] \circ \Pi^{-1},$$

with $\Pi := \pi\big|_{[0,\mathcal{T}) \times \mathbb{R}^n}$. The notion of Schwartz spaces can be extended to LCA groups (see [8] and [28]). The Schwartz-Bruhat space on G is given by

$$\mathscr{S}(G) := \{u \in C^\infty(G) \mid \forall(\alpha, \beta, \gamma) \in \mathbb{N}_0^n \times \mathbb{N}_0 \times \mathbb{N}_0^n : \rho_{\alpha,\beta,\gamma}(u) < \infty\},$$

where

$$\rho_{\alpha,\beta,\gamma}(u) := \sup_{(t,x) \in G} \left| x^\gamma \partial_t^\beta \partial_x^\alpha u(t, x) \right|.$$

$\mathscr{S}(G)$ becomes a topological vector space, if we equip $\mathscr{S}(G)$ with the seminorm topology of the family $\{\rho_{\alpha,\beta,\gamma} \mid (\alpha, \beta, \gamma) \in \mathbb{N}_0^n \times \mathbb{N}_0 \times \mathbb{N}_0^n\}$. The corresponding topological dual space $\mathscr{S}'(G)$ equipped with the weak* topology is referred to as the space of tempered distributions on G. Distributional derivatives for a tempered distribution u are defined by duality as in the classical case. The Schwartz-Bruhat space on $\widehat{G}$ is

$$\mathscr{S}(\widehat{G}) := \{u \in C^\infty(\widehat{G}) \mid \forall (\alpha, \beta, \gamma) \in \mathbb{N}_0^n \times \mathbb{N}_0^n \times \mathbb{N}_0 : \hat{\rho}_{\alpha,\beta,\gamma}(u) < \infty\},$$

with the generic semi-norms

$$\hat{\rho}_{\alpha,\beta,\gamma}(u) := \sup_{(k,\xi) \in \widehat{G}} \left| \xi^\alpha \partial_\xi^\beta k^\gamma u(k, \xi) \right|$$

inducing the topology.

2.3 Fourier Transform and Multiplier Theory

By $\mathscr{F}_G$ we denote the Fourier transform associated to the LCA group G equipped with the Haar measure introduced above:

$$\mathscr{F}_G : \mathscr{S}(G) \to \mathscr{S}(\widehat{G}),$$

$$\mathscr{F}_G[u](k, \xi) := \frac{1}{\mathcal{T}} \int_0^{\mathcal{T}} \int_{\mathbb{R}^n} u(t, x) \, e^{-ix \cdot \xi - ikt} \, dx \, dt.$$

Recall that $\mathscr{F}_G : \mathscr{S}(G) \to \mathscr{S}(\widehat{G})$ is a homeomorphism with inverse given by

$$\mathscr{F}_G^{-1} : \mathscr{S}(\widehat{G}) \to \mathscr{S}(G),$$

$$\mathscr{F}_G^{-1}[w](t, x) := \sum_{k \in \frac{2\pi}{\mathcal{T}}\mathbb{Z}} \int_{\mathbb{R}^n} w(k, \xi) \, e^{ix \cdot \xi + ikt} \, d\xi,$$

provided the Lebesgue measure $d\xi$ is normalized appropriately. By duality, $\mathscr{F}_G$ extends to a homeomorphism $\mathscr{F}_G : \mathscr{S}'(G) \to \mathscr{S}'(\widehat{G})$.

The Fourier symbol, with respect to $\mathscr{F}_G$, of the projection $\mathcal{P}$ introduced in (2.1.3) is the delta distribution $\delta_{\frac{2\pi}{\mathcal{T}}\mathbb{Z}}$ that is identified with the function

$$\delta_{\frac{2\pi}{\mathcal{T}}\mathbb{Z}} : \frac{2\pi}{\mathcal{T}}\mathbb{Z} \to \mathbb{C}, \qquad \delta_{\frac{2\pi}{\mathcal{T}}\mathbb{Z}}(k) := \begin{cases} 1 & \text{if } k = 0, \\ 0 & \text{if } k \neq 0. \end{cases} \qquad (2.3.1)$$

Via this symbol, the projections $\mathcal{P}$ and $\mathcal{P}_\perp$ extend to projections on $\mathscr{S}'(G)$:

$$\begin{aligned} \mathcal{P} : \mathscr{S}'(G) \to \mathscr{S}'(G), \quad \mathcal{P}u &:= \mathscr{F}_G^{-1}[\delta_{\frac{2\pi}{\mathcal{T}}\mathbb{Z}} \mathscr{F}_G[u]], \\ \mathcal{P}_\perp : \mathscr{S}'(G) \to \mathscr{S}'(G), \quad \mathcal{P}_\perp u &:= \mathscr{F}_G^{-1}[(1 - \delta_{\frac{2\pi}{\mathcal{T}}\mathbb{Z}}) \mathscr{F}_G[u]]. \end{aligned} \qquad (2.3.2)$$

At this point, we have introduced ample formalism to formulate the time-periodic problems that occur in this thesis as systems of partial differential equations in the time-space domain G. Moreover, the Fourier transform $\mathscr{F}_G$ enables us to investigate these systems in terms of Fourier multipliers. Due to the lack of a comprehensive L^q-multiplier theory in the general group setting, we shall utilize a so-called *transference principle* for this purpose. The transference principle goes back to DE LEEUW [22]. The theorem below is due to EDWARDS and GAUDRY [27].

Theorem 2.3.1 (DE LEEUW, EDWARDS and GAUDRY). *Let G and H be LCA groups. Moreover, let $\Phi : \widehat{G} \to \widehat{H}$ be a continuous homomorphism and $q \in [1, \infty]$. Assume that $m \in \mathrm{L}^\infty(\widehat{H}; \mathbb{C})$ is a continuous L^q-multiplier, i.e., there is a constant C such that*

$$\forall f \in \mathrm{L}^2(H) \cap \mathrm{L}^q(H) : \quad \|\mathscr{F}_H^{-1}[m\,\mathscr{F}_H[f]]\|_q \leq C\|f\|_q.$$

Then $m \circ \Phi \in \mathrm{L}^\infty(\widehat{G}; \mathbb{C})$ is also an L^q-multiplier with

$$\forall f \in \mathrm{L}^2(G) \cap \mathrm{L}^q(G) : \quad \|\mathscr{F}_G^{-1}[(m \circ \Phi)\,\mathscr{F}_G[f]]\|_q \leq C\|f\|_q.$$

Proof. See [27, Theorem B.2.1]. □

We shall make use of the following multiplier theorem of Marcinkiewicz type ([84, Chapter IV, §6]):

Theorem 2.3.2. *Let m be a bounded C^n function defined away from the coordinate axes on $\mathbb{R}^n$. Assume that there is a constant $A > 0$ such that*

$$\sup_{\varepsilon \in \{0,1\}^n} \operatorname*{ess\,sup}_{\xi \in \mathbb{R}^n} \left| \xi_1^{\varepsilon_1} \cdots \xi_n^{\varepsilon_n} \partial_{\xi_1}^{\varepsilon_1} \cdots \partial_{\xi_n}^{\varepsilon_n} m(\xi) \right| \leq A. \tag{2.3.3}$$

Then for any $q \in (1, \infty)$ there is a constant $C > 0$ such that

$$\forall f \in \mathrm{L}^2(\mathbb{R}^n) \cap \mathrm{L}^q(\mathbb{R}^n) : \|\mathscr{F}_{\mathbb{R}^n}^{-1}[m\mathscr{F}[f]]\|_q \leq CA\|f\|_q,$$

with $C = C(q)$.

Proof. See [44, Corollary 5.2.5]. □

2.4 Sobolev and Bessel Potential Spaces

This section is dedicated to introduce the function spaces appearing in this doctoral thesis. In Subsection 4.4 the Stokes system will be investigated. Since the methods used to prove the existence of a solution to this problem are highly based on Fourier multipliers and interpolation theory, we find it convenient to introduce all the relevant Sobolev spaces via Bessel potential spaces.

2.4.1 Functions on Spatial Domains

Classical (inhomogeneous) Bessel potential spaces are defined for $s \in \mathbb{R}$ and $q \in [1, \infty)$ by

$$\mathrm{H}^{s,q}(\mathbb{R}^n) := \big\{ u \in \mathscr{S}'(\mathbb{R}^n) \mid \mathscr{F}_{\mathbb{R}^n}^{-1}\big[(1 + |\xi|^2)^{\frac{s}{2}} \mathscr{F}_{\mathbb{R}^n}[u]\big] \in \mathrm{L}^q(\mathbb{R}^n) \big\},$$

$$\|u\|_{\mathrm{H}^{s,q}(\mathbb{R}^n)} := \|u\|_{s,q} := \big\|\mathscr{F}_{\mathbb{R}^n}^{-1}\big[(1 + |\xi|^2)^{\frac{s}{2}} \mathscr{F}_{\mathbb{R}^n}[u]\big]\big\|_q.$$

Classical Sobolev spaces on $\mathbb{R}^n$ are defined as Bessel potential spaces of integer order $k \in \mathbb{Z}$, and Sobolev spaces on $\Omega \subset \mathbb{R}^n$ via restriction:

$$\mathrm{W}^{k,q}(\mathbb{R}^n) := \mathrm{H}^{k,q}(\mathbb{R}^n), \qquad \mathrm{W}^{k,q}(\Omega) := \big\{ u_{|\Omega} \mid u \in \mathrm{W}^{k,q}(\mathbb{R}^n) \big\}.$$

Observe that for negative-order spaces, *i.e.*, when $k < 0$, the Sobolev space $\mathrm{W}^{k,q}(\Omega)$ coincides with the dual space $\big(\mathrm{W}^{-k,q'}(\Omega)\big)'$ and *not* with the dual space $(\mathrm{W}_0^{-k,q'}(\Omega))'$. Note, the zero in the subscript indicates that any $u \in \mathrm{W}_0^{-k,q'}(\Omega)$ vanishes at the boundary $\partial\Omega$. In the following, it is essential that the former meaning of $\mathrm{W}^{k,q}(\Omega)$ is used.

Homogeneous Bessel potential spaces are defined in accordance with [88] by introducing the subspace

$$Z(\mathbb{R}^n) := \big\{ \phi \in \mathscr{S}(\mathbb{R}^n) \mid \forall \alpha \in \mathbb{N}_0^n : \partial_\xi^\alpha \mathscr{F}_{\mathbb{R}^n}[\phi](0) = 0 \big\}$$

of $\mathscr{S}(\mathbb{R}^n)$, and for $s \in \mathbb{R}$ and $q \in [1, \infty)$ letting

$$\dot{\mathrm{H}}^{s,q}(\mathbb{R}^n) := \big\{ u \in Z'(\mathbb{R}^n) \mid \mathscr{F}_{\mathbb{R}^n}^{-1}\big[|\xi|^s \mathscr{F}_{\mathbb{R}^n}[u]\big] \in \mathrm{L}^q(\mathbb{R}^n) \big\},$$

$$\|u\|_{\dot{\mathrm{H}}^{s,q}(\mathbb{R}^n)} := \big\|\mathscr{F}_{\mathbb{R}^n}^{-1}\big[|\xi|^s \mathscr{F}_{\mathbb{R}^n}[u]\big]\big\|_q.$$

Due to the lack of regularity of $|\xi|^s$ at the origin, the above definition of $\dot{\mathrm{H}}^{s,q}(\mathbb{R}^n)$ is not meaningful as a subspace of $\mathscr{S}'(\mathbb{R}^n)$. Instead, $\dot{\mathrm{H}}^{s,q}(\mathbb{R}^n)$ is defined as a subspace of $Z'(\mathbb{R}^n)$. As such, $\dot{\mathrm{H}}^{s,q}(\mathbb{R}^n)$ is clearly a Banach space. As above, we define homogeneous Sobolev spaces on $\mathbb{R}^n$ as homogeneous Bessel potential spaces of integer order $k \in \mathbb{Z}$, and introduce homogeneous Sobolev spaces on $\Omega \subset \mathbb{R}^n$ via restriction:

$$\dot{\mathrm{W}}^{k,q}(\mathbb{R}^n) := \dot{\mathrm{H}}^{k,q}(\mathbb{R}^n), \qquad \dot{\mathrm{W}}^{k,q}(\Omega) := \big\{ u_{|\Omega} \mid u \in \dot{\mathrm{W}}^{k,q}(\mathbb{R}^n) \big\}. \quad (2.4.1)$$

By the Hahn-Banach Theorem, any functional in $Z'(\mathbb{R}^n)$ can be extended to a tempered distribution in $\mathscr{S}'(\mathbb{R}^n)$. If $s \leq 0$, the extension of an element in $\dot{\mathrm{H}}^{s,q}(\mathbb{R}^n)$ to $\mathscr{S}'(\mathbb{R}^n)$ is unique. In the case $s > 0$, one may verify that two extensions of an element in $\dot{\mathrm{H}}^{s,q}(\mathbb{R}^n)$ differ at most by addition of a polynomial of order strictly less than s. With this ambiguity in mind, one may consider $\dot{\mathrm{H}}^{s,q}(\mathbb{R}^n)$ as a normed ($s \le 0$) and semi-normed ($s > 0$) subspace of $\mathscr{S}'(\mathbb{R}^n)$.

Sobolev-Slobodeckiĭ spaces of both homogeneous and inhomogeneous type are defined via real interpolation in the usual way and equipped with the associated interpolation norms. That is, for $k \in \mathbb{Z}$ and $\alpha \in (0,1)$ we have

$$\mathrm{W}^{k+\alpha,q}(\Omega) := \big(\mathrm{W}^{k+1,q}(\Omega), \mathrm{W}^{k,q}(\Omega)\big)_{1-\alpha,q},$$
$$\dot{\mathrm{W}}^{k+\alpha,q}(\Omega) := \big(\dot{\mathrm{W}}^{k+1,q}(\Omega), \dot{\mathrm{W}}^{k,q}(\Omega)\big)_{1-\alpha,q}.$$

2.4.2 Functions on Time-Space Domains

Analogously, Bessel potential spaces with underlying time-space domain G are introduced via the Fourier transform $\mathscr{F}_G$ as

$$\mathrm{H}^{r,q}\big(\mathbb{T}; \mathrm{H}^{s,q}(\mathbb{R}^n)\big) :=$$
$$\big\{ u \in \mathscr{S}'(G) \mid \mathscr{F}_G^{-1}\big[(1+|k|^2)^{\frac{r}{2}}(1+|\xi|^2)^{\frac{s}{2}}\mathscr{F}_G[u]\big] \in \mathrm{L}^q\big(\mathbb{T}; \mathrm{L}^q(\mathbb{R}^n)\big) \big\}$$

equipped with the canonical norm. Again, we refer to Bessel potential spaces of integer order $k, l \in \mathbb{N}_0$ as Sobolev spaces:

$$\mathrm{W}^{k,q}\big(\mathbb{T}; \mathrm{W}^{l,q}(\mathbb{R}^n)\big) := \mathrm{H}^{k,q}\big(\mathbb{T}; \mathrm{H}^{l,q}(\mathbb{R}^n)\big).$$

Sobolev spaces on the time-space domain $\mathbb{T} \times \Omega$ are defined via restriction of the elements in the spaces above. In order to introduce homogeneous spaces, we let

$$Z(G) := \big\{ \phi \in \mathscr{S}(G) \mid \forall \alpha \in \mathbb{N}_0^n : \ \partial_\xi^\alpha \mathscr{F}_{\mathbb{R}^n}[\phi](t,0) = 0 \quad \forall t \in \mathbb{T} \big\}$$

and put

$$\mathrm{H}^{r,q}\big(\mathbb{T}; \dot{\mathrm{H}}^{s,q}(\mathbb{R}^n)\big) :=$$
$$\big\{ u \in Z'(G) \mid \mathscr{F}_G^{-1}\big[(1+|k|^2)^{\frac{r}{2}}|\xi|^s \mathscr{F}_G[u]\big] \in \mathrm{L}^q\big(\mathbb{T}; \mathrm{L}^q(\mathbb{R}^n)\big) \big\}.$$

As above, we may consider $\mathrm{H}^{r,q}\big(\mathbb{T}; \dot{\mathrm{H}}^{s,q}(\mathbb{R}^n)\big)$ as a subspace of $\mathscr{S}'(G)$ by extension.

Finally, Sobolev-Slobodeckiĭ spaces on the domain $\mathbb{T} \times \Omega$ are defined via real interpolation, *i.e.*,

$$\mathrm{W}^{k+\alpha,q}\big(\mathbb{T}; \mathrm{W}^{l,q}(\Omega)\big) := \big(\mathrm{W}^{k+1,q}\big(\mathbb{T}; \mathrm{W}^{l,q}(\Omega)\big), \mathrm{W}^{k,q}\big(\mathbb{T}; \mathrm{W}^{l,q}(\Omega)\big)\big)_{1-\alpha,q},$$
$$\mathrm{W}^{k,q}\big(\mathbb{T}; \mathrm{W}^{l+\alpha,q}(\Omega)\big) := \big(\mathrm{W}^{k,q}\big(\mathbb{T}; \mathrm{W}^{l+1,q}(\Omega)\big), \mathrm{W}^{k,q}\big(\mathbb{T}; \mathrm{W}^{l,q}(\Omega)\big)\big)_{1-\alpha,q},$$
$$(2.4.2)$$

for $k, l \in \mathbb{Z}$ and $\alpha \in (0, 1)$. In this way, all the function spaces appearing in this thesis attain rigorous definitions. It is easy to verify that these definitions coincide with a classical interpretation as Bochner spaces of vector-valued functions defined on the torus $\mathbb{T}$. That is, for any Lebesgue and Sobolev space $E(\Omega)$ they coincide with

$$\mathrm{L}^q\big(\mathbb{T}; E(\Omega)\big) := \overline{C^\infty\big(\mathbb{T}; E(\Omega)\big)}^{\|\cdot\|_{\mathrm{L}^q(\mathbb{T}; E(\Omega))}},$$

$$\mathrm{W}^{k,q}\big(\mathbb{T}; E(\Omega)\big) := \overline{C^\infty\big(\mathbb{T}; E(\Omega)\big)}^{\|\cdot\|_{\mathrm{W}^{k,q}(\mathbb{T}; E(\Omega))}},$$

where $C^\infty\big(\mathbb{T}; E(\Omega)\big)$ denotes the space of smooth vector-valued functions on the torus and is introduced by the same construction as in (2.2.1).

For simplicity, we omit the domain in the norms above and write $\|\cdot\|_q$ instead of $\|\cdot\|_{\mathrm{L}^q(\mathbb{T}; \mathrm{L}^q(\Omega))}$ and $\|\cdot\|_{\mathrm{L}^q(\Omega)}$ at some points in this thesis. Moreover, for stationary problems we make use of the notation $\|\cdot\|_{s,q}$ to express the norm $\|\cdot\|_{\mathrm{W}^{s,q}(\Omega)}$ as long as no confusion may arise, for $s \in \mathbb{R}$.

Remark 2.4.1. Observe, the torus $\mathbb{T}$ canonically inherits a topology and differentiable structure from $\mathbb{R}$ via the quotient mapping $\pi \colon \mathbb{R} \to \mathbb{T}$ in such a way that

$$C^\infty\big(\mathbb{T}; E(\Omega)\big) = \{f \colon \mathbb{T} \to E(\Omega) \mid f \circ \pi \in C^\infty(\mathbb{R}; E(\Omega))\}$$

for any generic Sobolev space $E(\Omega)$. The quotient map π employed as a lifting operator acts as an isometric isomorphism between $C^\infty\big(\mathbb{T}; E(\Omega)\big)$ and $C_{\mathrm{per}}^\infty\big(\mathbb{R}; E(\Omega)\big)$. Here, $C_{\mathrm{per}}^\infty\big(\mathbb{R}; E(\Omega)\big)$ is defined as the set of all smooth functions f such that $f(t + \mathcal{T}, x) = f(t, x)$ for all $(t, x) \in \mathbb{R} \times \Omega$. Consequently also the Sobolev spaces $\mathrm{W}^{k,p}\big(\mathbb{T}; E(\Omega)\big)$ and $\mathrm{W}_{\mathrm{per}}^{k,p}\big(\mathbb{R}; E(\Omega)\big)$ are isometrically isomorphic, where $\mathrm{W}_{\mathrm{per}}^{k,p}\big(\mathbb{R}; E(\Omega)\big)$ is given as the closure of $C_{\mathrm{per}}^\infty\big(\mathbb{R}; E(\Omega)\big)$ with respect to the Sobolev norm $\|\cdot\|_{\mathrm{W}^{k,q}(\mathbb{T}; E(\Omega))}$.

With this remark at hand, it is now possible to prove all the following theorems and claims in the group setting $G = \mathbb{T} \times \mathbb{R}^n$, where we can take advantage of the framework introduced in this chapter, and deduce the results also in the setting, where the time axis is the whole of $\mathbb{R}$.

Moreover, observe that $\mathcal{P}$ and $\mathcal{P}_\perp$ are complementary projections on the space $C^\infty\big(\mathbb{T}; E(\Omega)\big)$. We shall employ these projections to decompose the Lebesgue and Sobolev spaces introduced above. By continuity, $\mathcal{P}$ and $\mathcal{P}_\perp$ extend to bounded operators on $\mathrm{L}^q\big(\mathbb{T}; E(\Omega)\big)$ and $\mathrm{W}^{k,q}\big(\mathbb{T}; E(\Omega)\big)$.

Notation 2.4.2. Throughout this thesis we set

$$\mathrm{W}_\perp^{k,q}\big(\mathbb{T}; E(\Omega)\big) := \mathcal{P}_\perp \mathrm{W}^{k,q}\big(\mathbb{T}; E(\Omega)\big),$$

$$\mathrm{W}_\perp^{k,q}\big(\mathbb{T}; E(\Omega)\big) := \mathcal{P}_\perp \mathrm{W}^{k,q}\big(\mathbb{T}; E(\Omega)\big),$$

for any Sobolev space $E(\Omega)$, $q \in (1, \infty)$ and $k \in \mathbb{N}_0$. Moreover, we write $\mathrm{L}^q(\mathbb{T} \times \Omega)$ instead $\mathrm{L}^q\big(\mathbb{T}; \mathrm{L}^q(\Omega)\big)$.

2.4.3 Spaces of Solenoidal Functions

In the context of fluid mechanics one is interested in the existence of solenoidal vector fields. For this reason, we denote by $C_{0,\sigma}^\infty(\Omega)$ the space of all divergence free smooth vector valued fields, that is

$$C_{0,\sigma}^\infty(\Omega) := \{u \in C_0^\infty(\Omega)^n \mid \operatorname{div} u = 0\}.$$

Note that the subscript denotes that $C_0^\infty(\Omega)$ is the set of all smooth functions on Ω that are compactly supported. As usual, we define for $1 < q < \infty$

$$\mathrm{L}_\sigma^q(\Omega) := \overline{C_{0,\sigma}^\infty(\Omega)}^{\|\cdot\|_q} \qquad \text{and} \qquad \mathrm{W}_\sigma^{1,q}(\Omega) := \overline{C_{0,\sigma}^\infty(\Omega)}^{\|\cdot\|_{1,q}}$$

endowed with their natural norms. These spaces are usually introduced if the Stokes equations subject to no-slip boundary conditions are studied. Indeed, elements $u \in \mathrm{L}_\sigma^q(\Omega)$ satisfy $\nu \cdot u = 0$ on $\partial\Omega$, where ν denotes the outward unit normal to Ω. However, in the framework of fluid-structure interaction, especially in the context of free boundary problems, these boundary values cannot be prescribed, see for example Subsection 4.2. For this reason, we have to modify the functions spaces of solenoidal vector fields and introduce the spaces

$$\mathcal{L}_\sigma^q(\Omega) := \{u \in \mathrm{L}^q(\Omega)^n \mid \operatorname{div} u = 0\}$$

and

$$\mathcal{W}_\sigma^{1,q}(\Omega) := \{u \in \mathrm{W}^{1,q}(\Omega)^n \mid \operatorname{div} u = 0\}, \tag{2.4.3}$$

endowed with their natural norms. These spaces are usually considered if one is interested in Neumann-type boundary conditions, see for example [71]. Observe that these spaces do in general not coincide. Compared to $\mathrm{L}_\sigma^q(\Omega)$, elements u in $\mathcal{L}_\sigma^q(\Omega)$ only have a vanishing mean value of $\nu \cdot u$ on $\partial\Omega$, which is a weaker condition than $\nu \cdot u = 0$ on $\partial\Omega$.

2.5 Interpolation

In Section 4.4 of this doctoral thesis the Stokes system in the half space and the periodic half space will be investigated. Although the function spaces appearing in Section 4.4 can all be defined in terms of classical interpolation, however, proving existence of a solution to the Stokes equations relies on a somewhat more refined scale of interpolation spaces. More specifically, it is based on anisotropic Besov spaces with underlying time-space domain G, which we shall show coincide with the function spaces obtained by real interpolation of the Bessel potential spaces introduced above. Therefore, we shall carry out the identification of the interpolation spaces and their properties. This will be done by mimicking the proofs of similar results for classical isotropic Besov spaces. To this end, we fix an $m \in \mathbb{N}$ and introduce the parabolic length scale

$$|\eta, \xi| := (|\eta|^2 + |\xi|^{4m})^{\frac{1}{4m}} \quad \text{for } (\eta, \xi) \in \mathbb{R} \times \mathbb{R}^n. \tag{2.5.1}$$

The anisotropic Besov spaces defined in Definition 2.5.2 below pertain to time-periodic parabolic problems of order $2m$. Since the interpolation results deduced in this section are only required to investigate the Stokes system, a parabolic scale is sufficient and no further interpolation properties for the function spaces occurring in the context of nonlinear acoustics are necessary. For simplicity, we omit m in the notation for the function spaces below.

The anisotropic Besov spaces shall be based on the following anisotropic partition of unity:

Lemma 2.5.1. *Let $m \in \mathbb{N}$ and $|\eta, \xi|$ be given by (2.5.1). There is a $\phi \in C_0^\infty(\mathbb{R} \times \mathbb{R}^n)$ satisfying*

$$\text{supp } \phi = \{(\eta, \xi) \mid 2^{-1} \leq |\eta, \xi| \leq 2\}, \tag{2.5.2}$$

$$\phi(\eta, \xi) > 0 \quad \text{for} \quad 2^{-1} < |\eta, \xi| < 2, \tag{2.5.3}$$

$$\sum_{l=-\infty}^{\infty} \phi(2^{-2ml}\eta, 2^{-l}\xi) = 1 \quad \text{for} \quad |\eta, \xi| \neq 0. \tag{2.5.4}$$

Proof. Let $h \in C^\infty(\mathbb{R})$ with $\text{supp } h = \{y \in \mathbb{R} \mid 2^{-1} \leq |y| \leq 2\}$ and $h(y) > 0$ for $2^{-1} < |y| < 2$. Then $f : \mathbb{R} \times \mathbb{R}^n \to \mathbb{R}$, $f(\eta, \xi) := h(|\eta, \xi|)$ satisfies (2.5.2) and (2.5.3). Moreover, $f(2^{-2ml}\eta, 2^{-l}\xi) \neq 0$ iff $2^{l-1} < |\eta, \xi| < 2^{l+1}$. Thus $f(2^{-2ml}\eta, 2^{-l}\xi) \neq 0$ for at least one and at most two $l \in \mathbb{Z}$. Consequently,

$$\phi : \mathbb{R} \times \mathbb{R}^n \to \mathbb{R}, \quad \phi(\eta, \xi) := \begin{cases} \dfrac{f(\eta, \xi)}{\sum_{l=-\infty}^{\infty} f(2^{-2ml}\eta, 2^{-l}\xi)} & \text{if } |\eta, \xi| \neq 0, \\ 0 & \text{if } |\eta, \xi| = 0 \end{cases}$$

is well-defined. It is easy to verify that ϕ satisfies (2.5.2)–(2.5.4). $\qquad \square$

Definition 2.5.2 (Anisotropic Besov and Bessel Potential Spaces). *Let $\phi \in C_0^\infty(\mathbb{R} \times \mathbb{R}^n)$ be as in Lemma 2.5.1, $s \in \mathbb{R}$ and $p, q \in [1, \infty)$. We define anisotropic Besov spaces*

$$\mathrm{B}_{pq,\perp}^s(G) := \{f \in \mathscr{S}_\perp'(G) \mid \|f\|_{\mathrm{B}_{pq,\perp}^s} < \infty\},$$

$$\|f\|_{\mathrm{B}_{pq,\perp}^s} := \bigg(\sum_{l=0}^\infty \big(2^{sl} \|\mathscr{F}_G^{-1}[\phi(2^{-2ml}k, 2^{-l}\xi)\mathscr{F}_G[f]]\|_p \big)^q \bigg)^{\frac{1}{q}}, \tag{2.5.5}$$

and anisotropic Bessel potential spaces

$$H_{p,\perp}^s(G) := \{f \in \mathscr{S}_\perp'(G) \mid \|f\|_{H_{p,\perp}^s} < \infty\},$$

$$\|f\|_{H_{p,\perp}^s} := \|\mathscr{F}_G^{-1}[|k, \xi|^s \mathscr{F}_G[f]]\|_p. \tag{2.5.6}$$

Observe that $\mathrm{B}_{pq,\perp}^s(G)$ and $H_{p,\perp}^s(G)$ are defined as subspaces of $\mathscr{S}_\perp'(G)$ rather than $\mathscr{S}'(G)$. Recalling (2.3.2), it is easy to verify that $\|\cdot\|_{\mathrm{B}_{pq,\perp}^s}$ and $\|\cdot\|_{H_{p,\perp}^s}$ are therefore norms (rather than mere semi-norms), and $\mathrm{B}_{pq,\perp}^s(G)$ and $H_{p,\perp}^s(G)$ Banach spaces. Observe further that Definition 2.5.2 does not coincide with the classical definition of a Besov and Bessel potential spaces. Note that real interpolation of anisotropic Bessel potential spaces yields anisotropic Besov spaces (as in the case of classical (isotropic) spaces). Moreover, throughout this thesis the Besov spaces are only used for real interpolation of anisotropic Bessel potential spaces of purely oscillatory functions, that is, functions f with $\mathscr{F}_G[f](0, \cdot) = 0$. Therefore, we find it convenient to define the Besov and Bessel potential norms $\|\cdot\|_{\mathrm{B}_{pq,\perp}^s}$ and $\|\cdot\|_{H_{p,\perp}^s}$ as in (2.5.5) and (2.5.6), respectively, where the zeroth Fourier mode on $\mathbb{T}$ is left out.

Lemma 2.5.3. *Let $p, q \in (1, \infty)$, $\theta \in (0, 1)$, $s_0, s_1 \in \mathbb{R}$ and $s := (1-\theta)s_0 + \theta s_1$. If $s_0 \neq s_1$, then $\big(H_{p,\perp}^{s_0}(G), H_{p,\perp}^{s_1}(G)\big)_{\theta,q} = \mathrm{B}_{pq,\perp}^s(G)$ with equivalent norms.*

Proof. For $l \in \mathbb{N}_0$ and $r \in \mathbb{R}$ let

$$\mathfrak{m}_l^r : \mathbb{R} \times \mathbb{R}^n \to \mathbb{C}, \quad \mathfrak{m}_l^r(\eta, \xi) := \phi\big(2^{-2ml}\eta, 2^{-l}\xi\big)|\eta, \xi|^{-r}.$$

We claim that $\mathfrak{m}_l^r$ is an $L^p(\mathbb{R}; L^p(\mathbb{R}^n))$-multiplier, which we verify by showing that $\mathfrak{m}_l^r$ meets the condition of Marcinkiewicz's multiplier theorem (see Theorem 2.3.2). For this purpose, we utilize only

$$\operatorname{supp}\phi\big(2^{-2ml}\cdot, 2^{-l}\cdot\big) = \{(\eta, \xi) \in \mathbb{R} \times \mathbb{R}^n \mid 2^{l-1} \leq |\eta, \xi| \leq 2^{l+1}\}, \tag{2.5.7}$$

and that $g(\eta, \xi) := |\eta, \xi|^{-r}$ is parabolically $(-r)$-homogeneous, that is,

$$\forall \lambda > 0 : \quad g(\eta, \xi) = \lambda^{-r} g(\lambda^{-2m}\eta, \lambda^{-1}\xi). \tag{2.5.8}$$

From (2.5.7) we immediately obtain $\|\mathfrak{m}_l^r\|_\infty \leq c_0 \|\phi\|_\infty 2^{-lr}$, with c_0 independent on l. By (2.5.8), we further observe that

$$\eta \, \partial_\eta \mathfrak{m}_l^r(\eta, \xi) = 2^{-2ml} \eta \, (\partial_\eta \phi) \big(2^{-2ml}\eta, 2^{-l}\xi\big) \, g(\eta, \xi)$$
$$+ \phi\big(2^{-2ml}\eta, 2^{-l}\xi\big) \, \lambda^{-r} \, (\partial_\eta g)(\lambda^{-2m}\eta, \lambda^{-1}\xi) \, \lambda^{-2m}\eta.$$

Choosing $\lambda := |\eta, \xi|$ and recalling (2.5.7), we thus deduce $\|\eta \, \partial_\eta \mathfrak{m}_l^r\|_\infty \leq c_1 \|\phi\|_\infty 2^{-lr}$, with c_1 independent on l. Similarly, we obtain

$$\sum_{\alpha \in \{0,1\}^{n+1}} \|\xi_1^{\alpha_1} \cdots \xi_n^{\alpha_n} \eta^{\alpha_{n+1}} \partial_{\xi_1}^{\alpha_1} \cdots \partial_{\xi_n}^{\alpha_n} \partial_\eta^{\alpha_{n+1}} \mathfrak{m}_l^r\|_\infty \leq c_2 \|\phi\|_\infty 2^{-lr}$$

with c_2 independent on l. It follows from Marcinkiewicz's multiplier theorem (Theorem 2.3.2) that $\mathfrak{m}_l^r$ is an $L^p(\mathbb{R}; L^p(\mathbb{R}^n))$-multiplier. Consequently, the Transference Principle (Theorem 2.3.1) implies that $\mathfrak{m}_{l \mid \frac{2\pi}{T}\mathbb{Z} \times \mathbb{R}^n}^r$ is an $L^p(\mathbb{T}; L^p(\mathbb{R}^n))$-multiplier with

$$\left\| \phi \mapsto \mathscr{F}_G^{-1} \big[\mathfrak{m}_l^r(k, \xi) \mathscr{F}_G[\phi] \big] \right\|_{\mathscr{L}(L^p(\mathbb{T}; L^p(\mathbb{R}^n)), L^p(\mathbb{T}; L^p(\mathbb{R}^n)))} < c_3 \|\phi\|_\infty 2^{-lr}.$$

Let $f \in \big(H^{s_0}_{p,\perp}(G), H^{s_1}_{p,\perp}(G)\big)_{\theta,q}$. Consider a decomposition $f = f_0 + f_1$ with $f_0 \in H^{s_0}_{p,\perp}(G)$ and $f_1 \in H^{s_1}_{p,\perp}(G)$. We deduce

$$\|\mathscr{F}_G^{-1} [\phi(2^{-2ml}k, 2^{-l}\xi) \mathscr{F}_G[f]] \|_p$$
$$\leq \|\mathscr{F}_G^{-1} \big[\mathfrak{m}_l^{s_0} \mathscr{F}_G \big[\mathscr{F}_G^{-1} \big[|k, \xi|^{s_0} \mathscr{F}_G[f_0] \big] \big] \big] \|_p$$
$$+ \|\mathscr{F}_G^{-1} \big[\mathfrak{m}_l^{s_1} \mathscr{F}_G \big[\mathscr{F}_G^{-1} \big[|k, \xi|^{s_1} \mathscr{F}_G[f_1] \big] \big] \big] \|_p$$
$$\leq c_4 \big(2^{-ls_0} \|f_0\|_{H^{s_0}_{p,\perp}} + 2^{-ls_1} \|f_1\|_{H^{s_1}_{p,\perp}} \big).$$

We now employ the K-method (see for example [4, Chapter 3.1]) to characterize the interpolation space $\big(H^{s_0}_{p,\perp}(G), H^{s_1}_{p,\perp}(G)\big)_{\theta,q}$. Taking infimum over all decompositions f_0, f_1 in the inequality above, we find that

$$\|\mathscr{F}_G^{-1} [\phi(2^{-2ml}k, 2^{-l}\xi) \mathscr{F}_G[f]] \|_p \leq c_5 \, 2^{-ls_0} \, K\big(2^{l(s_0 - s_1)}, f, H^{s_0}_{p,\perp}, H^{s_1}_{p,\perp}\big),$$

which implies

$$\|f\|_{\mathrm{B}^s_{pq,\perp}} \leq c_6 \bigg(\sum_{l=0}^\infty \big(2^{\theta l(s_1 - s_0)} \, K(2^{l(s_0 - s_1)}, f, H^{s_0}_{p,\perp}, H^{s_1}_{p,\perp})\big)^q \bigg)^{\frac{1}{q}}$$
$$\leq c_7 \, \|f\|_{\big(H^{s_0}_{p,\perp}, H^{s_1}_{p,\perp}\big)_{\theta,q}},$$

where the last inequality above is valid since $s_0 \neq s_1$.

Now consider $f \in B^s_{pq,\perp}(G)$ and let $l \in \mathbb{N}_0$. Choose $\psi \in C_0^\infty(\mathbb{R} \times \mathbb{R}^n)$ with $\psi(\eta, \xi) = 1$ for $2^{-1} \leq |\eta, \xi| \leq 2$ and $\operatorname{supp} \psi = \{(\eta, \xi) \in \mathbb{R} \times \mathbb{R}^n \mid 4^{-1} \leq |\eta, \xi| \leq 4\}$. Using the same technique as above, this time utilizing the multiplier

$$\widetilde{\mathfrak{m}}^r_l : \mathbb{R} \times \mathbb{R}^n \to \mathbb{C}, \quad \widetilde{\mathfrak{m}}^r_l(\eta, \xi) := \psi\big(2^{-2ml}\eta, 2^{-l}\xi\big)|\eta, \xi|^{-r},$$

we can estimate

$$\|\mathscr{F}_G^{-1}\big[\phi(2^{-2ml}k, 2^{-l}\xi)\mathscr{F}_G[f]\big]\|_{H^{s_1}_{p,\perp}}$$
$$= \|\mathscr{F}_G^{-1}\big[\psi(2^{-2ml}k, 2^{-l}\xi)\,\phi(2^{-2ml}k, 2^{-l}\xi)\mathscr{F}_G[f]\big]\|_{H^{s_1}_{p,\perp}}$$
$$\leq c_8\, 2^{ls_1}\|\mathscr{F}_G^{-1}\big[\phi(2^{-2ml}k, 2^{-l}\xi)\mathscr{F}_G[f]\big]\|_p,$$

and similarly

$$\|\mathscr{F}_G^{-1}\big[\phi(2^{-2ml}k, 2^{-l}\xi)\mathscr{F}_G[f]\big]\|_{H^{s_0}_{p,\perp}}$$
$$\leq c_9\, 2^{ls_0}\|\mathscr{F}_G^{-1}\big[\phi(2^{-2ml}k, 2^{-l}\xi)\mathscr{F}_G[f]\big]\|_p.$$

We thus obtain

$$2^{-l\theta(s_0-s_1)}2^{l(s_0-s_1)}\|\mathscr{F}_G^{-1}\big[\phi(2^{-2ml}k, 2^{-l}\xi)\mathscr{F}_G[f]\big]\|_{H^{s_1}_{p,\perp}}$$
$$= 2^{ls}2^{-ls_1}\|\mathscr{F}_G^{-1}\big[\phi(2^{-2ml}k, 2^{-l}\xi)\mathscr{F}_G[f]\big]\|_{H^{s_1}_{p,\perp}}$$
$$\leq c_{10}2^{ls}\|\mathscr{F}_G^{-1}\big[\phi(2^{-2ml}k, 2^{-l}\xi)\mathscr{F}_G[f]\big]\|_p$$

and

$$2^{-l\theta(s_0-s_1)}\|\mathscr{F}_G^{-1}\big[\phi(2^{-2ml}k, 2^{-l}\xi)\mathscr{F}_G[f]\big]\|_{H^{s_0}_{p,\perp}}$$
$$= 2^{ls}2^{-ls_0}\|\mathscr{F}_G^{-1}\big[\phi(2^{-2ml}k, 2^{-l}\xi)\mathscr{F}_G[f]\big]\|_{H^{s_0}_{p,\perp}}$$
$$\leq c_{11}2^{ls}\|\mathscr{F}_G^{-1}\big[\phi(2^{-2ml}k, 2^{-l}\xi)\mathscr{F}_G[f]\big]\|_p.$$

We now employ the J-method (see for example [4, Chapter 3.2]) to characterize the interpolation space $\big(H^{s_0}_{p,\perp}(G), H^{s_1}_{p,\perp}(G)\big)_{\theta,q}$. By the last two estimates above, we see that

$$2^{-l\theta(s_0-s_1)}\,J\big(2^{l(s_0-s_1)}, \mathscr{F}_G^{-1}\big[\phi(2^{-2ml}k, 2^{-l}\xi)\mathscr{F}_G[f]\big]\big)$$
$$\leq c_{12}\,2^{ls}\|\mathscr{F}_G^{-1}\big[\phi(2^{-2ml}k, 2^{-l}\xi)\mathscr{F}_G[f]\big]\|_p.$$

Since $\mathcal{P}_\perp f = f$, we find that $f = \sum_{l=0}^{\infty} \mathscr{F}_G^{-1}\big[\phi(2^{-2ml}k, 2^{-l}\xi)\mathscr{F}_G[f]\big]$ with convergence in the space $H_{p,\perp}^{s_0}(G) + H_{p,\perp}^{s_1}(G)$. Recalling that $s_0 \neq s_1$, we thus conclude that

$$\|f\|_{\left(H_{p,\perp}^{s_0}, H_{p,\perp}^{s_1}\right)_{\theta,q}} \leq c_{13}\|f\|_{\mathrm{B}_{pq,\perp}^s},$$

and thereby the lemma. $\qquad\qquad\qquad\qquad\qquad\qquad\qquad\qquad\square$

2.6 Embedding and Trace properties of Time-Periodic Sobolev spaces

Subject of this subsection is to introduce some embedding and trace properties of anisotropic time-periodic Sobolev spaces which will be used to determine L^q estimates in the nonlinear case.

2.6.1 Embedding Properties

Before considering the trace spaces and the corresponding operators that appear in this doctoral thesis, we shall begin by proving some embedding properties. For reasons of simplicity we first introduce the following notation:

Notation 2.6.1. Throughout this thesis we let

$$\mathbb{X}^q(\mathbb{T} \times \Omega) := \mathrm{W}^{2,q}\big(\mathbb{T}; \mathrm{L}^q(\Omega)\big) \cap \mathrm{W}^{1,q}\big(\mathbb{T}; \mathrm{W}^{2,q}(\Omega)\big),$$
$$\mathbb{Y}^q(\mathbb{T} \times \Omega) := \mathrm{W}^{3,q}\big(\mathbb{T}; \mathrm{L}^q(\Omega)\big) \cap \mathrm{W}^{1,q}\big(\mathbb{T}; \mathrm{W}^{4,q}(\Omega)\big),$$

equipped with the canonical norms

$$\|\cdot\|_{\mathbb{X}^q(\mathbb{T}\times\Omega)} := \big(\|\cdot\|_{\mathrm{W}^{2,q}(\mathbb{T};\mathrm{L}^q(\Omega))}^q + \|\cdot\|_{\mathrm{W}^{1,q}(\mathbb{T};\mathrm{W}^{2,q}(\Omega))}^q\big)^{\frac{1}{q}},$$
$$\|\cdot\|_{\mathbb{Y}^q(\mathbb{T}\times\Omega)} := \big(\|\cdot\|_{\mathrm{W}^{3,q}(\mathbb{T};\mathrm{L}^q(\Omega))}^q + \|\cdot\|_{\mathrm{W}^{1,q}(\mathbb{T};\mathrm{W}^{4,q}(\Omega))}^q\big)^{\frac{1}{q}},$$

respectively. The projected function spaces $\mathbb{X}_\perp^q$ and $\mathbb{Y}_\perp^q$ are defined in a similar manner.

The following lemma concerning embedding properties utilized in the examination of the Blackstock-Crighton equation in Section 3.4 shall be proven:

Lemma 2.6.2. *Let $q \in (1, \infty)$, $n \geq 2$ and $\Omega \subset \mathbb{R}^n$ be a bounded domain of class C^4. The embeddings*

$$\mathbb{Y}^q(\mathbb{T} \times \Omega) \hookrightarrow W^{2,q}\big(\mathbb{T}; W^{2,q}(\Omega)\big), \tag{2.6.1}$$

$$W^{2,q}\big(\mathbb{T}; L^q(\Omega)\big) \cap L^q\big(\mathbb{T}; W^{4,q}(\Omega)\big) \hookrightarrow W^{1,q}\big(\mathbb{T}; W^{2,q}(\Omega)\big), \tag{2.6.2}$$

$$L^q\big(\mathbb{R}_+; \mathbb{X}^q(\mathbb{T} \times \mathbb{R}^{n-1})\big) \cap W^{2,q}\big(\mathbb{R}_+; W^{1,q}\big(\mathbb{T}; L^q(\mathbb{R}^{n-1})\big)\big)$$
$$\hookrightarrow W^{1,q}\big(\mathbb{R}_+; W^{1,q}\big(\mathbb{T}; W^{1,q}(\mathbb{R}^{n-1})\big)\big), \tag{2.6.3}$$

and ($l \in \{1, 2, 3\}$)

$$L^q\big(\mathbb{R}_+; \mathbb{Y}^q(\mathbb{T} \times \mathbb{R}^{n-1})\big) \cap W^{4,q}\big(\mathbb{R}_+; W^{1,q}\big(\mathbb{T}; L^q(\mathbb{R}^{n-1})\big)\big)$$
$$\hookrightarrow W^{l,q}\big(\mathbb{R}_+; W^{1,q}\big(\mathbb{T}; W^{4-l,q}(\mathbb{R}^{n-1})\big)\big) \tag{2.6.4}$$

are continuous.

Proof. Here, we only prove the embedding (2.6.1), as the others can be shown analogously. But first, observe that the regularity of Ω suffices to ensure the existence of a continuous extension operator $E \colon \mathbb{Y}^q(\mathbb{T} \times \Omega) \to \mathbb{Y}^q(\mathbb{T} \times \mathbb{R}^n)$ and therefore it suffices to prove Lemma 2.6.2 for $\Omega = \mathbb{R}^n$. In this setting, we can employ the Fourier transform $\mathscr{F}_G$ in time and space to characterize the Sobolev spaces

$$W^{2,q}\big(\mathbb{T}; W^{2,q}(\mathbb{R}^n)\big)$$
$$= \Big\{ f \in \mathscr{S}'(G) \; \Big| \; \mathscr{F}_G^{-1}\big[(1 + |k|^2)(1 + |\xi|^2)\mathscr{F}_G[f]\big] \in L^q(G) \Big\}$$

and

$$\mathbb{Y}^q(G) = W^{3,q}\big(\mathbb{T}; L^q(\mathbb{R}^n)\big) \cap W^{1,q}\big(\mathbb{T}; W^{4,q}(\mathbb{R}^n)\big)$$
$$= \big\{ f \in \mathscr{S}'(G) \mid \mathscr{F}_{\mathbb{T} \times \mathbb{R}^n}^{-1}\big[m_{\mathbb{Y}} \mathscr{F}_{\mathbb{T} \times \mathbb{R}^n}[f]\big] \in L^q(\mathbb{T} \times \mathbb{R}^n) \big\}$$

where $(k, \xi) \in \mathbb{Z} \times \mathbb{R}^n$ is an element of the corresponding dual group, and $m_{\mathbb{Y}}$ is given by

$$m_{\mathbb{Y}}(k, \xi) := (1 + |k|^2)^{\frac{3}{2}} + (1 + |k|^2)^{\frac{1}{2}}(1 + |\xi|^2)^2.$$

Now observe that

$$\|f\|_{W^{2,q}(\mathbb{T}; W^{2,q}(\mathbb{R}^n))} \leq c_0 \Big\| \mathscr{F}_G^{-1}\Big[m \, \mathscr{F}_G\big[\mathscr{F}_G^{-1}[m_{\mathbb{Y}}(k, \xi)\mathscr{F}_G[f]]\big]\Big]\Big\|_q, \tag{2.6.5}$$

where the multiplier

$$m\colon \mathbb{R} \times \mathbb{R}^n \to \mathbb{C}, \quad m(\eta, \xi) := \frac{(1 + |k|^2)^{\frac{1}{2}}(1 + |\xi|^2)}{(1 + |k|^2) + (1 + |\xi|^2)^2}$$

satisfies the condition of Marcinkiewicz's Multiplier Theorem (Theorem 2.3.2). Indeed, by Young's inequality $|\eta||\xi|^2 \leq \frac{1}{2}|\eta|^2 + \frac{1}{2}|\eta||\xi|^4$, whence $\|m\|_\infty < \infty$. Similarly, one may verify that

$$\max_{\varepsilon \in \{0,1\}^{n+1}} \|\xi_1^{\varepsilon_1} \cdots \xi_n^{\varepsilon_n} \eta^{\varepsilon_{n+1}} \partial_{\xi_1}^{\varepsilon_1} \cdots \partial_{\xi_n}^{\varepsilon_n} \partial_\eta^{\varepsilon_{n+1}} m(\eta, \xi)\|_\infty < \infty.$$

Consequently, m is an $L^q(\mathbb{R} \times \mathbb{R}^n)$-multiplier. By de Leeuw's transference principle for Fourier multipliers on LCA groups (Theorem 2.3.1), it follows that the restriction $m_{|\mathbb{Z} \times \mathbb{R}^n}$ is an $L^q(G)$-multiplier. From (2.6.5) we thus deduce

$$\|f\|_{W^{2,q}(\mathbb{T}; W^{2,q}(\mathbb{R}^n))} \leq c_1 \left\| \mathscr{F}_G^{-1}[m_\mathbb{Y}(k, \xi)\mathscr{F}_G[f]] \right\|_{L^q(G)} \leq c_2 \|f\|_{\mathbb{Y}^q(G)},$$

and we conclude (2.6.1). The embeddings (2.6.2)–(2.6.4) can be established in a completely similar manner. ☐

Finally, embedding properties of $\mathcal{T}$-time-periodic anisotropic Sobolev spaces shall be established. Since these properties are employed to different problems, we consider anisotropic Sobolev spaces of parabolic type in the most general way as possible. For this reason we propose an embedding theorem concerning the embedding properties of anisotropic Sobolev spaces of parabolic type which are of the form

$$W_\perp^{m, 2m, q}(\mathbb{T} \times \Omega) := W_\perp^{m,q}(\mathbb{T}; L^q(\Omega)) \cap L_\perp^q(\mathbb{T}; W^{2m,q}(\Omega)), \qquad (2.6.6)$$

with $m \in \mathbb{N}$. The following embedding theorem is later used to establish L^q estimates for the nonlinear terms appearing in this doctoral thesis. In what follows, we only consider purely oscillatory functions, that is, functions that have a vanishing mean value in time.

Theorem 2.6.3 (Embedding Theorem). *Let $\Omega \subset \mathbb{R}^n$ ($n \geq 2$) be the whole space $\mathbb{R}^n$, the half space $\mathbb{R}_+^n$ or a bounded domain with a Lipschitz boundary and $q \in (1, \infty)$. Moreover, let $m \in \mathbb{N}$, $M_t \in \mathbb{N}_0$ and $m_x \in \mathbb{N}_0^n$ such that*

$$0 \leq M_x + 2M_t \leq 2m,$$

with $M_x := |m_x|$, and $\alpha, \beta \in [0, 2(m - M_t) - M_x]$ such that $\beta := 2(m - M_t) - M_x - \alpha$. Assume that $p, r \in [q, \infty]$ satisfy

$$
\begin{cases}
r \leq \dfrac{2q}{2 - \alpha q} & \text{if } \alpha q < 2, \\[2mm]
r < \infty & \text{if } \alpha q = 2, \\[2mm]
r \leq \infty & \text{if } \alpha q > 2,
\end{cases}
\qquad
\begin{cases}
p \leq \dfrac{nq}{n - \beta q} & \text{if } \beta q < n, \\[2mm]
p < \infty & \text{if } \beta q = n, \\[2mm]
p \leq \infty & \text{if } \beta q > n.
\end{cases}
$$

Then there exists a constant $C_1 = C_1(\mathcal{T}, \Omega, p, q, r) > 0$ such that

$$
\|\partial_x^{m_x} \partial_t^{M_t} u\|_{\mathrm{L}_\perp^r(\mathbb{T}; \mathrm{L}^p(\Omega))} \leq C_1 \|u\|_{\mathrm{W}_\perp^{m, 2m, q}(\mathbb{T} \times \Omega)} \tag{2.6.7}
$$

holds for all $u \in \mathrm{W}_\perp^{m, 2m, q}(\mathbb{T} \times \Omega)$.

In order to prove Theorem 2.6.3 it is necessary to introduce several interpolation properties. But first observe that the regularity of Ω suffices to ensure the existence of a continuous extension operator $E \colon \mathrm{W}_\perp^{m, 2m, q}(\mathbb{T} \times \Omega) \to \mathrm{W}_\perp^{m, 2m, q}(\mathbb{T} \times \mathbb{R}^n)$ as in the case of classical Sobolev spaces. Note further that in this framework we can write $\partial_x^{m_x} \partial_t^{M_t} u$ as

$$
\partial_x^{m_x} \partial_t^{M_t} u = \mathcal{F}_\alpha *_{\mathbb{T}} \mathcal{F}_\beta *_{\mathbb{R}^n} \mathcal{F}, \tag{2.6.8}
$$

with

$$
\mathcal{F}_\alpha := \mathscr{F}_{\mathbb{T}}^{-1}\left[\left(1 - \delta_{\frac{2\pi}{\mathcal{T}}\mathbb{Z}}(k)\right)|k|^{-\frac{\alpha}{2}}\right],
$$

$$
\mathcal{F}_\beta := \mathscr{F}_{\mathbb{R}^n}^{-1}\left[\left(1 + |\xi|^2\right)^{-\frac{\beta}{2}}\right],
$$

$$
\mathcal{F} := \mathscr{F}_G^{-1}\left[\frac{\left(1 - \delta_{\frac{2\pi}{\mathcal{T}}\mathbb{Z}}(k)\right)i^{M_t + M_x}k^{M_t}|k|^{\frac{\alpha}{2}}\xi^{m_x}\left(1 + |\xi|^2\right)^{\frac{\beta}{2}}}{1 + |k|^m + |\xi|^{2m}}\right.
$$

$$
\left. \mathscr{F}_G\left[(1 + |k|^m + |\xi|^{2m})u\right]\right]. \tag{2.6.9}
$$

Here, $*_{\mathbb{T}}$ is to be understood as the convolution with respect to the torus $\mathbb{T}$ and $*_{\mathbb{R}^n}$ as the convolution with respect to $\mathbb{R}^n$. The estimate (2.6.7) follows if we can prove that the fraction appearing in the representation of $\mathcal{F}$ is a Fourier multiplier, and the mappings $\phi \mapsto \mathcal{F}_\alpha *_{\mathbb{T}} \phi$ and $\phi \mapsto \mathcal{F}_\beta *_{\mathbb{R}^n} \phi$ define continuous linear operators from $\mathrm{L}_\perp^q(\mathbb{T})$ into $\mathrm{L}_\perp^r(\mathbb{T})$ and $\mathrm{L}^q(\mathbb{R}^n)$ into $\mathrm{L}^p(\mathbb{R}^n)$, respectively. For this purpose we calculate the inverse Fourier transforms $\mathcal{F}_\alpha$ and $\mathcal{F}_\beta$, and we begin by considering $\mathcal{F}_\alpha$ in the case $\frac{\alpha}{2} < 1$. Choosing for example $[-\frac{1}{2}\mathcal{T}, \frac{1}{2}\mathcal{T})$ as a realization of $\mathbb{T}$, we obtain that

$\mathcal{F}_\alpha(t) = C_2 t^{\frac{\alpha}{2}-1} + h_\alpha(t)$ with $h_\alpha \in C_\perp^\infty(\mathbb{T})$ and $C_2 > 0$, see [44, Example 3.1.19]. For $\frac{\alpha}{2} = 1$ a simple calculation shows that

$$\mathcal{F}_\alpha(t) = \sum_{k \in \frac{2\pi}{\mathcal{T}}\mathbb{Z}\setminus\{0\}} \frac{e^{ikt}}{|k|} = \frac{\mathcal{T}}{2\pi}\left(\sum_{k=1}^\infty \frac{e^{i\frac{2\pi}{\mathcal{T}}kt}}{k} + \sum_{k=1}^\infty \frac{e^{-i\frac{2\pi}{\mathcal{T}}kt}}{k}\right)$$

$$= \frac{\mathcal{T}}{2\pi}\left(-\sum_{k=1}^\infty (-1)^{k+1}\frac{(-e^{i\frac{2\pi}{\mathcal{T}}t})^k}{k} - \sum_{k=1}^\infty (-1)^{k+1}\frac{(-e^{-i\frac{2\pi}{\mathcal{T}}t})^k}{k}\right).$$

In order to calculate both series on the right-hand side of the equality above, we make use of the identity

$$\log(1+x) = \sum_{k=1}^\infty (-1)^{k+1}\frac{x^k}{k}$$

for all $x \in (-1, 1]$. Therefore, we obtain

$$\mathcal{F}_\alpha(t) = \frac{\mathcal{T}}{2\pi}\left(-\log(1 - e^{i\frac{2\pi}{\mathcal{T}}t}) - \log(1 - e^{-i\frac{2\pi}{\mathcal{T}}t})\right)$$

$$= -\frac{\mathcal{T}}{2\pi}\log\left((1 - e^{i\frac{2\pi}{\mathcal{T}}t})(1 - e^{-i\frac{2\pi}{\mathcal{T}}t})\right) = -\frac{\mathcal{T}}{2\pi}\log\left(2 - 2\cos(\frac{2\pi}{\mathcal{T}}t)\right),$$

for all $t \neq c\mathcal{T}$ with $c \in \mathbb{Z}$. In the last case where $\frac{\alpha}{2} > 1$, the series $\sum_{k \in \frac{2\pi}{\mathcal{T}}\mathbb{Z}\setminus\{0\}} |k|^{-\frac{\alpha}{2}}$ converges and therefore $k^{-\frac{\alpha}{2}} \in \ell^1(\frac{2\pi}{\mathcal{T}}\mathbb{Z}\setminus\{0\})$. Moreover, this series is a convergent majorant of $\mathcal{F}_\alpha$ and we deduce that

$$g_\alpha(t) := \mathcal{F}_\alpha(t) = \sum_{k \in \frac{2\pi}{\mathcal{T}}\mathbb{Z}\setminus\{0\}} \frac{e^{ikt}}{|k|^{\frac{\alpha}{2}}} \in \mathscr{F}_\mathbb{T}^{-1}\left[\ell^1\left(\frac{2\pi}{\mathcal{T}}\mathbb{Z}\right)\right] \subset C_\perp(\mathbb{T}).$$

All in all the inverse Fourier transform $\mathcal{F}_\alpha$ is given by

$$\mathcal{F}_\alpha(t) = \begin{cases} C_2|t|^{\frac{\alpha}{2}-1} + h_\alpha(t) & \text{if } \frac{\alpha}{2} < 1, \\ -\dfrac{\mathcal{T}}{2\pi}\log\left(2 - 2\cos(\frac{2\pi}{\mathcal{T}}t)\right) & \text{if } \frac{\alpha}{2} = 1, \\ g_\alpha(t) & \text{if } \frac{\alpha}{2} > 1, \end{cases} \qquad (2.6.10)$$

where $h_\alpha \in C_\perp^\infty(\mathbb{T})$ and $g_\alpha \in L_\perp^\infty(\mathbb{T})$.

Lemma 2.6.4. *In the situation of Theorem 2.6.3 the mapping $\phi \mapsto \mathcal{F}_\alpha *_\mathbb{T} \phi$ defines a continuous linear operator from $L_\perp^q(\mathbb{T})$ into $L_\perp^r(\mathbb{T})$ for $p, r \in (1, \infty)$ satisfying*

$$1 + \frac{1}{r} = \frac{1}{p} + \frac{1}{q},$$

and either $\mathcal{F}_\alpha \in L_\perp^{p,\infty}(\mathbb{T})$ and $r < \infty$ or $\mathcal{F}_\alpha \in L_\perp^p(\mathbb{T})$ and $r \leq \infty$.

Proof. Here we use the standard convention $\frac{1}{0} = \infty$. In order to prove the lemma we consider the three cases $\alpha q < 2$, $\alpha q = 2$ and $\alpha q > 2$, and employ Young's inequality to the representation formula deduced in (2.6.10). If $\alpha q < 2$, we have that $\mathcal{F}_\alpha(t) = C_2|t|^{\frac{\alpha}{2}-1} + h_\alpha(t)$, since $0 < \frac{\alpha}{2} < \frac{1}{q} < 1$ for all $q \in (1,\infty)$. Clearly, $\mathcal{F}_\alpha \in \mathrm{L}_\perp^{\frac{1}{1-\frac{\alpha}{2}},\infty}(\mathbb{T})$ (see Proposition 1.1.6. and the corresponding example in [44]) and Young's inequality (see [44, Theorem 1.4.24]) implies

$$\|\mathcal{F}_\alpha *_\mathrm{T} \phi\|_{\mathrm{L}_\perp^r(\mathbb{T})} \leq c_0 \|\mathcal{F}_\alpha\|_{\mathrm{L}_\perp^{\frac{1}{1-\frac{\alpha}{2}},\infty}(\mathbb{T})} \|\phi\|_{\mathrm{L}_\perp^q(\mathbb{T})},$$

for all $q, r \in (1,\infty)$ satisfying

$$1 + \frac{1}{r} = \frac{1}{\frac{1}{1-\frac{\alpha}{2}}} + \frac{1}{q} \Leftrightarrow r = \frac{2q}{2-\alpha q}.$$

Note that $\mathcal{F}_\alpha \in \mathrm{L}_\perp^p(\mathbb{T})$ for all $p < \frac{1}{1-\frac{\alpha}{2}}$ and therefore Young's inequality yields

$$\|\mathcal{F}_\alpha *_\mathrm{T} \phi\|_{\mathrm{L}_\perp^r(\mathbb{T})} \leq c_1 \|\mathcal{F}_\alpha\|_{\mathrm{L}_\perp^p(\mathbb{T})} \|\phi\|_{\mathrm{L}_\perp^q(\mathbb{T})},$$

for all $p, q, r \in (1,\infty)$ satisfying

$$r < \frac{2q}{2-\alpha q} < \infty \qquad \text{and} \qquad p < \frac{1}{1-\frac{\alpha}{2}}.$$

In the case $\alpha q = 2$ we also have that $0 < \frac{\alpha}{2} = \frac{1}{q} < 1$ for all $q \in (1,\infty)$, hence we have again $\mathcal{F}_\alpha(t) = C_2|t|^{\frac{\alpha}{2}-1} + h_\alpha(t)$ and similarly it follows that $\mathcal{F}_\alpha \in \mathrm{L}_\perp^{p,\infty}(\mathbb{T})$ and $r < \infty$.

Let us now consider $\alpha q > 2$. Here we have to distinguish again the three cases

$$0 < \frac{1}{q} < \frac{\alpha}{2} < 1, \quad \frac{\alpha}{2} = 1 \quad \text{and} \quad \frac{\alpha}{2} > 1.$$

In the first case we have again that $\mathcal{F}_\alpha \in \mathrm{L}_\perp^p(\mathbb{T})$ for any $p \leq \frac{1}{1-\frac{1}{q}} < \frac{1}{1-\frac{\alpha}{2}}$, hence Young's inequality implies that $\mathcal{F}_\alpha$ is a continuous linear operator from $\mathrm{L}_\perp^q(\mathbb{T})$ into $\mathrm{L}_\perp^r(\mathbb{T})$ with $p, q, r \in (1,\infty)$ satisfying

$$1 + \frac{1}{r} = \frac{1}{q} + \frac{1}{p} \geq \frac{1}{q} + \left(1 - \frac{1}{q}\right) \quad \Rightarrow \quad r \leq \infty.$$

If $\frac{\alpha}{2} = 1$, we have $\mathcal{F}_\alpha(t) = -\frac{\mathcal{T}}{2\pi}\big(\log(2) + \log(1 - \cos(\frac{2\pi}{\mathcal{T}}t))\big)$ and observe that the critical point is only $t = 0$. Hence, for an arbitrary $\varepsilon > 0$ and $t > \varepsilon > 0$ we have $\mathcal{F}_\alpha \in \mathrm{L}^p_\perp\big([-\frac{1}{2}\mathcal{T}, \frac{1}{2}\mathcal{T}) \setminus [-\varepsilon, \varepsilon]\big)$ for all $p \in [1, \infty)$ with

$$\|\mathcal{F}_\alpha\|_{\mathrm{L}^p_\perp([-\frac{1}{2}\mathcal{T}, \frac{1}{2}\mathcal{T})\setminus[-\varepsilon,\varepsilon])} \leq c_1.$$

In the case $t \in (-\varepsilon, \varepsilon)$ the cos in the representation formula of $\mathcal{F}_\alpha$ can be expressed as a Taylor series at $t = 0$, that is,

$$1 - \cos\left(\frac{2\pi}{\mathcal{T}}t\right) = \frac{2\pi^2 t^2}{\mathcal{T}^2}\left(1 - 2\sum_{s=2}^{\infty} \frac{(-1)^s}{(2s)!}\left(\frac{2\pi t}{\mathcal{T}}\right)^{2s-2}\right).$$

Hence, by choosing ε sufficiently small, such that

$$\frac{1}{2} < 1 - 2\sum_{s=2}^{\infty} \frac{(-1)^s}{(2s)!}\left(\frac{2\pi t}{\mathcal{T}}\right)^{2s-2} < 1,$$

we thus have for all $p \in [1, \infty)$ that

$$\|\mathcal{F}_\alpha\|_{\mathrm{L}^p((-\varepsilon,\varepsilon))} \leq c_2(\varepsilon)\Big(|\log(2)| + \Big\|\log\Big(\frac{2\pi^2 t^2}{\mathcal{T}^2}\Big)\Big\|_{\mathrm{L}^p((-\varepsilon,\varepsilon))} + |\log(\tfrac{1}{2})|\Big) < \infty,$$

as $\log(\frac{2\pi^2 t^2}{\mathcal{T}^2}) \in \mathrm{L}^p((-\varepsilon, \varepsilon))$. Consequently, it follows that $\mathcal{F}_\alpha \in \mathrm{L}^p_\perp(\mathbb{T})$ for all $p \in [1, \infty)$ and with Young's inequality

$$\|\mathcal{F}_\alpha *_\mathbb{T} \phi\|_{\mathrm{L}^r_\perp(\mathbb{T})} \leq c_3 \|\mathcal{F}_\alpha\|_{\mathrm{L}^p_\perp(\mathbb{T})} \|\phi\|_{\mathrm{L}^q_\perp(\mathbb{T})}$$

follows, with $p = \frac{rq}{q+r(q-1)} \in [1, \infty)$. Furthermore, we also get

$$\|\mathcal{F}_\alpha *_\mathbb{T} \phi\|_{\mathrm{L}^\infty_\perp(\mathbb{T})} \leq c_4 \|\mathcal{F}_\alpha\|_{\mathrm{L}^{\frac{q}{q-1}}_\perp(\mathbb{T})} \|\phi\|_{\mathrm{L}^q_\perp(\mathbb{T})}.$$

Finally, for $\frac{\alpha}{2} > 1$ we have that $\mathcal{F}_\alpha = g_\alpha \in \mathrm{L}^\infty_\perp(\mathbb{T})$ and therefore $\mathcal{F}_\alpha \in \mathrm{L}^p_\perp(\mathbb{T})$ for $p \in [1, \infty)$. Similarly to the second case, we deduce from Young's inequality for $r \leq \infty$ that

$$\|\mathcal{F}_\alpha *_\mathbb{T} \phi\|_{\mathrm{L}^r_\perp(\mathbb{T})} \leq c_5 \|\mathcal{F}_\alpha\|_{\mathrm{L}^p_\perp(\mathbb{T})} \|\phi\|_{\mathrm{L}^q_\perp(\mathbb{T})}.$$

$\square$

For $\mathcal{F}_\beta$ we have

$$|\mathcal{F}_\beta| \leq \begin{cases} c_6(\beta, n)e^{-\frac{|x|}{2}} & \text{if } |x| \geq 2, \\ c_7(\beta, n)H_\beta(x) & \text{if } |x| < 2, \end{cases} \tag{2.6.11}$$

where H_β is given by

$$H_\beta(x) := \begin{cases} |x|^{\beta-n} + 1 & \text{if } 0 < \beta < n, \\ \log(\frac{2}{|x|}) + 1 & \text{if } \beta = n, \\ 1 & \text{if } \beta > n, \end{cases}$$

see [45, Proposition 6.1.5].

Lemma 2.6.5. *In the situation of Theorem 2.6.3 we have that the mapping $\phi \mapsto \mathcal{F}_\beta *_{\mathbb{R}^n} \phi$ defines a continuous linear operator from $L^q(\mathbb{R}^n)$ into $L^p(\mathbb{R}^n)$ for $p, q, r \in (1, \infty)$ satisfying*

$$1 + \frac{1}{r} = \frac{1}{p} + \frac{1}{q}.$$

Proof. We will proceed similarly to the proof of Lemma 2.6.4 and distinguish the three cases $0 < \beta < n$, $\beta = n$ and $\beta > n$. But before we start, note that for any $s \in [1, \infty)$ we have

$$\|e^{-\frac{|x|}{2}}\|^s_{L^s(\mathbb{R}^n)} = \int_{\mathbb{R}^n} e^{-\frac{|x|s}{2}}\,\mathrm{d}x = \left(\frac{s}{2}\right)^n \int_{\mathbb{R}^n} e^{-|x|}\,\mathrm{d}x$$

$$= \left(\frac{s}{2}\right)^n \omega_n \int_0^\infty r^{n-1}e^{-r}\mathrm{d}r = \left(\frac{s}{2}\right)^n \omega_n \Gamma(n) < \infty,$$

with ω_n denotes the area of the surface of the n-dimensional unit ball. Hence, $e^{-\frac{|x|}{2}} \in L^s(\mathbb{R}^n)$ for all $s \in [1, \infty]$, and therefore we deduce from (2.6.11) that it suffices to study $\mathcal{H}_\beta$ in order to prove the assertion.

If $\beta q < n$ we have that $\mathcal{H}_\beta(x) = |x|^{\beta-n} + 1 \in L^{\frac{n}{n-\beta}, \infty}(B_2)$ and also in $L^s(B_2)$ for all $s < \frac{n}{n-\beta}$. Similarly to the proof of Lemma 2.6.4 we obtain the assertion for all $p \le \frac{nq}{n-\beta q}$ via an application of Young's inequality. In the case $\beta q = n$ we have again that $\beta < n$ and Young's inequality yields

$$\|\mathcal{H}_\beta *_{\mathbb{R}^n} \phi\|_{L^p(\mathbb{R}^n)} \le c_0 \|\mathcal{H}_\beta\|_{L^s(\mathbb{R}^n)}\|\phi\|_{L^q(\mathbb{R}^n)}$$

for any $p < \infty$. For $\beta q > n$ we have to consider the three cases $\beta < n$, $\beta = n$ and $\beta > n$ and deduce analogously to the proof of Lemma 2.6.4 that the mapping $\phi \mapsto \mathcal{F}_\beta *_{\mathbb{R}^n} \phi$ defines a continuous linear operator from $L^q(\mathbb{R}^n)$ into $L^p(\mathbb{R}^n)$ for $p \le \infty$. $\qquad\square$

Finally, we show that $\mathcal{F} \in L^q_\perp(\mathbb{T}; L^q(\mathbb{R}^n))$. For this purpose we consider

$$M_{\mathbb{T}}(k,\xi) := \frac{\left(1 - \delta_{\frac{2\pi}{\mathcal{T}}\mathbb{Z}}(k)\right)i^{M_t+M_x}k^{M_t}|k|^{\frac{\alpha}{2}}\xi^{m_x}(1+|\xi|^2)^{\frac{\beta}{2}}}{1+|k|^m+|\xi|^{2m}} \qquad (2.6.12)$$

and prove that $M_{\mathbb{T}}$ is an $L^q(\mathbb{T} \times \mathbb{R}^n)$ multiplier via an application of the multiplier theorem of Marcinkiewicz and the transference principle. This is summarized in the next lemma.

Lemma 2.6.6. *Let α, β, m, M_t, M_x and q be as in Theorem 2.6.3. Then $M_{\mathbb{T}}$ is an $L^q(\mathbb{T}; L^q(\mathbb{R}^n))$-multiplier.*

Proof. Let $\chi \in C_0^\infty(\mathbb{R};\mathbb{R})$ be a cut-off function with

$$\chi(\eta) = 0 \text{ for } |\eta| \le \frac{\pi}{\mathcal{T}}, \quad \chi(\eta) = 1 \text{ for } |\eta| \ge \frac{2\pi}{\mathcal{T}},$$

and define

$$M_{\mathbb{R}}(\eta,\xi) := \frac{\chi(\eta)i^{M_t+M_x}\eta^{M_t}|\eta|^{\frac{\alpha}{2}}\xi^{m_x}(1+|\xi|^2)^{\frac{\beta}{2}}}{1+|\eta|^m+|\xi|^{2m}}.$$

Lemma A.2.2 yields that $M_{\mathbb{R}}$ is an $L^q(\mathbb{R}\times\mathbb{R}^n)$-multiplier, hence an application of the transference principle (Theorem 2.3.1) completes the proof. $\quad\square$

Proof of Theorem 2.6.3. The theorem extends similar embeddings established in [37, Theorem 4.1] and can be shown by a similar technique. As mentioned above it is sufficient to prove the assertion in the whole space, since the regularity of Ω suffices to ensure the existence of an extension operator $E\colon W^{m,2m,q}_\perp(\mathbb{T} \times \Omega) \to W^{m,2m,q}_\perp(\mathbb{T} \times \mathbb{R}^n)$. In this setting, we can employ the Fourier transform $\mathscr{F}_{\mathbb{T}\times\mathbb{R}^n}$ in time and space to present the partial derivatives of u as

$$\partial_x^{m_x}\partial_t^{M_t}u = \mathcal{F}_\alpha *_{\mathbb{T}} \mathcal{F}_\beta *_{\mathbb{R}^n} \mathcal{F}.$$

Recall (2.6.9) that $\mathcal{F}_\alpha$, $\mathcal{F}_\beta$ and $\mathcal{F}$ are given by

$$\mathcal{F}_\alpha := \mathscr{F}_{\mathbb{T}}^{-1}\left[\left(1 - \delta_{\frac{2\pi}{\mathcal{T}}\mathbb{Z}}(k)\right)|k|^{-\frac{\alpha}{2}}\right],$$

$$\mathcal{F}_\beta := \mathscr{F}_{\mathbb{R}^n}^{-1}\left[(1+|\xi|^2)^{-\frac{\beta}{2}}\right],$$

$$\mathcal{F} := \mathscr{F}_G^{-1}\left[M_{\mathbb{T}}(k,\xi)\mathscr{F}_G[(1+\partial_t^m + \Delta^m)u]\right],$$

whit $M_{\mathbb{T}}$ given as in (2.6.12). Furthermore, observe that $\partial_x^{m_x}\partial_t^{M_t}u = \mathscr{F}_{\mathbb{T}}^{-1}\left[\left(1 - \delta_{\frac{2\pi}{\mathcal{T}}\mathbb{Z}}(k)\right)\mathscr{F}_{\mathbb{T}}[\partial_x^{m_x}\partial_t^{M_t}u]\right]$. Under the condition imposed on α and

β the mappings $\phi \mapsto \mathcal{F}_\alpha *_\mathbb{T} \phi$ and $\phi \mapsto \mathcal{F}_\beta *_{\mathbb{R}^n} \phi$ are continuous linear operators from $L_\perp^q(\mathbb{T})$ into $L_\perp^r(\mathbb{T})$ and $L^q(\mathbb{R}^n)$ into $L^p(\mathbb{R}^n)$ due to Lemma 2.6.4 and Lemma 2.6.5, respectively. Thus, these lemmas yield

$$\|\partial_x^{m_x} \partial_t^{M_t} u\|_{L_\perp^r(\mathbb{T};L^p(\mathbb{R}^n))} = \left(\int_\mathbb{T} \|\mathcal{F}_\beta *_{\mathbb{R}^n} \mathcal{F}_\alpha *_\mathbb{T} \mathcal{F}\|_{L^p(\mathbb{R}^n)}^r \, \mathrm{d}t \right)^{\frac{1}{r}}$$

$$\leq c_0 \left(\int_\mathbb{T} \|\mathcal{F}_\alpha *_\mathbb{T} \mathcal{F}\|_{L^q(\mathbb{R}^n)}^r \, \mathrm{d}t \right)^{\frac{1}{r}}$$

$$\leq c_1 \left(\int_{\mathbb{R}^n} \|\mathcal{F}_\alpha *_\mathbb{T} \mathcal{F}\|_{L_\perp^r(\mathbb{T})}^q \, \mathrm{d}x \right)^{\frac{1}{q}} \leq c_2 \|\mathcal{F}\|_q,$$

where the second inequality above follows from Minkowski's integral inequality. Moreover, we obtain from Lemma 2.6.6 that $M_\mathbb{T}$ is an $L^q(\mathbb{T} \times \mathbb{R}^n)$-multiplier and therefore

$$\|\partial_x^{m_x} \partial_t^{M_t} u\|_{L_\perp^r(\mathbb{T};L^p(\mathbb{R}^n))} \leq c_3 \|\mathcal{F}\|_q \leq c_4 \|u\|_{W_\perp^{m,2m,q}(\mathbb{T} \times \mathbb{R}^n)}$$

holds for all $u \in W_\perp^{m,2m,q}(\mathbb{T} \times \mathbb{R}^n)$. $\qquad\square$

2.6.2 Trace Space Operators

Four types of trace operators are employed in the following:

$$\mathrm{Tr}_0 \colon C^\infty(\mathbb{T} \times \overline{\Omega}) \to C^\infty(\mathbb{T} \times \partial\Omega), \qquad \mathrm{Tr}_0(u) := u_{|\mathbb{T} \times \partial\Omega},$$

$$\mathrm{Tr}_1 \colon C^\infty(\mathbb{T} \times \overline{\Omega}) \to C^\infty(\mathbb{T} \times \partial\Omega), \qquad \mathrm{Tr}_1(u) := \partial_\nu u_{|\mathbb{T} \times \partial\Omega},$$

$$\mathrm{Tr}_D \colon C^\infty(\mathbb{T} \times \overline{\Omega}) \to C^\infty(\mathbb{T} \times \partial\Omega)^2, \qquad \mathrm{Tr}_D(u) := (u, \Delta u)_{|\mathbb{T} \times \partial\Omega},$$

$$\mathrm{Tr}_N \colon C^\infty(\mathbb{T} \times \overline{\Omega}) \to C^\infty(\mathbb{T} \times \partial\Omega)^2, \qquad \mathrm{Tr}_N(u) := (\partial_\nu u, \partial_\nu \Delta u)_{|\mathbb{T} \times \partial\Omega},$$

where ν is the outer normal vector on $\partial\Omega$ and $\partial_\nu u$ denotes the normal derivative $\frac{\partial u}{\partial \nu}$ of u. In order to characterize appropriate trace spaces, we introduce for $q \in [1, \infty)$

$$\mathrm{T}_{D_1}^q(\mathbb{T} \times \partial\Omega) := \mathrm{W}^{2-\frac{1}{2q},q}\big(\mathbb{T}; L^q(\partial\Omega)\big) \cap \mathrm{W}^{1,q}\big(\mathbb{T}; \mathrm{W}^{2-\frac{1}{q},q}(\partial\Omega)\big),$$

$$\mathrm{T}_{D_2}^q(\mathbb{T} \times \partial\Omega) := \mathrm{W}^{3-\frac{1}{2q},q}\big(\mathbb{T}; L^q(\partial\Omega)\big) \cap \mathrm{W}^{1,q}\big(\mathbb{T}; \mathrm{W}^{4-\frac{1}{q},q}(\partial\Omega)\big),$$

$$\mathrm{T}_{N_1}^q(\mathbb{T} \times \partial\Omega) := \mathrm{W}^{\frac{3}{2}-\frac{1}{2q},q}\big(\mathbb{T}; L^q(\partial\Omega)\big) \cap \mathrm{W}^{1,q}\big(\mathbb{T}; \mathrm{W}^{1-\frac{1}{q},q}(\partial\Omega)\big),$$

$$\mathrm{T}_{N_2}^q(\mathbb{T} \times \partial\Omega) := \mathrm{W}^{\frac{5}{2}-\frac{1}{2q},q}\big(\mathbb{T}; L^q(\partial\Omega)\big) \cap \mathrm{W}^{1,q}\big(\mathbb{T}; \mathrm{W}^{3-\frac{1}{q},q}(\partial\Omega)\big).$$

These spaces can be identified as trace spaces in the following sense:

Lemma 2.6.7. *Let $\Omega \subset \mathbb{R}^n$, $n \geq 2$, be a domain with a $C^{1,1}$-smooth boundary. The trace operators Tr_0 and Tr_1 extend to bounded operators:*

$$\mathrm{Tr}_0 : \mathbb{X}^q(\mathbb{T} \times \Omega) \to \mathrm{T}^q_{D_1}(\mathbb{T} \times \partial\Omega), \qquad (2.6.13)$$

$$\mathrm{Tr}_1 : \mathbb{X}^q(\mathbb{T} \times \Omega) \to \mathrm{T}^q_{N_1}(\mathbb{T} \times \partial\Omega). \qquad (2.6.14)$$

Moreover, the operators above possess a continuous right-inverse. If we further assume that Ω has a $C^{3,1}$-smooth boundary, the following trace operators extend to bounded operators which have a continuous right-inverse:

$$\mathrm{Tr}_0 : \mathbb{Y}^q(\mathbb{T} \times \Omega) \to \mathrm{T}^q_{D_2}(\mathbb{T} \times \partial\Omega), \qquad (2.6.15)$$

$$\mathrm{Tr}_0 : \mathrm{W}^{2,4,q}(\mathbb{T} \times \Omega) \to \mathrm{W}^{2-\frac{1}{2q},4-\frac{1}{q},q}(\mathbb{T} \times \partial\Omega), \qquad (2.6.16)$$

$$\mathrm{Tr}_D : \mathbb{Y}^q(\mathbb{T} \times \Omega) \to \mathrm{T}^q_{D_2}(\mathbb{T} \times \partial\Omega) \times \mathrm{T}^q_{D_1}(\mathbb{T} \times \partial\Omega), \qquad (2.6.17)$$

$$\mathrm{Tr}_N : \mathbb{Y}^q(\mathbb{T} \times \Omega) \to \mathrm{T}^q_{N_2}(\mathbb{T} \times \partial\Omega) \times \mathrm{T}^q_{N_1}(\mathbb{T} \times \partial\Omega). \qquad (2.6.18)$$

Proof. It suffices to verify the assertions in the half space case $\Omega := \mathbb{R}^n_+$. The general case of a bounded domain Ω with a Lipschitz boundary then follows via localization. Due to this localization argument, higher boundary regularity is needed for the trace operators (2.6.15) - (2.6.18). Observe that

$$
\begin{aligned}
\mathbb{Y}^q(\mathbb{T} \times \mathbb{R}^n_+) &= \mathrm{L}^q\big(\mathbb{R}_+; \mathbb{Y}^q(\mathbb{T} \times \mathbb{R}^{n-1})\big) \cap \mathrm{W}^{4,q}\big(\mathbb{R}_+; \mathrm{W}^{1,q}\big(\mathbb{T}; \mathrm{L}^q(\mathbb{R}^{n-1})\big)\big) \\
&\quad \cap \mathrm{W}^{3,q}\big(\mathbb{R}_+; \mathrm{W}^{1,q}\big(\mathbb{T}; \mathrm{W}^{1,q}(\mathbb{R}^{n-1})\big)\big) \\
&\quad \cap \mathrm{W}^{2,q}\big(\mathbb{R}_+; \mathrm{W}^{1,q}\big(\mathbb{T}; \mathrm{W}^{2,q}(\mathbb{R}^{n-1})\big)\big) \\
&\quad \cap \mathrm{W}^{1,q}\big(\mathbb{R}_+; \mathrm{W}^{1,q}\big(\mathbb{T}; \mathrm{W}^{3,q}(\mathbb{R}^{n-1})\big)\big) \\
&= \mathrm{L}^q\big(\mathbb{R}_+; \mathbb{Y}^q(\mathbb{T} \times \mathbb{R}^{n-1})\big) \cap \mathrm{W}^{4,q}\big(\mathbb{R}_+; \mathrm{W}^{1,q}\big(\mathbb{T}; \mathrm{L}^q(\mathbb{R}^{n-1})\big)\big),
\end{aligned}
$$

where the last equality is due to the embeddings (2.6.4). From [87, Theorem 1.8.3] we obtain that Tr_0 extends to a continuous operator

$$\mathrm{Tr}_0 : \mathbb{Y}^q(\mathbb{T} \times \mathbb{R}^n_+) \to \big(\mathrm{W}^{1,q}(\mathbb{T}; \mathrm{L}^q(\mathbb{R}^{n-1})), \mathbb{Y}^q(\mathbb{T} \times \mathbb{R}^{n-1})\big)_{1-1/4q,\,q}.$$

One may verify that $\mathbb{Y}^q(\mathbb{T} \times \mathbb{R}^{n-1})$ and $\mathrm{W}^{1,q}\big(\mathbb{T}; \mathrm{L}^q(\mathbb{R}^{n-1})\big)$ form a quasi-linearizable interpolation couple; see [87, Definition 1.8.4]. Indeed, an admissible operator in the sense of [87, Definition 1.8.4] is given by $V_1(\mu) := \mu^{-1}\big(\mu^{-1} - \partial_t^2 + \Delta^2\big)^{-1}$, where invertibility of $\big(\mu^{-1} - \partial_t^2 + \Delta^2\big) : \mathbb{Y}^q(\mathbb{T} \times \mathbb{R}^{n-1}) \to \mathrm{W}^{1,q}\big(\mathbb{T}; \mathrm{L}^q(\mathbb{R}^{n-1})\big)$ can be established by an analysis of the multiplier $(\eta, \xi) \to \big(\mu^{-1} + \eta^2 + |\xi|^4\big)^{-1}$ and an application of de Leeuw's transference principle as in the proof of Lemma 2.6.2. Consequently, one obtains even better properties of the trace operator, namely that it possesses a continuous right inverse; see [87, Theorem 1.8.5]. Moreover, we can utilize the property stated in [87, Theorem 1.12.1] concerning interpolation of intersections of spaces that form quasilinearizable interpolation couples to conclude

$$\left(W^{1,q}(\mathbb{T}; L^q(\mathbb{R}^{n-1})), \mathbb{Y}^q(\mathbb{T} \times \mathbb{R}^{n-1})\right)_{1-1/4q,q} = T^q_{D_2}(\mathbb{T} \times \mathbb{R}^{n-1}),$$

which verifies (2.6.15). Alternatively, one may utilize the theorem of GRIS-VARD [48] on intersection of interpolation spaces, which implies the interpolation identity above. The assertions (2.6.13)–(2.6.18) follow in a similar way. □

Lemma 2.6.8. *Let $\Omega \subset \mathbb{R}^n$, $n \geq 2$, be a bounded domain with a C^4-smooth boundary. The embedding*

$$W^{2-\frac{1}{2q}, 4-\frac{1}{q}, q}(\mathbb{T} \times \partial\Omega) \hookrightarrow T^q_{D_1}(\mathbb{T} \times \partial\Omega) \qquad (2.6.19)$$

is continuous.

Proof. Denote the embedding (2.6.2) by ι. Recalling (2.6.13) and (2.6.16), we find that $\mathrm{Tr}_0 \circ \iota \circ R_0$ yields the embedding (2.6.19), with R_0 being a continuous right-inverse to Tr_0. □

2.7 Mathematical Tools from Fluid Mechanics

Here we are going to introduce some useful tools which are used in the framework of fluid mechanics. Moreover, we will introduce some tools to estimate the L^q-norms by the corresponding gradient norm, namely we introduce Poincaré's inequality on layer-like domains.

2.7.1 Poincaré's Inequality in Layer Domains

This subsection is dedicated to the introduction of Poincaré's inequality on

$$\Omega_\ell = \{x \in \mathbb{R}^n \mid x' \in \mathbb{R}^{n-1}, 0 < x_n < \ell\}, \qquad (2.7.1)$$

if on one part of the boundary a (homogeneous) Dirichlet boundary condition is achieved. Here, $\ell \in (0, \infty)$ and q' denotes the Hölder conjugated corresponding to q.

Lemma 2.7.1 (Poincaré's inequality). *Let $q \in (1, \infty)$, $\ell \in (0, \infty)$ and Ω_ℓ the infinite layer defined in (2.7.1). Then Poincaré's inequality*

$$\|u\|_{L^q(\Omega_\ell)} \le C_3 \|\nabla u\|_{L^q(\Omega_\ell)} \tag{2.7.2}$$

holds for $C_3 = \ell > 0$ and any $u \in C^\infty(\overline{\Omega}_\ell)$, such that either $u(x', 0) = 0$ for all $x' \in \mathbb{R}^{n-1}$ or $u(x', \ell) = 0$ for all $x' \in \mathbb{R}^{n-1}$.

Proof. Without loss of generality we assume that u vanishes at the lower boundary, that is $u(x', 0) = 0$. In order to prove (2.7.2) we first observe that for any $q \in (1, \infty)$ the inequality

$$|u(x', x_n)|^q \le \left(\int_0^{x_n} |\partial_{x_n} u(x', s)| \, \mathrm{d}s \right)^q \le \left(\int_0^\ell 1 \cdot |\partial_{x_n} u(x', s)| \, \mathrm{d}s \right)^q$$

$$\le \left[\left(\int_0^\ell |1|^{q'} \, \mathrm{d}s \right)^{\frac{1}{q'}} \left(\int_0^\ell |\partial_{x_n} u(x', s)|^q \, \mathrm{d}s \right)^{\frac{1}{q}} \right]^q$$

$$= \ell^{\frac{q}{q'}} \int_0^\ell |\partial_{x_n} u(x', s)|^q \, \mathrm{d}s \le \ell^{\frac{q}{q'}} \int_0^\ell |\nabla u(x', s)|^q \, \mathrm{d}s$$

holds. Observe, since $u(x', 0)$ vanishes, we can write u as an integral and deduce the first inequality above. Moreover, this implies

$$\int_{\Omega_\ell} |u|^q \, \mathrm{d}x \le \ell^{\frac{q}{q'}} \int_{\mathbb{R}^{n-1}} \int_0^\ell \int_0^\ell |\nabla u(x', s)|^q \, \mathrm{d}s \, \mathrm{d}x_n \, \mathrm{d}x'$$

$$= \ell^{\frac{q}{q'}+1} \int_{\mathbb{R}^{n-1}} \int_0^\ell |\nabla u(x', s)|^q \, \mathrm{d}s \, \mathrm{d}x' = \ell^{\frac{q}{q'}+1} \int_{\Omega_\ell} |\nabla u(x)|^q \, \mathrm{d}x,$$

that is the Poincaré inequality (2.7.2) with $C_3 = \ell^{\frac{1}{q} + \frac{1}{q'}} = \ell$. $\qquad\qquad\square$

Remark 2.7.2. Observe that in the case of $\ell = 1$ we obtain $C_3 = 1$. Moreover, Lemma 2.7.1 stays valid if the infinite layer is replaced by the periodic layer $\Omega_{\ell,\mathrm{per}}$ of arbitrary (finite) thickness $\ell \in (0, \infty)$. That is, let

$$\Omega_{\ell,\mathrm{per}} := \{x \in \mathbb{R}^n \mid x' \in \mathbb{T}_0^{n-1}, \, 0 < x_n < \ell\},$$

then for any $u \in C^\infty(\overline{\Omega}_{\ell,\mathrm{per}})$, such that u vanishes at one component of the boundary, the estimate

$$\|u\|_{\mathrm{L}^q(\Omega_{\ell,\mathrm{per}})} \leq C_3 \|\nabla u\|_{\mathrm{L}^q(\Omega_{\ell,\mathrm{per}})}$$

holds. Here, $\mathbb{T}_0$ denotes the torus $\mathbb{R}/L\mathbb{Z}$ with fixed $L > 0$.

Remark 2.7.3. Let $m \in \mathbb{N}_0$ and $q \in (1, \infty)$. By density of $C^\infty(\overline{\Omega}_\ell)$ and $C^\infty(\overline{\Omega}_{\ell,\mathrm{per}})$ in $\mathrm{W}^{m,q}(\Omega_\ell)$ and $\mathrm{W}^{m,q}(\Omega_{\ell,\mathrm{per}})$, the assertion of Lemma 2.7.1 stays valid if we consider $u \in \mathrm{W}^{m,q}(\Omega_\ell)$ and $u \in \mathrm{W}^{m,q}(\Omega_{\ell,\mathrm{per}})$, respectively.

2.7.2 Solenoidal Vector Fields and the Divergence Problem

In fluid mechanics usually solenoidal vector fields are considered. However, in the linearization of such problems the Stokes system with fully inhomogeneous data occur (see for example Chapter 4), and in order to utilize classical methods and results for the Stokes system, we have to deal with the inhomogeneous divergence. For this reason, the investigation of

$$\begin{cases} \operatorname{div} u = f & \text{in } \Omega, \\ u = 0 & \text{on } \partial\Omega, \end{cases} \tag{2.7.3}$$

with $f \in \mathrm{L}^2(\Omega)^3$ on $\Omega := \mathbb{T}_0^2 \times (0,1) \subset \mathbb{R}^3$ is subject of this subsection. This divergence problem is later used to "lift" the inhomogeneous divergence. That is, we subtract the solution u to (2.7.3) from the solution to the considered problem from fluid mechanics to obtain a divergence free vector field. Observe that due to the Gaussian theorem, the assumption

$$\int_\Omega f \, \mathrm{d}x = 0 \tag{2.7.4}$$

is necessary to find a solution to (2.7.3). Here, we want to construct a solution to the divergence problem such that the following inequality holds

$$\|u\|_{\mathrm{L}^q(\Omega)} \leq C |f|_{-1,q}^*, \tag{2.7.5}$$

where the right-hand side is given by

$$|f|_{-1,q}^* = \sup_{\phi \in \dot{\mathrm{W}}^{1,q'}(\Omega);\, |\phi|_{1,q'}=1} |(f, \phi)|, \tag{2.7.6}$$

and q' is the Hölder conjugated of q. It is well known that in the case of a bounded domain, this problem is resolved by the Bogovskiĭ operator $\mathcal{B}\colon C_0^\infty(\Omega) \to C_0^\infty(\Omega)^n$, if f satisfies (2.7.4), see for example [36, Section III.3]. Clearly, $\mathcal{B}f$ is a highly non-unique solution to (2.7.3), since one can always add a solenoidal function $v \in W_0^{1,q}(\Omega)$ to $\mathcal{B}f$ and still has a solution to the divergence problem. This operator is called the *Bogovskiĭ operator*, and there exists a continuous extension to a linear operator from $W_0^{m,q}$ to $W_0^{m+1,q}$, for $m \in \mathbb{N}_0$.

Theorem 2.7.4 (Bogovskiĭ Operator). *Let $q \in (1,\infty)$. Then the Bogovskiĭ operator $\mathcal{B}\colon C_0^\infty(\Omega) \to C_0^\infty(\Omega)^n$ has a continuous (linear) extension $\mathcal{B}\colon W_0^{m,q}(\Omega) \to W_0^{m+1,q}(\Omega)^n$, with $m \in \mathbb{N}_0$. Given $f \in W_0^{m,q}(\Omega)$ satisfying (2.7.4), then $u = \mathcal{B}f$ is a solution to (2.7.3) satisfying*

$$\|\nabla u\|_{l,q} \leq C_4 \|f\|_{l,q}, \tag{2.7.7}$$

for all $l = 0, \ldots, m$. Moreover, there exists a constant $C_5 = C_5(n,q,\Omega) > 0$ such that (2.7.5) holds for all $f \in L^q(\Omega)$, with $|\cdot|^*_{-1,q}$ defined as in (2.7.6).

Proof. The first part of the Theorem can be proven as in [36, Theorem III.3.3].To see that (2.7.5) holds, we mimic the steps of the proofs of [36, Theorem III.3.5] and [42, Theorem 2.5]. □

3 Nonlinear Acoustics

In the field of nonlinear acoustics, the propagation of acoustic waves of large amplitudes through a viscous medium is studied. The description of this process is governed by the equations of fluid dynamics which are basically nonlinear equations. More precisely, the motion of a viscous, heat-conducting fluid is governed by mass conservation, momentum conservation, energy conservation, and a thermodynamic equation of state. The compressible Navier-Stokes equations describe the conservation of mass and momentum when viscous effects are taken into account. When heat-conducting effects are assumed, conservation of energy is described by the Kirchhoff-Fourier equations. Using the equation of state of an ideal fluid, and assuming the flow to be irrotational, KUZNETSOV [57] eliminated all but one dependent variable from the Navier-Stokes and Kirchhoff-Fourier equations to obtain the so-called Kuznetsov equation

$$\partial_t^2 u - \Delta u - \frac{b}{c^2}\partial_t\Delta u - \partial_t\left(\frac{1}{\rho_0 c^4}\frac{B}{2A}(\partial_t u)^2 + |\nabla u|^2\right) = f. \qquad (3.0.1)$$

BLACKSTOCK [7] derived from these equations in a similar way the following model for a viscous, heat-conducting fluid:

$$(a\Delta - \partial_t)\left(\partial_t^2 u - c^2\Delta u - b\partial_t\Delta u\right) - \partial_t^2\left(\frac{1}{c^2}\frac{B}{2A}(\partial_t u)^2 + |\nabla u|^2\right) = f. \qquad (3.0.2)$$

In these models, u denotes the potential, also referred to as the acoustic potential, of the Eulerian velocity field, and f a forcing term acting on the fluid. The constant a is the heat conductivity of the fluid, c the speed of sound, and ρ_0 the mass density. The diffusivity of sound b is a measure of energy dissipation due to viscosity and heat conduction in the fluid. Finally, the so-called (acoustic) parameter of nonlinearity B/A, introduced by BEYER [6] (see also [30]), is the quotient of the second and first coefficient in the Taylor expansion of the pressure-density relationship.

Equation (3.0.2) is called the Blackstock-Crighton-Kuznetsov equation, and it is used as a model for acoustic wave propagation in a medium in which both nonlinear and dissipative effects are taken into account. Assuming that $c^2|\nabla u|^2 \approx (\partial_t u)^2$, one obtains the simplified Blackstock-Crighton-Westervelt equation:

$$(a\Delta - \partial_t)\left(\partial_t^2 u - c^2\Delta u - b\partial_t\Delta u\right) - \partial_t^2\left(\frac{1}{c^2}\left(1 + \frac{B}{2A}\right)(\partial_t u)^2\right) = f. \quad (3.0.3)$$

If both nonlinear and dissipative terms are neglected in (3.0.2) (the latter is obtained by setting $b = 0$), the model reduces to the classical wave equation.

In the present chapter of this doctoral thesis we investigate if the dissipative effects occurring in (3.0.1)–(3.0.3) are sufficient to avoid resonance, that is, the occurrence of an unbounded solution when the system is excited by a periodic force. For undamped linear hyperbolic systems such as the wave equation, it is easy to show existence of unbounded solutions when a periodic force of a certain frequency (the eigenfrequency) is applied to it. In physical terms, energy conservation in such systems means that even a minor amount of work by the force over each periodic can accumulate in the system and lead to an unbounded response, *i.e.*, resonance. The systems (3.0.1) – (3.0.3) are damped by the term $\partial_t\Delta u$, which introduces a source of energy dissipation. The question thus arises whether this dissipation dampens the system sufficiently to avoid resonance. By showing existence of a periodic, and hence bounded, solution for periodic forces of arbitrary frequency, we establish that this is indeed the case; at least when the magnitude of the force is sufficiently restricted. We may therefore conclude that the dissipative effects of viscosity and heat conduction in these models brings about a sufficient energy absorption mechanism to avoid resonance.

As our main result of this chapter we show existence of $\mathcal{T}$-time-periodic solutions to (3.0.1), (3.0.2) and (3.0.3) under periodic forcing. We shall treat non-homogeneous boundary values of both Dirichlet and Neumann type. Before we begin the investigation of these systems, we introduce in the first part of this chapter the necessary conservation laws which describe the motion of a viscous, heat-conducting fluid and briefly explain how KUZNETSOV and BLACKSTOCK derived these nonlinear systems. Such nonlinear problems are usually treated via a fixed-point argument, for example the contraction mapping principle, which again is strongly based on the L^q estimates of the solution to the corresponding linearizations

$$\partial_t^2 u - \Delta u - \frac{b}{c^2}\Delta\partial_t u = f, \quad (3.0.4)$$

and

$$(a\Delta - \partial_t)\left(\partial_t^2 v - c^2 \Delta v - b\partial_t \Delta v\right) = f, \qquad (3.0.5)$$

respectively. Therefore, a careful investigation of these linear problems is of high significance. Note that (3.0.5) is in particular the damped wave equation coupled with the heat equation. Hence, the investigation hereof is based on the results we obtain for (3.0.4). For this reason a careful investigation of (3.0.4) is necessary. The second part of this chapter is dedicated to the investigation of the damped wave equation (3.0.4) in the whole space $\mathbb{R}^n$, the half space $\mathbb{R}_+^n$ and on a bounded domain $\Omega \subset \mathbb{R}^n$ of dimension greater or equal to 2. We shall prove existence of a non-resonant solution to (3.0.4) and establish the necessary L^q estimates.

The Kuznetsov equation (3.0.1) will be studied in Section 3.3. Here we utilize the contraction mapping principle and the embedding properties of Sobolev spaces of time-periodic functions collected in Section 2.6 to examine the nonlinear problem.

Finally, the Blackstock-Crighton-Kuznetsov equation (3.0.2) and the Blackstock-Crighton-Westervelt (3.0.3) are treated. Existence of solutions to these systems is obtained in a similar way as for the Kuznetsov equation. Since the linearized Blackstock-Crighton is basically a damped wave equation, most of the work is done in Section 3.2 and we will study the linear and nonlinear case together in Section 3.4.

The results presented in Section 3.2–3.4 were already published in [13, 14].

3.1 Models

The description of the propagation of an acoustic wave through a viscous medium is based on the theory of the motion of the fluid, which we assume to be diffusive. For this reason, viscosity and heat conduction effects appear in the description of the medium. In order to identify the state of the fluid at a point x at time t, six variables are needed, *i.e.*, we have to find six equations that rule the propagation of an acoustic wave. More precisely the six variables are the pressure p, density ρ, temperature T and the three components of the velocity v. The equations used to determine this state are the conservation laws of mass, momentum and energy, and a thermodynamic equation of state.

The law of conservation of mass states that mass in an isolated system $V(t) \subset \mathbb{R}^n$ remains constant over time. Since the mass of the fluid domain

$V(t)$ is given by

$$m(V(t)) = \int_{V(t)} \rho(y)\,\mathrm{d}y,$$

the conservation law of mass yields that

$$\frac{\mathrm{d}}{\mathrm{d}t} m(V(t)) = \frac{\mathrm{d}}{\mathrm{d}t} \int_{V(t)} \rho(y)\,\mathrm{d}y = 0.$$

Let $V_0 \subset \mathbb{R}^n$ be an arbitrary (time independent) fluid domain, such that at time $t > 0$, $V(t)$ results from V_0. Then, utilizing integration by a change of variable and transferring the integral above to an integral defined on V_0, we deduce that

$$0 = \frac{\mathrm{d}}{\mathrm{d}t} \int_{V(t)} \rho(y)\,\mathrm{d}y = \int_{V_0} \partial_t \rho + \mathrm{div}(\rho v)\,\mathrm{d}x,$$

and thus

$$\frac{\mathrm{D}\rho}{\mathrm{D}t} + \rho \nabla \cdot v = 0, \tag{3.1.1}$$

where $\frac{\mathrm{D}}{\mathrm{D}t}$ denotes the material derivative and is defined as

$$\frac{\mathrm{D}}{\mathrm{D}t} := \frac{\partial}{\partial t} + v_i \frac{\partial}{\partial x_i} = \frac{\partial}{\partial t} + v \cdot \nabla.$$

In a mechanical system the total momentum, which is independent of external forces, is time-independent, that is, for any time t the total momentum is constant. This conservation of momentum is governed by the compressible Navier-Stokes equations

$$\rho \left[\frac{\partial v}{\partial t} + (v \cdot \nabla)v \right] = -\nabla p + \eta \Delta v + \left(\zeta + \frac{\eta}{3} \right) \nabla (\nabla \cdot v), \tag{3.1.2}$$

whenever viscous effects are taken into account, see [73, Subsection 2.1.2]. Here ζ and η are the bulk and shear viscosity, respectively.

The fifth and last conservation law we need is the energy conservation which states that the reduction per time of energy contents in a fixed volume is equal to the energy inserted into the volume. Similar to the first laws we find the Kirchhoff-Fourier equation

$$\rho\,\mathrm{T} \left[\frac{\partial s}{\partial t} + (v \cdot \nabla)s \right] = \kappa \Delta\,\mathrm{T} + \zeta(\nabla \cdot v)^2 + \frac{1}{2}(\partial_i v_j + \partial_j v_i - \frac{2}{3}\nabla \cdot v \delta_{ij})^2, \tag{3.1.3}$$

which describes the conservation of energy when heat-conducting effects appear, see [73, Subsection 2.1.3]. Here s is the entropy per mass, κ the heat conduction number and δ_{ij} the Kronecker tensor.

Finally, we further assume the fluid to satisfy the equation of an ideal fluid, which is given by

$$\frac{p}{\rho} = (c_p - c_v)\, \mathrm{T}, \tag{3.1.4}$$

where c_p and c_v are the capacities per unit mass of the fluid of constant pressure and volume, respectively. In order to simplify the analysis we assume that the flow is irrotational, that is

$$\nabla \times v = 0 \qquad \Rightarrow \qquad v = -\nabla u,$$

where u is the velocity potential. With this equations at hand we are now able to derive two model equations in nonlinear acoustics, namely the Kuznetsov equation and the Blackstock-Crighton equation. Therefore, we eliminate all but one dependent variable (namely the velocity potential u) from the system (3.1.1) $-$ (3.1.4).

First we eliminate the velocity. For this purpose we replace v in the Navier-Stokes equations (3.1.2) with the velocity potential u, then utilizing the identities

$$\operatorname{curl} \operatorname{curl} = \operatorname{grad} \operatorname{div} - \Delta,$$

and

$$\frac{1}{2}\nabla v^2 = (v \cdot \nabla)v + v \times (\nabla \times v),$$

which yields

$$\nabla\left[-\frac{\partial u}{\partial t} + \frac{1}{2}(\nabla u)^2 \right] = -\frac{1}{\rho}\nabla p - \frac{1}{\rho}\left(\frac{4}{3}\eta + \zeta \right)\nabla(\Delta u).$$

Next we have to eliminate the remaining terms by employing the other conservation laws. For this reason we introduce some notation which was first used by LIGHTHILL and later by BLACKSTOCK in [7] to derive the Blackstock-Crighton equation. In what follows, all nondissipative linear terms are regarded as first-order terms. All nonlinear terms that are quadratic and do not involve viscosity or heat conduction are referred to as second-order terms. Moreover, the linear viscosity and heat conduction terms are also terms of second-order. The remaining terms are referred to as higher-order terms. Only retaining the second order of the small quantities in all equations, (3.1.3) simplifies to

$$\rho_0 \, T_0 \, \frac{\partial s}{\partial t} = \kappa \Delta \, T \, .$$

Here ρ_0 and T_0 are the equilibrium values of ρ and T, respectively. After a short calculation we deduce the Kuznetsov equation (3.0.1). For more details on the derivation of (3.0.1) and on the constants occurring herein we refer to [57] or Chapter 2 in [73].

To derive the Blackstock-Crighton equation we follow the approach of BLACKSTOCK in [7], which was first introduced by LIGHTHILL, see [67]. Here, we utilize the same notation as above. In this setting it is convenient to work with scalar and vector potentials, *i.e.*, with u and w from the Helmholtz decomposition $v = \nabla \times u + \nabla w$ of the velocity v, see [7]. Retaining only the first-order and second-order terms and neglecting the higher-order terms, BLACKSTOCK eliminated all but one dependent variable from the equations (3.1.1) – (3.1.4) to obtain (3.0.2). To be more precisely, BLACKSTOCK derived (3.0.2) without the damping term $\Delta^2 \partial_t u$. In contrast to [7] we follow the idea in [9] and retain the damping mechanism. For more details on how to derive (3.0.2) we refer to [7, 9, 18].

The Blackstock-Crighton-Westervelt equation (3.0.3) is obtained from the Blackstock-Crighton-Kuznetsov model (3.0.2) by neglecting local nonlinear effects in the sense that $c^2 |\nabla u|^2 \approx (\partial_t u)^2$ herein.

3.2 The Damped Wave Equation

This section is dedicated to the study of the linear time-periodic damped wave equation with inhomogeneous Dirichlet and Neumann boundary conditions. We are interested in the existence of $\mathcal{T}$-time-periodic solutions to

$$\begin{cases} \partial_t^2 u - \Delta u - \dfrac{b}{c^2} \partial_t \Delta u = f & \text{in } \mathbb{T} \times \Omega, \\ \\ \qquad\qquad\qquad u = g & \text{on } \mathbb{T} \times \partial\Omega, \end{cases} \qquad (3.2.1)$$

and

$$\begin{cases} \partial_t^2 u - \Delta u - \dfrac{b}{c^2} \partial_t \Delta u = f & \text{in } \mathbb{T} \times \Omega, \\ \\ \qquad\qquad\qquad \dfrac{\partial u}{\partial \nu} = g & \text{on } \mathbb{T} \times \partial\Omega, \end{cases} \qquad (3.2.2)$$

that are sufficiently regular. Due to proving the absence of resonance in this system, existence of such periodic solutions would be a first clue. Without loss of generality, let $b = c = 1$. Throughout this section we will always denote by ν the outer unit normal vector at $\partial\Omega$. We shall consider both data and solutions that are time-periodic with the same period $\mathcal{T} > 0$ and study (3.2.1) and (3.2.2) in the whole space, the half space and on a bounded domain. Instead of relying on a Poincaré map, which is the standard procedure in the investigation of time-periodic problems like (3.0.4), and also the approach used in [56], we obtain the estimates for the solution directly via a representation formula for the solution. We hereby circumvent completely the theory for the corresponding initial-value problem, which is needed to construct a Poincaré map, and develop a much more direct approach. Additionally, the representation formula we establish also seems interesting in the context of resonance, or rather the avoidance hereof, since it exposes the way different modes of the solution are damped in relation to the modes of the forcing term. We shall briefly outline the method in the whole-space case $\Omega = \mathbb{R}^n$, $n \geq 2$. The main advantage in the framework where the time axis is the torus $\mathbb{T}$, is that it is possible to use the Fourier transform $\mathscr{F}_G$ in combination with the space of tempered distributions $\mathscr{S}'(G)$ and derive from (3.0.4), that is,

$$\partial_t^2 u - \Delta u - \frac{b}{c^2}\partial_t\Delta u = f \qquad \text{in } \mathbb{T} \times \Omega, \tag{3.2.3}$$

the representation formula

$$u = \mathscr{F}_G^{-1}\left[\frac{1}{|\xi|^2 - k^2 + ik|\xi|^2}\mathscr{F}_G[f]\right], \tag{3.2.4}$$

when $f \in \mathscr{S}(G)$. The term $ik|\xi|^2$ in the denominator of the Fourier multiplier in (3.2.4) stems from the damping. For modes $k \neq 0$, the multiplier is bounded due to the damping term, whereas the mode $k = 0$ of the multiplier is not "damped" at all. To obtain the desired estimates of u, we shall therefore split the "damped" and "non-damped" modes, in this case

$$u = \mathscr{F}_G^{-1}\left[\frac{1}{|\xi|^2}\mathscr{F}_G[f]\right] + \mathscr{F}_G^{-1}\left[\frac{(1 - \delta_{\frac{2\pi}{\mathcal{T}}\mathbb{Z}}(k))}{|\xi|^2 - k^2 + ik|\xi|^2}\mathscr{F}_G[f]\right] = u_{\mathrm{s}} + u_{\mathrm{tp}}, \tag{3.2.5}$$

where $\delta_{\frac{2\pi}{\mathcal{T}}\mathbb{Z}}$ denotes the Dirac measure on $\frac{2\pi}{\mathcal{T}}\mathbb{Z}$, see (2.3.1). The main advantage of the decomposition is that the bounded multiplier in the representation of u_{tp} leads to a better L^q estimate than can be obtained for the full solution u. The necessity of the decomposition becomes obvious when addressing the question of regularity of a solution. We shall establish the estimate by invoking a transference principle (Theorem 2.3.1) for group multipliers which allows us to transfer the investigation of the multiplier into a Euclidean setting. This principle was originally established by DE LEEUW [22] and later generalized by EDWARDS and GAUDRY [27]. Observe, the representation formula for the solution to the damped wave equation is only available in the whole space. However, the half space problem is investigated via a reflection principle and a lifting argument. In combination, these two techniques help to deduce the existence and L^q estimates for time-periodic solutions to (3.2.3) with inhomogeneous purely oscillatory data f_{tp} and g_{tp}. Since the operator corresponding to the damped wave equation on a bounded domain is an isomorphism, the existence and regularity assertions follow by a localization argument. The estimate of u_{s} can be obtained by standard methods. For this reason, the focus in the current section lies on the study of the purely oscillatory problem and we briefly outline the proofs for the steady state case.

3.2.1 The Steady State Damped Wave Equation

This subsection is dedicated to the investigation of the stationary damped wave equation, that is,

$$\begin{cases} -\Delta u = f & \text{in } \Omega, \\ \quad\; u = g & \text{on } \partial\Omega, \end{cases} \tag{3.2.6}$$

for boundary values of Dirichlet type, and the corresponding Neumann boundary value problem

$$\begin{cases} -\Delta u = f & \text{in } \Omega, \\ \dfrac{\partial u}{\partial \nu} = g & \text{on } \partial\Omega. \end{cases} \tag{3.2.7}$$

Note that (3.2.6) and (3.2.7) are the Laplace equation or rather Poisson's equation with inhomogeneous Dirichlet and Neumann boundary values, respectively. Hence, a large collection of theory concerning existence of a solution is available and these systems can be investigated with standard methods. In the whole space and the half space we obtain the following existence result in case of Dirichlet boundary conditions.

Lemma 3.2.1 (Dirichlet Laplace in the Whole and Half Space). *Let $q \in (1, \infty)$ and $n \geq 2$. Let further $\Omega = \mathbb{R}^n$ or $\Omega = \mathbb{R}_+^n$. Then for any $f \in L^q(\Omega)$ and $g \in W^{2-\frac{1}{q},q}(\partial\Omega)$ there is a solution $u \in \dot{W}^{2,q}(\Omega)$ to (3.2.6) such that*

$$\|\nabla^2 u\|_{L^q(\Omega)} \leq C_6 \big(\|f\|_{L^q(\Omega)} + \|g\|_{W^{2-\frac{1}{q},q}(\partial\Omega)} \big), \qquad (3.2.8)$$

holds, with $C_6 = C_6(n, q, \Omega) > 0$.

Proof. Let us first consider (3.2.6) in the whole space with $f = \delta$, where $\delta \in \mathscr{S}'(\Omega)$ is the Dirac delta distribution. For this problem a fundamental solution is available, which is given by

$$\Gamma(x) := \begin{cases} -\frac{1}{2\pi} \ln |x|, & n = 2, \\ \frac{1}{\omega_n} \frac{1}{|x|^{n-2}}, & n > 2, \end{cases}$$

where ω_n denotes the area of the surface of the n-dimensional unit ball. Therefore, we obtain the existence of a solution to (3.2.6) by convolution, *i.e.*, $u := \Gamma * f$. Moreover, for the second order (mixed) derivatives of u it holds that

$$\|\partial_{x_i} \partial_{x_j} u\|_{L^q(\mathbb{R}^n)} = \|\mathscr{F}_{\mathbb{R}^n}^{-1} [\xi_i \xi_j \mathscr{F}_{\mathbb{R}^n}[\Gamma] \mathscr{F}_{\mathbb{R}^n}[f]]\|_{L^q(\mathbb{R}^n)}$$
$$= \left\| \mathscr{F}_{\mathbb{R}^n}^{-1} \left[\frac{\xi_i}{|\xi|} \frac{\xi_j}{|\xi|} \mathscr{F}_{\mathbb{R}^n}[f] \right] \right\|_{L^q(\mathbb{R}^n)},$$

with $\frac{\xi_i}{|\xi|}$ denoting the Riesz transform (see [44, Definition 4.1.13]). Since the Riesz transform is bounded in $L^q(\mathbb{R}^n)$ for any $q \in (1, \infty)$, it follows that $\frac{\xi_i}{|\xi|}$ is an $L^q(\mathbb{R}^n)$-multiplier, and thus the *a priori* L^q estimate (3.2.8) follows.

Existence of a solution to Poisson's equation with homogeneous boundary values in the half space is obtained via an odd reflection. For more details we refer to the proof of Lemma 3.2.6, where this principle is demonstrated for the purely oscillatory half space problem corresponding to the damped wave equation (3.2.1). Since the trace operators corresponding to the Dirichlet boundary value problem is continuous and has a continuous right inverse (see [87, Theorem 1.8.3]), existence of a solution to (3.2.6) satisfying (3.2.8) follows with a lifting argument, that is, we utilize the fact, that for any $g \in W^{2-\frac{1}{q},q}(\Omega)$ there is a $v \in W^{2,q}(\Omega)$ such that $v_{|\partial\Omega} = g$. By subtracting v from the solution u, it suffices to study (3.2.6) with $g = 0$. $\qquad \square$

In the case of a bounded domain we do not only get existence of a solution, we further obtain that the solution to the Dirichlet boundary value problem (3.2.6) is unique and satisfies better L^q estimates.

Lemma 3.2.2 (Dirichlet Laplace in a Bounded Domain). *Let q and n be as in Lemma 3.2.1, and let $\Omega \subset \mathbb{R}^n$ be a bounded domain with a $C^{1,1}$- smooth boundary. Then for any $f \in L^q(\Omega)$ and $g \in W^{2-\frac{1}{q},q}(\partial\Omega)$ there is a unique solution $u \in W^{2,q}(\Omega)$ to (3.2.6) such that*

$$\|u\|_{W^{2,q}(\Omega)} \leq C_7\big(\|f\|_{L^q(\Omega)} + \|g\|_{W^{2-\frac{1}{q},q}(\partial\Omega)}\big),$$

holds, with $C_7 = C_7(n,q,\Omega) > 0$.

Proof. Employing the theorem of Lax-Milgram existence of a weak solution is established, which by regularity theory for elliptic equations (see for example [89, Chapter 3, §20]) is as regular as the data allows for, *i.e.*, we have $u \in W^{2,q}(\Omega)$. Moreover, a priori L^q estimates for the solution u follow from the theorem of Agmon, Douglas and Nirenberg [3]. $\square$

Remark 3.2.3. For more details on this approach we refer to Subsection 4.5 and Subsection 4.6 in this doctoral thesis, where a hyperbolic-parabolic coupled system is investigated by Galerkin approximation and a localization argument to deduce existence of a solution and the corresponding L^q estimates, respectively. Note, we could have employed Galerkin approximation instead of Lax-Milgram to establish a solution to (3.2.6).

Observe, in the case of Neumann type boundary conditions a further compatibility condition is necessary. If u is a solution to the damped wave equation (3.2.7) with right-hand side (f, g) on a bounded domain, Gauß's theorem yields

$$\int_\Omega f \, \mathrm{d}x = -\int_\Omega \Delta u \, \mathrm{d}x = -\int_{\partial\Omega} \frac{\partial u}{\partial \nu} \, \mathrm{d}S = -\int_{\partial\Omega} g \, \mathrm{d}S,$$

and thus

$$\int_\Omega f \, \mathrm{d}x + \int_{\partial\Omega} g \, \mathrm{d}S = 0. \qquad (3.2.9)$$

For Neumann boundary values the existence and regularity results in the half space or in a bounded domain are collected in the following lemma.

Lemma 3.2.4 (Neumann Boundary Value Problem). *Let $q \in (1, \infty)$, $n \geq 2$ and Ω be either the half space $\mathbb{R}^n_+$ or a bounded domain $\Omega \subset \mathbb{R}^n$ with a $C^{1,1}$-smooth boundary. Furthermore, let $f \in \mathrm{L}^q(\Omega)$ and $g \in \mathrm{W}^{1-\frac{1}{q},q}(\partial\Omega)$. If Ω is a bounded domain, assume further that f and g satisfy (3.2.9). Then there exists a solution $u \in \dot{\mathrm{W}}^{2,q}(\Omega)$ to (3.2.7) and a constant $C_8 = C_9(n, q, \Omega) > 0$ such that*

$$\|\nabla^2 u\|_{\mathrm{L}^q(\Omega)} \leq C_9 \big(\|f\|_{\mathrm{L}^q(\Omega)} + \|g\|_{\mathrm{W}^{1-\frac{1}{q},q}(\partial\Omega)} \big), \qquad (3.2.10)$$

holds.

Proof. Let us first consider the steady state Neumann problem (3.2.7) in the half space $\Omega = \mathbb{R}^n_+$. Existence of a solution follows similarly as in the proof of Lemma 3.2.1 by utilizing a reflection principle. Instead of the odd reflection we used for the Dirichlet boundary value problem, we employ an even reflection (see for example the proof of Lemma 3.2.7 below). The assertion for inhomogeneous boundary data follow via the same lifting argument as for Dirichlet type conditions. On a bounded domain standard theory on elliptic equations as in the proof of Lemma 3.2.2 implies the existence of a solution $u \in \mathrm{W}^{2,q}(\Omega)$ to (3.2.7) satisfying the L^q estimate (3.2.10). More precisely, we further obtain that u is unique and is subject to

$$\|u\|_{\mathrm{W}^{2,q}(\Omega)} \leq C_9 \big(\|f\|_{\mathrm{L}^q(\Omega)} + \|g\|_{\mathrm{W}^{1-\frac{1}{q},q}(\partial\Omega)} \big).$$

$\square$

Since the stationary case can be treated with standard theory, the challenge will be the investigation of the purely oscillatory part, which is outlined in the following three subsections.

3.2.2 The Purely Oscillatory Damped Wave Equation in the Whole Space

In what follows, we are going to consider purely oscillatory data f, *i.e.*, data with a mean-zero condition in time, and study the damped wave equation (3.2.3) with purely oscillatory data. Hereinafter, we always consider the space dimension $n \geq 2$ and we study

$$\partial_t^2 u - \Delta u - \partial_t \Delta u = f \quad \text{in } G. \qquad (3.2.11)$$

Before starting our investigation we want to recall Notation 2.6.1 which is used throughout this chapter. As we will see, the strong solutions to the damped wave equation (3.2.11) are elements in the anisotropic Sobolev spaces of type

$$\mathbb{X}_\perp^q(\mathbb{T} \times \Omega) := \mathcal{P}_\perp W^{2,q}(\mathbb{T}; L^q(\Omega)) \cap \mathcal{P}_\perp W^{1,q}(\mathbb{T}; W^{2,q}(\Omega))$$
$$:= W_\perp^{2,q}(\mathbb{T}; L^q(\Omega)) \cap W_\perp^{1,q}(\mathbb{T}; W^{2,q}(\Omega))$$

The function space $\mathbb{X}_\perp^q$ is equipped with the canonical norm

$$\|f\|_{\mathbb{X}_\perp^q(\mathbb{T}\times\Omega)} := \left(\|f\|_{W_\perp^{2,q}(\mathbb{T};L^q(\Omega))}^q + \|f\|_{W_\perp^{1,q}(\mathbb{T};W^{2,q}(\Omega))}^q \right)^{\frac{1}{q}}.$$

Recall, the occurring trace spaces related to the Dirichlet and Neumann trace corresponding to $\mathbb{X}_\perp^q$ are given by

$$T_{\perp,D_1}^q(\mathbb{T} \times \partial\Omega) = W_\perp^{2-\frac{1}{2q},q}(\mathbb{T}; L^q(\partial\Omega)) \cap W_\perp^{1,q}(\mathbb{T}; W^{2-\frac{1}{q},q}(\partial\Omega)),$$
$$T_{\perp,N_1}^q(\mathbb{T} \times \partial\Omega) = W_\perp^{\frac{3}{2}-\frac{1}{2q},q}(\mathbb{T}; L^q(\partial\Omega)) \cap W_\perp^{1,q}(\mathbb{T}; W^{1-\frac{1}{q},q}(\partial\Omega)),$$

equipped with trace space norms. These spaces can be identified as trace spaces in the sense of Lemma 2.6.7.

Equation (3.2.11) can be investigated in the framework introduced in Chapter 2. Utilizing these tools, we obtain the following lemma.

Lemma 3.2.5 (Purely-Periodic Whole Space Problem). *Let $q \in (1,\infty)$. For any $f \in L_\perp^q(G)$ there exists a solution $u \in \mathbb{X}_\perp^q(G)$ to the damped wave equation (3.2.11) satisfying*

$$\|u\|_{\mathbb{X}_\perp^q(G)} \le C_{10}\|f\|_{L_\perp^q(G)}, \tag{3.2.12}$$

where $C_{10} = C_{10}(n,q,\mathcal{T}) > 0$. The solution is unique in $\mathscr{S}_\perp'(G)$.

Proof. Formally applying the Fourier transform $\mathscr{F}_G$ in (3.2.11), we obtain from (2.3.2), *i.e.*, from $\mathscr{F}_G[f] = \mathscr{F}_G[\mathcal{P}_\perp f] = \left(1 - \delta_{\frac{2\pi}{\mathcal{T}}\mathbb{Z}}(k)\right)\mathscr{F}_G[f]$, that

$$u = \mathscr{F}_G^{-1}\big[M(k,\xi)\mathscr{F}_G[f]\big] := \mathscr{F}_G^{-1}\left[\frac{\left(1 - \delta_{\frac{2\pi}{\mathcal{T}}\mathbb{Z}}(k)\right)}{|\xi|^2 - k^2 + ik|\xi|^2}\mathscr{F}_G[f]\right],$$

where $M \in L^\infty(\widehat{G})$ is bounded. Hence, u is well-defined as an element in $\mathscr{S}_\perp'(G)$. Note, the multiplier theory we are going to utilize in order to establish the L^q estimates is only available in the whole space $\mathbb{R} \times \mathbb{R}^n$, however the transference principle (Theorem 2.3.1) allows us to make use of this theory, if a continuous homomorphism $\Phi \colon \widehat{G} \to \mathbb{R} \times \mathbb{R}^n$ exists, such that $M = m \circ \Phi$ where m is an $L^q(\mathbb{R} \times \mathbb{R}^n)$-multiplier. For this purpose, let $\chi \in C_0^\infty(\mathbb{R}; \mathbb{R})$ be a "cut-off" function with

$$\chi(\eta) = \begin{cases} 1, & \text{for } |\eta| \leq \frac{\pi}{\mathcal{T}}, \\ 0, & \text{for } |\eta| \geq \frac{2\pi}{\mathcal{T}}, \end{cases}$$

and define

$$m : \mathbb{R} \times \mathbb{R}^n \to \mathbb{C}, \quad m(\eta, \xi) := \frac{1 - \chi(\eta)}{|\xi|^2 - \eta^2 + i\eta|\xi|^2}.$$

Clearly, the mapping Φ given by

$$\Phi \colon \widehat{G} \to \mathbb{R} \times \mathbb{R}^n, \quad \Phi(k, \xi) := (k, \xi),$$

is a continuous homomorphism and $M = m \circ \Phi$. Consequently, it suffices to show that m is a continuous $L^q(\mathbb{R} \times \mathbb{R}^n)$-multiplier to obtain from Theorem 2.3.1 that M is a $L^q(G)$-multiplier. This will be done by employing the multiplier theorem of Marcinkiewicz, see Theorem 2.3.2. Therefore, observe that for all $(\eta, \xi) \in \mathbb{R} \times \mathbb{R}^n$ with $|\eta| > \frac{\pi}{\mathcal{T}}$ the following inequalities hold

$$\begin{aligned} I_1 &:= \frac{1}{(|\xi|^2 - \eta^2)^2 + \eta^2|\xi|^4} \leq 4 \max\left(\frac{\mathcal{T}^4}{\pi^4}, \frac{\mathcal{T}^6}{\pi^6}\right) < \infty, \\ I_2 &:= \frac{\eta^2}{(|\xi|^2 - \eta^2)^2 + \eta^2|\xi|^4} < 4 \max\left(\frac{\mathcal{T}^2}{\pi^2}, \frac{\mathcal{T}^4}{\pi^4}\right) < \infty, \\ I_3 &:= \frac{\eta^2|\xi|^4}{(|\xi|^2 - \eta^2)^2 + \eta^2|\xi|^4} \leq 1 < \infty, \\ I_4 &:= \frac{|\xi|^4}{(|\xi|^2 - \eta^2)^2 + \eta^2|\xi|^4} < \frac{\mathcal{T}^2}{\pi^2} < \infty, \end{aligned} \tag{3.2.13}$$

see Section A.1.1 of the appendix. For the first order derivatives of m we have

$$\begin{aligned} \partial_\eta m(\eta, \xi) &= \frac{-\chi'(\eta)}{|\xi|^2 - \eta^2 + i\eta|\xi|^2} - \frac{(1 - \chi(\eta))(i|\xi|^2 - 2\eta)}{(|\xi|^2 - \eta^2 + i\eta|\xi|^2)^2}, \\ \partial_{\xi_j} m(\eta, \xi) &= -2\frac{(1 - \chi(\eta))(\xi_j + i\eta\xi_j)}{(|\xi|^2 - \eta^2 + i\eta|\xi|^2)^2}, \end{aligned} \tag{3.2.14}$$

and by iteration we see that for pairwise disjoint $j_1, \ldots, j_r \in \{1, \ldots, n\}$

$$\partial_{\xi_{j_1}} \cdots \partial_{\xi_{j_r}} m(\eta, \xi) = (-1)^r \cdot 2^r \cdot r! \cdot (1 - \chi(\eta)) \frac{\prod_{l=1}^r (\xi_{j_l} + i\eta\xi_{j_l})}{(|\xi|^2 - \eta^2 + i\eta|\xi|^2)^{r+1}},$$

(3.2.15)

holds, as the numerator of $\partial_{\xi_{j_1}} \cdots \partial_{\xi_{j_r}} m$ does not depend on ξ_i for $i \neq j_s$ with $s \in \{1, \ldots, r\}$, see $(3.2.14)_2$ above. With this representation formula at hand we now take derivative of (3.2.15) with respect to η and deduce for the mixed derivatives of m that

$$\partial_\eta \partial_{\xi_{j_1}} \cdots \partial_{\xi_{j_r}} m(\eta, \xi) = (-1)^r \cdot 2^r \cdot r! \cdot (S_1 + S_2 + S_3)$$

(3.2.16)

with

$$S_1 := -\frac{\chi'(\eta) \prod_{l=1}^r (\xi_{j_l} + i\eta\xi_{j_l})}{(|\xi|^2 - \eta^2 + i\eta|\xi|^2)^{r+1}},$$

$$S_2 := \sum_{s=1}^r \frac{(1 - \chi(\eta))i\xi_{j_s} \prod_{l=1, l \neq s}^r (\xi_{j_l} + i\eta\xi_{j_l})}{(|\xi|^2 - \eta^2 + i\eta|\xi|^2)^{r+1}},$$

and

$$S_3 := -(r+1)\frac{(1 - \chi(\eta))(i|\xi|^2 - 2\eta) \prod_{l=1}^r (\xi_{j_l} + i\eta\xi_{j_l})}{(|\xi|^2 - \eta^2 + i\eta|\xi|^2)^{r+2}}.$$

To verify the condition (2.3.3) of Marcinkiewicz multiplier theorem, we let $\gamma \in \{0, 1\}$, $\varepsilon \in \{0, 1\}^n$ and $r = |\varepsilon|$. Employing the estimate derived in (3.2.13) to the representation (3.2.16) we find

$$|\eta^\gamma \xi^\varepsilon \partial_\eta^\gamma \partial_\xi^\varepsilon m(k, \xi)|^2 \leq 2^{2r}((r+1)!)^2 \left(\frac{\pi^2}{\mathcal{T}^2} |\chi'(\eta)|^2 I_1 (I_3 + I_4)^2 \right.$$

$$\left. + |1 - \chi(\eta)|^2 (I_3 + I_4)^{r-1} \big(I_1 I_3 + (I_1 I_3 + 4I_2^2)(I_3 + I_4) \big) \right) < \infty.$$

(3.2.17)

But this is already (2.3.3). Hence, Theorem 2.3.2 implies that m is an $L^q(\mathbb{R} \times \mathbb{R}^n)$-multiplier that satisfies

$$\|u\|_{L^q(\mathbb{R} \times \mathbb{R}^n)} \leq c_0 \|f\|_{L^q(\mathbb{R} \times \mathbb{R}^n)}.$$

Utilizing the transference principle (Theorem 2.3.1), we also obtain that M is an $L^q(G)$-multiplier that obeys the estimate

$$\|u\|_{L^q(G)} \leq c_1 \|f\|_{L^q(G)}.$$

In order to prove that the solution is as regular as claimed, we have to differentiate u with respect to time and space and repeat the argumentation above, *i.e.*, we have to consider the functions

$$\partial_t^\alpha u = \mathscr{F}_G^{-1}\big[(ik)^\alpha M(k,\xi)\mathscr{F}_G[f]\big],$$
$$\partial_x^\beta u = \mathscr{F}_G^{-1}\big[i^{|\beta|}\xi^\beta M(k,\xi)\mathscr{F}_G[f]\big],$$
$$\partial_t\partial_x^\beta u = \mathscr{F}_G^{-1}\big[i^{|\beta|+1}k\xi^\beta M(k,\xi)\mathscr{F}_G[f]\big],$$

and consider $(ik)^\alpha M$, $i^{|\beta|}\xi^\beta M$ and $i^{|\beta|+1}k\xi^\beta M$ instead of M, and $(i\eta)^\alpha m$, $i^{|\beta|}\xi^\beta m$ and $i^{|\beta|+1}\eta\xi^\beta m$ instead of m. Here, $\alpha \in \{1,2\}$ and $\beta \in \mathbb{R}^n$, with $1 \leq |\beta| \leq 2$. Taking advantage of the estimates and representation formulas deduced in (3.2.13) and (3.2.14) – (3.2.16), as well as

$$I_5 := \frac{\eta^4}{(|\xi|^2 - \eta^2)^2 + \eta^2|\xi|^4} \leq 4\max\left(\frac{\mathcal{T}^2}{\pi^2},1\right) < \infty, \qquad \text{for } |\eta| > \frac{\pi}{\mathcal{T}}, \tag{3.2.18}$$

a similar calculation as in (3.2.17) implies the assertion. To elaborate one term, let us consider the time derivative $\partial_t u$. We have that

$$|i\eta\, m|^2 = |1 - \chi(\eta)|^2 I_2 < \infty.$$

The additional η in the numerator only implies that by deriving $i\eta\, m$ we get a further term in the representation formulas of $\partial_\eta(i\eta\, m)$, $\partial_\xi^\varepsilon(i\eta\, m)$ and $\partial_\eta\partial_\xi^\varepsilon(i\eta\, m)$. More precisely, we deduce for the mixed derivatives of $i\eta\, m$ from the representation formulas (3.2.14) – (3.2.16) that

$$\partial_\eta\partial_{\xi_{j_1}}\cdots\partial_{\xi_{j_r}}(i\eta m) = i\partial_{\xi_{j_1}}\cdots\partial_{\xi_{j_r}}m + i\eta\partial_\eta\partial_{\xi_{j_1}}\cdots\partial_{\xi_{j_r}}m.$$

Utilizing the fact that

$$|\xi^\varepsilon\partial_\xi^\varepsilon m(k,\xi)| \leq 2^{2r}(r!)^2|1 - \chi(\eta)|^2 I_1(I_3 + I_4^k) < \infty$$

holds for all $\eta \in \mathbb{R}$ with $|\eta| > \frac{\pi}{\mathcal{T}}$, we find

$$|\eta^\gamma\xi^\varepsilon\partial_\eta^\gamma\partial_\xi^\varepsilon(i\eta m)| \leq 2^{2r}((r+1)!)^2\left(\frac{\pi^2}{\mathcal{T}^2}|\chi'(\eta)|^2 I_2(I_3+I_4)^2 \right.$$
$$\left. + |1-\chi(\eta)|^2(I_3+I_4)^{r-1}\big(I_2I_3 + (I_2I_3 + 4I_2I_5)(I_3+I_4)\big)\right) < \infty.$$

Hence $i\eta \cdot m$ is an $L^q(\mathbb{R} \times \mathbb{R}^n)$-multiplier, and by Theorem 2.3.1 even ikM is an $L^q(\mathbb{T} \times \mathbb{R}^n)$-multiplier. Observe that with the representation formulas (3.2.14) – (3.2.16) at hand we can examine $\partial_t^2 u$, $\partial_x^\beta u$ and $\partial_t \partial_x^\beta u$ in a similarly way by employing I_1 - I_5. Therefore, we obtain that $u \in \mathbb{X}_\perp^q(G)$ is a solution to (3.2.11) and satisfies (3.2.12).

It remains to show uniqueness. Assume that $v \in \mathscr{S}_\perp'(G)$ is another solution with $\mathcal{P}v = 0$. Then

$$\partial_t^2 (u - v) - \Delta (u - v) - \partial_t \Delta (u - v) = 0.$$

Formally applying the Fourier transform $\mathscr{F}_G$ to this identity, we deduce $(|\xi|^2 - k^2 + ik|\xi|^2) \mathscr{F}_G[u - v] = 0$ and therefore $\operatorname{supp} \mathscr{F}_G[u-v] \subset \{(0,0)\}$. Since $\mathcal{P}(u - v) = 0$, it follows that $\delta_{\frac{2\pi}{T}\mathbb{Z}}(k)\mathscr{F}_G[u - v] = \mathscr{F}_G[\mathcal{P}(u - v)] = 0$, and thus we must have $(0,0) \notin \operatorname{supp} \mathscr{F}_G[u - v]$. Consequently, $\operatorname{supp} \mathscr{F}_G[u - v] = \emptyset$ and $u = v$. $\qquad\square$

3.2.3 The Purely Oscillatory Damped Wave Equation in the Half Space

This subsection is dedicated to the investigation of the damped wave equation with purely oscillatory data and both Dirichlet and Neumann type boundary conditions in the half space $\mathbb{R}_+^n$. We will begin by considering the fully inhomogeneous Dirichlet boundary value problem

$$\begin{cases} \partial_t^2 u - \Delta u - \partial_t \Delta u = f & \text{in } \mathbb{T} \times \mathbb{R}_+^n, \\ \qquad\qquad\qquad u = g & \text{on } \mathbb{T} \times \partial\mathbb{R}_+^n, \end{cases} \qquad (3.2.19)$$

with data (f, g) in the class

$$\begin{aligned} f &\in \mathrm{L}_\perp^q(\mathbb{T} \times \mathbb{R}_+^n), \\ g &\in \mathrm{T}_{\perp,D_1}^q(\mathbb{T} \times \partial\mathbb{R}_+^n). \end{aligned} \qquad (3.2.20)$$

Recall that

$$\mathrm{T}_{\perp,D_1}^q(\mathbb{T} \times \partial\mathbb{R}_+^n) = \mathrm{W}_\perp^{2 - \frac{1}{2q}, q}(\mathbb{T}; \mathrm{L}^q(\partial\mathbb{R}_+^n)) \cap \mathrm{W}_\perp^{1,q}(\mathbb{T}; \mathrm{W}^{2 - \frac{1}{q}, q}(\partial\mathbb{R}_+^n)).$$

Combining a reflection principle and Lemma 3.2.5, existence of a solution to the damped wave equation and the L^q estimate follow. The results are stated in the following lemma.

Lemma 3.2.6 (Dirichet Half Space Problem). *Let $n \geq 2$ and $q \in (1, \infty)$. For any (f, g) given in (3.2.20) there is a unique solution $u \in \mathbb{X}_\perp^q(\mathbb{T} \times \mathbb{R}_+^n)$ to (3.2.19) satisfying*

$$\|u\|_{\mathbb{X}_\perp^q(\mathbb{T} \times \mathbb{R}_+^n)} \leq C_{11}\big(\|f\|_{\mathrm{L}_\perp^q(\mathbb{T} \times \mathbb{R}_+^n)} + \|g\|_{\mathrm{T}_{\perp,D_1}^q(\mathbb{T} \times \mathbb{R}^{n-1})}\big), \qquad (3.2.21)$$

with $C_{11} = C_{11}(n, q, \mathcal{T}) > 0$. *If additionally* $f \in \mathrm{L}_{\perp}^{s}(\mathbb{T} \times \mathbb{R}_{+}^{n})$ *and* $g \in \mathrm{T}_{\perp,D_1}^{s}(\mathbb{T} \times \partial\mathbb{R}_{+}^{n})$ *for some* $s \in (1, \infty)$, *then also* $u \in \mathbb{X}_{\perp}^{s}(\mathbb{T} \times \mathbb{R}_{+}^{n})$.

Proof. Let us first consider the damped wave equation with homogeneous Dirichlet boundary values, *i.e.*, $g = 0$. Here, existence of a solution $u \in \mathbb{X}_{\perp}^{q}(\mathbb{T} \times \mathbb{R}_{+}^{n})$ to (3.2.19) follows from a reflection principle in combination with the results deduced for the corresponding whole space problem stated in Lemma 3.2.5. Therefore, we set

$$\tilde{f}(t, x) := \begin{cases} f(t, x', x_n) & \text{if } x_n \geq 0, \\ -f(t, x', -x_n) & \text{if } x_n < 0, \end{cases}$$

with $x' = (x_1, \ldots, x_{n-1})$ and consider the whole space problem

$$\partial_t^2 \tilde{u} - \Delta \tilde{u} - \partial_t \Delta \tilde{u} = \tilde{f} \quad \text{in } \mathbb{T} \times \mathbb{R}^n. \tag{3.2.22}$$

Existence of a solution $\tilde{u} \in \mathbb{X}_{\perp}^{q}(\mathbb{T} \times \mathbb{R}^{n})$ to this problem is obtained by utilizing Lemma 3.2.5. Moreover, $\tilde{u}$ and thus also $u := \tilde{u}_{|\mathbb{T} \times \mathbb{R}_{+}^{n}}$ satisfy (3.2.21) with $g = 0$. To classify u as a solution to (3.2.19) it remains to verify that u satisfies the boundary condition, *i.e.*, $\mathrm{Tr}_0 \left[\tilde{u}_{|\mathbb{T} \times \mathbb{R}_{+}^{n}} \right] = 0$. But since Lemma 3.2.5 yields that the solution to (3.2.22) is unique and $-\tilde{u}(t, x', -x_n)$ satisfies

$$\left(\partial_t^2 - \Delta - \partial_t \Delta \right) \left[-\tilde{u}(t, x', -x_n) \right] = -\tilde{f}(t, x', -x_n) = \tilde{f}(t, x),$$

we obtain that $\tilde{u}(t, x', x_n) = -\tilde{u}(t, x', -x_n)$, and therefore $\mathrm{Tr}_0 \left[\tilde{u}_{|\mathbb{T} \times \mathbb{R}_{+}^{n}} \right] = -\mathrm{Tr}_0 \left[\tilde{u}_{|\mathbb{T} \times \mathbb{R}_{+}^{n}} \right]$. Consequently we have $\mathrm{Tr}_0 \left[\tilde{u}_{|\mathbb{T} \times \mathbb{R}_{+}^{n}} \right] = 0$, and conclude that $u := \tilde{u}_{|\mathbb{T} \times \mathbb{R}_{+}^{n}}$ is a solution to the half space problem (3.2.19) which satisfies the L^q estimate (3.2.21). As the trace operator $\mathrm{Tr}_0 : \mathbb{X}_{\perp}^{q}(\mathbb{T} \times \mathbb{R}_{+}^{n}) \to \mathrm{T}_{\perp,D_1}^{q}(\mathbb{T} \times \partial\mathbb{R}_{+}^{n})$ is continuous and surjective, see Lemma 2.6.7, we find for any $g \in \mathrm{T}_{\perp,D_1}^{q}(\mathbb{T} \times \partial\mathbb{R}_{+}^{n})$ a $v \in \mathbb{X}_{\perp}^{q}(\mathbb{T} \times \mathbb{R}_{+}^{n})$ such that $\mathrm{Tr}_0[v] = g$. By setting $u := v + w$ where $w \in \mathbb{X}_{\perp}^{q}(\mathbb{T} \times \mathbb{R}_{+}^{n})$ is a solution to (3.2.19) with right-hand side $(f - \partial_t^2 v + \Delta v + \partial_t \Delta v, 0)$ and v as above, existence of a solution for the fully inhomogeneous problem (3.2.19) satisfying the L^q estimate (3.2.21) follows.

It remains to prove the uniqueness assertion. For this purpose, let $u \in \mathbb{X}_{\perp}^{q}(\mathbb{T} \times \mathbb{R}_{+}^{n})$ be a solution to (3.2.19) with $f = 0$ and $g = 0$. Let $\psi \in \mathrm{L}_{\perp}^{q'}(\mathbb{T} \times \mathbb{R}_{+}^{n})$ be arbitrary. By the argument above there exists a function $\phi \in \mathbb{X}_{\perp}^{q'}(\mathbb{T} \times \mathbb{R}_{+}^{n})$ satisfying $\partial_t^2 \phi - \Delta \phi - \partial_t \Delta \phi = \psi$ and $\phi_{|\mathbb{T} \times \partial\mathbb{R}_{+}^{n}} = 0$. Put $\tilde{\phi}(t, x) := \phi(-t, x)$ and observe that the integrability properties of u

are sufficient to carry out the following integration by parts:

$$\frac{1}{\mathcal{T}} \int\limits_0^{\mathcal{T}} \int\limits_{\mathbb{R}_+^n} u\,\psi \,\mathrm{d}x\,\mathrm{d}t = \frac{1}{\mathcal{T}} \int\limits_0^{\mathcal{T}} \int\limits_{\mathbb{R}_+^n} u \left(\partial_t^2 \tilde{\phi} - \Delta \tilde{\phi} + \partial_t \Delta \tilde{\phi} \right) \,\mathrm{d}x\,\mathrm{d}t$$

$$= \frac{1}{\mathcal{T}} \int\limits_0^{\mathcal{T}} \int\limits_{\mathbb{R}_+^n} \left(\partial_t^2 u - \Delta u - \partial_t \Delta u \right) \tilde{\phi}\,\mathrm{d}x\,\mathrm{d}t = 0.$$

Since ψ was arbitrary, it follows that $u = 0$.

Now assume in addition that $f \in \mathrm{L}_\perp^s(\mathbb{T} \times \mathbb{R}_+^n)$ for some $s \in (1, \infty)$ and $g = 0$. A unique solution $\tilde{U} \in \mathbb{X}_\perp^s(\mathbb{T} \times \mathbb{R}^n)$ to the damped wave equation (3.2.19), which is unique in $\mathscr{S}'(G)$, is now obtained in the same way as above by an application of the reflection principle and Lemma 3.2.5. Thus, $\tilde{U} = \tilde{u}$ and $u \in \mathbb{X}_\perp^s(\mathbb{T} \times \mathbb{R}_+^n)$. By the same lifting argument as above, the assertion also holds for inhomogeneous boundary values $g \in \mathrm{T}_{\perp, D_1}^s(\mathbb{T} \times \partial\mathbb{R}_+^n)$. $\qquad\square$

Before we continue our investigation of the Dirichlet boundary value problem, we will consider the corresponding Neumann problem and prove similar results as in Lemma 3.2.6, *i.e.*, we study the half space problem

$$\begin{cases} \partial_t^2 u - \Delta u - \partial_t \Delta u = f & \text{in } \mathbb{T} \times \mathbb{R}_+^n, \\ \dfrac{\partial u}{\partial \nu} = g & \text{on } \mathbb{T} \times \partial\mathbb{R}_+^n, \end{cases} \qquad (3.2.23)$$

and prove existence of a time-periodic solution u which is as regular as the data allow. Since the approach for the Neumann boundary value problem is similar to that we employed for (3.2.19), we will briefly sketch the proof. For the Neumann boundary value problem another trace space is necessary in order to get the optimal regularity result for the solution to (3.2.23). As pointed out in the previous subsection, the "correct" trace space for this Neumann problem is

$$\mathrm{T}_{\perp, N_1}^q(\mathbb{T} \times \partial\mathbb{R}_+^n) := \mathrm{W}_\perp^{\frac{3}{2} - \frac{1}{2q}, q}\left(\mathbb{T}; \mathrm{L}^q(\partial\mathbb{R}_+^n)\right) \cap \mathrm{W}_\perp^{1, q}\left(\mathbb{T}; \mathrm{W}^{1 - \frac{1}{q}, q}(\partial\mathbb{R}_+^n)\right).$$

Altogether we obtain for the fully inhomogeneous problem the following existence result.

Lemma 3.2.7 (Neumann Half Space Problem). *Let* $q \in (1, \infty)$, $n \geq 2$ *and consider*

$$(f, g) \in \mathrm{L}_\perp^q(\mathbb{T} \times \mathbb{R}_+^n) \times \mathrm{T}_{\perp, N_1}^q(\mathbb{T} \times \partial\mathbb{R}_+^n).$$

Then the damped wave equation (3.2.23) admits a unique solution $u \in$ $\mathbb{X}_{\perp}^q(\mathbb{T} \times \mathbb{R}_+^n)$ such that

$$\|u\|_{\mathbb{X}_{\perp}^q(\mathbb{T} \times \mathbb{R}_+^n)} \leq C_{12}\big(\|f\|_{\mathrm{L}_{\perp}^q(\mathbb{T} \times \mathbb{R}_+^n)} + \|g\|_{\mathrm{T}_{\perp,N_1}^q(\mathbb{T} \times \mathbb{R}^{n-1})}\big), \qquad (3.2.24)$$

holds with $C_{12} = C_{12}(n, p, \mathcal{T}) > 0$. If additionally $f \in \mathrm{L}_{\perp}^s(\mathbb{T} \times \mathbb{R}_+^n)$ and $g \in \mathrm{T}_{\perp,N_1}^s(\mathbb{T} \times \partial \mathbb{R}_+^n)$ for some $s \in (1, \infty)$, then also $u \in \mathbb{X}_{\perp}^s(\mathbb{T} \times \mathbb{R}_+^n)$.

Proof. Similarly as in the proof of Lemma 3.2.6 we utilize a reflection principle, but instead of an odd reflection we will use an even reflection. For this purpose, let us consider the half space problem (3.2.23) with homogeneous boundary value $g = 0$ and inhomogeneous data f. By setting

$$\tilde{f}(t, x', x_n) := \begin{cases} f(t, x', x_n) & \text{if } x_n \geq 0, \\ f(t, x', -x_n) & \text{if } x_n < 0, \end{cases}$$

and considering the damped wave equation with right-hand side $\tilde{f}$, existence of a solution $u \in \mathbb{X}_{\perp}^q(\mathbb{T} \times \mathbb{R}_+^n)$ to (3.2.23) follows again by employing Lemma 3.2.5 and restricting the resulting whole space solution $\tilde{u}$ to the half space $\mathbb{T} \times \mathbb{R}_+^n$. Observe that due to the reflection properties of $\tilde{f}$ we have that $\tilde{u}(t, x', -x_n)$ and $\tilde{u}(t, x)$ are solutions to the same whole space problem with right-hand side $\tilde{f}(t, x)$, hence the uniqueness property of $\tilde{u}$ yields the accuracy of the boundary condition. For inhomogeneous boundary data g existence of a solution follows again by the same lifting argument as in the proof of Lemma 3.2.6. Uniqueness of the solution in the space $\mathbb{X}_{\perp}^q(\mathbb{T} \times \mathbb{R}_+^n)$ follows by a similar argumentation as in Lemma 3.2.6.

If additionally that $(f, g) \in \mathrm{L}_{\perp}^s(\mathbb{T} \times \mathbb{R}_+^n) \times \mathrm{T}_{\perp,N_1}^s(\mathbb{T} \times \partial \mathbb{R}_+^n)$, it follows as in the proof of Lemma 3.2.6 that $u \in \mathbb{X}_{\perp}^s(\mathbb{T} \times \mathbb{R}_+^n)$ $\qquad \square$

3.2.4 The Purely Oscillatory Damped Wave Equation in a Bounded Domain

In this subsection we are going to consider the damped wave equation in a bounded domain $\Omega \subset \mathbb{R}^n$ with both Dirichlet and Neumann boundary conditions. We will prove similar results as in the previous two subsections. In order to obtain the L^q estimates, we make use of a localization argument, which is based on an appropriate choice of finitely many balls $B_j \subset \mathbb{R}^n$ covering Ω, where each $j \in \{1, \ldots, m\}$ is of one of the two types:

1. type $\mathbb{R}^n$: if $\overline{B}_j \subset \Omega$,

2. type $\mathbb{R}^n_{\omega_j}$: if $\overline{B}_j \cap \partial\Omega \neq \emptyset$.

In the second case, $\omega_j \colon \mathbb{R}^{n-1} \to \mathbb{R}$ denotes a Lipschitz function with $\overline{B}_j \cap \partial\Omega \subset \text{graph}(\omega_j)$ in the respective local coordinates, and we further refer to $\mathbb{R}^n_{\omega_j} := \{x \in \mathbb{R}^n \mid x_n > \omega_j(x')\}$ as the bent half space. Note that the bent half space $\mathbb{R}^n_{\omega_j}$ stems from the half space $\mathbb{R}^n_+$ by perturbing it with a "small" perturbation term ω_j. Here, the expression "small" means that ω_j satisfies $\|\nabla'\omega_j\|_\infty < \delta$ for some $\delta > 0$ that can be chosen small. Recall that by ∇' we denote the gradient $(\partial_{x_1}, \dots, \partial_{x_{n-1}})^\top$.

The key lemma for bounded domains Ω with a boundary of class $C^{1,1}$ reads as follows.

Lemma 3.2.8. *Let $q \in (1, \infty)$, $n \geq 2$ and $\Omega \subset \mathbb{R}^n$ be a bounded domain with boundary of class $C^{1,1}$. The operator*

$$\mathcal{K} \colon \mathbb{X}^q_\perp(\mathbb{T} \times \Omega) \to \mathrm{L}^q_\perp(\mathbb{T} \times \Omega) \times \mathrm{T}^q_{\perp, D_1}(\mathbb{T} \times \partial\Omega),$$
$$\mathcal{K}(u) := \left(\partial_t^2 u - \Delta u - \partial_t \Delta u, \mathrm{Tr}_0\, u \right)$$

is injective and has a dense range. Moreover, for all $u \in \mathbb{X}^q_\perp(\mathbb{T} \times \Omega)$ it holds

$$\|u\|_{\mathbb{X}^q_\perp(\mathbb{T}\times\Omega)} \leq C_{13} \big(\|(\partial_t^2 - \Delta - \partial_t\Delta)u\|_{\mathrm{L}^q_\perp(\mathbb{T}\times\Omega)} + \|u\|_{\mathrm{L}^q_\perp(\mathbb{T}\times\Omega)}$$
$$+ \|\mathrm{Tr}_0\, u\|_{\mathrm{T}^q_{\perp,D_1}(\mathbb{T}\times\partial\Omega)} \big),$$
$$(3.2.25)$$

where $C_{13} = C_{13}(n, q, \Omega, \mathcal{T}) > 0$.

Proof. For $k \in \frac{2\pi}{\mathcal{T}}\mathbb{Z} \setminus \{0\}$ consider the equation

$$\begin{cases} -k^2 v - (1 + ik)\Delta v = h & \text{in } \Omega, \\ v = 0 & \text{on } \partial\Omega. \end{cases} \qquad (3.2.26)$$

Standard theory for elliptic boundary value problems (see the proof of Lemma 3.2.2) yields for every $h \in \mathrm{L}^q(\Omega)$ a unique solution $v \in \mathrm{W}^{2,q}(\Omega)$ to (3.2.26). If $u \in \mathbb{X}^q_\perp(\mathbb{T} \times \Omega)$ satisfies $\mathcal{K}(u) = 0$, then $\mathscr{F}_\mathbb{T}[u](k, \cdot) \in \mathrm{W}^{2,q}(\Omega)$ solves (3.2.26) with a homogeneous right-hand side. Consequently $\mathscr{F}_\mathbb{T}[u](k, \cdot) = 0$. Since $k \in \frac{2\pi}{\mathcal{T}}\mathbb{Z} \setminus \{0\}$ was arbitrary and $\mathscr{F}_\mathbb{T}[u](0, \cdot) = 0$ by the assumption $\mathcal{P}u = 0$, it follows that $u = 0$. Consequently, $\mathcal{K}$ is injective.

To show that $\mathcal{K}$ has a dense range, consider $(f,g) \in \mathrm{L}^q_\perp(\mathbb{T} \times \Omega) \times \mathrm{T}^q_{\perp,D_1}(\mathbb{T} \times \partial\Omega)$. Choose $\mathrm{G} \in \mathbb{X}^q_\perp(\mathbb{T} \times \Omega)$ with $\mathrm{Tr}_0\,\mathrm{G} = g$. Since trigonometric polynomials with coefficients in $\mathrm{L}^q(\Omega)$ are dense in $\mathrm{L}^q_\perp(\mathbb{T} \times \Omega)$, there is a sequence $\{p_\ell\}_{\ell=1}^\infty \subset \mathrm{L}^q(\mathbb{T} \times \Omega)$ of trigonometric polynomials with $p_\ell \to f - (\partial_t^2\mathrm{G} - \Delta\mathrm{G} - \partial_t\Delta\mathrm{G})$ in $\mathrm{L}^q(\mathbb{T} \times \Omega)$ as $\ell \to \infty$. If we can find a solution $\tilde{u}_\ell$ to $\mathcal{K}(\tilde{u}_\ell) = (p_\ell, 0)$, then $\mathcal{K}(\tilde{u}_\ell + \mathrm{G}) \to (f,g)$, and we may conclude that $\mathcal{K}$ has a dense range. To show existence of $\tilde{u}_\ell$, it clearly suffices to solve $\mathcal{K}(\tilde{u}_\ell) = (p_\ell, 0)$ for a simple trigonometric polynomial $p_\ell := h_\ell\,\mathrm{e}^{ikt}$ with arbitrary $h_\ell \in \mathrm{L}^q(\Omega)$ and $k \in \frac{2\pi}{\mathcal{T}}\mathbb{Z} \setminus \{0\}$. A solution to this problem is given by $\tilde{u}_\ell := v_{k,\ell}\,\mathrm{e}^{ikt}$, where $v_{k,\ell}$ is the solution to (3.2.26).

Finally, we show (3.2.25) by a localization method as described above. For this purpose, we choose finitely many balls $B_j \subset \mathbb{R}^n$, $j \in \{1, \ldots, m\}$ covering Ω, where either $\overline{B}_j \subset \Omega$ or $\overline{B}_j \cap \partial\Omega \neq \emptyset$. Moreover, these balls are chosen sufficiently small, such that the boundary of $B_j \cap \Omega$ can be expressed by the graph of a function $\omega_j \in C^{1,1}(\mathbb{R}^{n-1})$ which satisfies the property

$$\|\nabla'\omega_j\|_\infty < \delta < \infty, \qquad \text{and} \qquad \|(\nabla')^2\omega_j\|_\infty < \mathrm{d} < \infty, \qquad (3.2.27)$$

with $\delta, \mathrm{d} > 0$. Compared to d the constant δ can be chosen arbitrary small. Let $\psi_j \in C_0^\infty(\mathbb{R}^n)$ be a partition of unity satisfying

$$\mathrm{supp}\,\psi_j \subset B_j \qquad \text{and} \qquad \sum_{j=1}^m \psi_j = 1 \quad \text{in } \Omega.$$

We obtain for $j \in \{1, \ldots, m\}$ and $v_j := \psi_j u$

$$\partial_t^2 v_j - \Delta v_j - \partial_t\Delta v_j = f_j \quad \text{in } \mathbb{T} \times (\Omega \cap B_j), \qquad (3.2.28)$$

where

$$\begin{aligned} f_j &:= \psi_j(\partial_t^2 - \Delta - \partial_t\Delta)u - (\Delta\psi_j)(u + \partial_t u) \\ &\quad - 2(\nabla\psi_j)(\nabla u + \partial_t\nabla u). \end{aligned}$$

If B_j is completely contained in Ω, (3.2.28) can be interpreted as a problem in the whole space $\mathbb{T} \times \mathbb{R}^n$, and Lemma 3.2.5 yields

$$\|v_j\|_{\mathbb{X}^q_\perp(\mathbb{T}\times\mathbb{R}^n)} \leq c_0\|f_j\|_{\mathrm{L}^q_\perp(\mathbb{T}\times\mathbb{R}^n)},$$

and therefore

$$\begin{aligned} \|v_j\|_{\mathbb{X}^q_\perp(\mathbb{T}\times B_j)} \leq c_1\big(&\|(\partial_t^2 - \Delta - \partial_t\Delta)u\|_{\mathrm{L}^q_\perp(\mathbb{T}\times\Omega)} + \|u\|_{\mathrm{L}^q_\perp(\mathbb{T}\times\Omega)} \\ &+ \|\nabla u\|_{\mathrm{L}^q_\perp(\mathbb{T}\times\Omega)} + \|\partial_t u\|_{\mathrm{L}^q_\perp(\mathbb{T}\times\Omega)} + \|\partial_t\nabla u\|_{\mathrm{L}^q_\perp(\mathbb{T}\times\Omega)}\big), \end{aligned}$$

with $c_1 = c_1(\psi_j) > 0$. In the second case, where $\overline{B}_j \cap \partial\Omega \neq \emptyset$, we interpret (3.2.28) as a problem in the bent half space $\mathbb{T} \times \mathbb{R}^n_{\omega_j}$. To verify (3.2.25), we first have to "flatten" the boundary and find an equivalent formulation in the half space. Therefore, let

$$\phi_{\omega_j} : \mathbb{R}^n_{\omega_j} \to \mathbb{R}^n_+, \quad \phi_{\omega_j}(x) := \tilde{x} := \big(x', x_n - \omega_j(x')\big). \qquad (3.2.29)$$

For a function v_j defined on $\mathbb{T} \times \mathbb{R}^n_{\omega_j}$, we set $\Phi[v_j](t, \tilde{x}) := \tilde{v}_j(t, \tilde{x}) := v_j(t, \phi^{-1}_{\omega_j}(\tilde{x}))$, where $(t, \tilde{x}) \in \mathbb{T} \times \mathbb{R}^n_+$. Observe that

$$\Phi\big[(\partial_t^2 - \Delta - \partial_t\Delta)v_j\big] = \big(\partial_t^2 - \Delta - \partial_t\Delta + \tilde{R}\big)\tilde{v}_j, \qquad (3.2.30)$$

where $\tilde{R}_j : \mathbb{X}^q_\perp(\mathbb{T} \times \mathbb{R}^n_+) \to \mathrm{L}^q_\perp(\mathbb{T} \times \mathbb{R}^n_+)$ is given by

$$\begin{aligned}
\tilde{R}_j\tilde{v}_j := &- |\nabla'\omega_j|^2\partial^2_{x_n}\tilde{v}_j + 2\nabla\omega_j \cdot \nabla\partial_{x_n}\tilde{v}_j + \Delta'\omega_j\, \partial_{x_n}\tilde{v}_j \\
&- |\nabla'\omega_j|^2\partial_t\partial^2_{x_n}\tilde{v}_j + 2\nabla\omega_j \cdot \nabla\partial_t\partial_{x_n}\tilde{v}_j + \Delta'\omega_j\, \partial_t\partial_{x_n}\tilde{v}_j
\end{aligned} \qquad (3.2.31)$$

with $\nabla\omega_j = \begin{pmatrix} \nabla'\omega_j \\ 0 \end{pmatrix}$. With view to (3.2.27), we can estimate

$$\|\tilde{R}_j\tilde{v}_j\|_{\mathrm{L}^q_\perp(\mathbb{T}\times\mathbb{R}^n_+)} \leq C_{14}(\delta)\|\tilde{v}_j\|_{\mathbb{X}^q_\perp(\mathbb{T}\times\mathbb{R}^n_+)} + C_{15}\|\partial_t\nabla\tilde{v}_j\|_{\mathrm{L}^q_\perp(\mathbb{T}\times\mathbb{R}^n_+)} \qquad (3.2.32)$$

with $C_{14}(\delta) = \delta(2 + \delta)$ and $C_{15} = C_{15}(\mathrm{d})$. It is standard to verify that

$$\begin{aligned}
\Phi &: \mathrm{L}^q_\perp(\mathbb{T} \times \mathbb{R}^n_{\omega_j}) \to \mathrm{L}^q_\perp(\mathbb{T} \times \mathbb{R}^n_+), \\
\Phi &: \mathbb{X}^q_\perp(\mathbb{T} \times \mathbb{R}^n_{\omega_j}) \to \mathbb{X}^q_\perp(\mathbb{T} \times \mathbb{R}^n_+), \\
\Phi &: \mathrm{T}^q_{\perp,D_1}(\mathbb{T} \times \partial\mathbb{R}^n_{\omega_j}) \to \mathrm{T}^q_{\perp,D_1}(\mathbb{T} \times \partial\mathbb{R}^n_+)
\end{aligned} \qquad (3.2.33)$$

are homeomorphisms. Hence, we can equivalently transform (3.2.28) into the half space problem

$$(\partial_t^2 - \Delta - \partial_t\Delta)\tilde{v}_j = \tilde{F}_j \qquad \text{in } \mathbb{T} \times \mathbb{R}^n_+,$$

with $\tilde{F}_j := \tilde{f}_j - \tilde{R}_j\tilde{v}_j$, and recalling Lemma 3.2.6, especially (3.2.21), and employing (3.2.32) we deduce

$$\begin{aligned}
\|\tilde{v}_j\|_{\mathbb{X}^q_\perp(\mathbb{T}\times\mathbb{R}^n_+)} \leq\ & c_2\big(\|\tilde{F}_j\|_{\mathrm{L}^q_\perp(\mathbb{T}\times\mathbb{R}^n_+)} + \|\mathrm{Tr}_0\,\tilde{v}_j\|_{\mathrm{T}^q_{\perp,D_1}(\mathbb{T}\times\partial\mathbb{R}^n_+)}\big) \\
\leq\ & c_2\big(\|\tilde{f}_j\|_{\mathrm{L}^q_\perp(\mathbb{T}\times\mathbb{R}^n_+)} + C_{14}(\delta)\|\tilde{v}_j\|_{\mathbb{X}^q_\perp(\mathbb{T}\times\mathbb{R}^n_+)} \\
& + C_{15}\|\partial_t\nabla\tilde{v}_j\|_{\mathrm{L}^q_\perp(\mathbb{T}\times\mathbb{R}^n_+)} + \|\mathrm{Tr}_0\,\tilde{v}_j\|_{\mathrm{T}^q_{\perp,D_1}(\mathbb{T}\times\partial\mathbb{R}^n_+)}\big).
\end{aligned}$$

Hence, by choosing δ sufficiently small, we find

$$\|\tilde{v}_j\|_{\mathbb{X}_\perp^q(\mathbb{T}\times\mathbb{R}_+^n)} \leq c_3\big(\|\tilde{f}_j\|_{L_\perp^q(\mathbb{T}\times\mathbb{R}_+^n)} + C_{15}\|\partial_t\nabla\tilde{v}_j\|_{L_\perp^q(\mathbb{T}\times\mathbb{R}_+^n)}$$
$$+ \|\mathrm{Tr}_0\,\tilde{v}_j\|_{T_{\perp,D_1}^q(\mathbb{T}\times\partial\mathbb{R}_+^n)}\big),$$

which in view of the homeomorphism properties of Φ stated in (3.2.33) and the definition of $\tilde{v}_j$ leads to the estimates

$$\|v_j\|_{\mathbb{X}_\perp^q(\mathbb{T}\times(B_j\cap\Omega))} \leq \|v_j\|_{\mathbb{X}_\perp^q(\mathbb{T}\times\mathbb{R}_{\omega_j}^n)}$$
$$\leq c_4\big(\|\tilde{f}_j\|_{L_\perp^q(\mathbb{T}\times\mathbb{R}_+^n)} + C_{15}\|\partial_t\nabla\tilde{v}_j\|_{L_\perp^q(\mathbb{T}\times\mathbb{R}_+^n)} + \|\mathrm{Tr}_0\,\tilde{v}_j\|_{T_{\perp,D_1}^q(\mathbb{T}\times\partial\mathbb{R}_+^n)}\big)$$
$$\leq c_5(\psi_j,\mathrm{d})\big(\|(\partial_t^2-\Delta-\partial_t\Delta)u\|_{L_\perp^q(\mathbb{T}\times\Omega)} + \|u\|_{L_\perp^q(\mathbb{T}\times\Omega)} + \|\nabla u\|_{L_\perp^q(\mathbb{T}\times\Omega)}$$
$$+ \|\partial_t u\|_{L_\perp^q(\mathbb{T}\times\Omega)} + \|\partial_t\nabla u\|_{L_\perp^q(\mathbb{T}\times\Omega)} + \|\mathrm{Tr}_0\,u\|_{T_{\perp,D_1}^q(\mathbb{T}\times\partial\Omega)}\big).$$

Summing up over $j \in \{1,\ldots,m\}$, we get

$$\|u\|_{\mathbb{X}_\perp^q(\mathbb{T}\times\Omega)} \leq c_6\big(\|(\partial_t^2-\Delta-\partial_t\Delta)u\|_{L_\perp^q(\mathbb{T}\times\Omega)} + \|u\|_{L_\perp^q(\mathbb{T}\times\Omega)} + \|\nabla u\|_{L_\perp^q(\mathbb{T}\times\Omega)}$$
$$+ \|\partial_t u\|_{L_\perp^q(\mathbb{T}\times\Omega)} + \|\partial_t\nabla u\|_{L_\perp^q(\mathbb{T}\times\Omega)} + \|\mathrm{Tr}_0\,u\|_{T_{\perp,D_1}^q(\mathbb{T}\times\partial\Omega)}\big),$$

and by exploiting Ehrling's Lemma, see for example [2, Chapter 5],

$$\|u\|_{\mathbb{X}_\perp^q(\mathbb{T}\times\Omega)} \leq c_7\big(\|(\partial_t^2-\Delta-\partial_t\Delta)u\|_{L_\perp^q(\mathbb{T}\times\Omega)} + \|u\|_{L_\perp^q(\mathbb{T}\times\Omega)}$$
$$+ \|\mathrm{Tr}_0\,u\|_{T_{\perp,D_1}^q(\mathbb{T}\times\partial\Omega)}\big)$$

follows. $\qquad\qquad\qquad\qquad\qquad\qquad\qquad\qquad\qquad\qquad\qquad\qquad\square$

In the next step we show that the term $\|u\|_q$ on the right-hand side in (3.2.25) can be dropped.

Lemma 3.2.9. *Let $q \in (1,\infty)$, $n \geq 2$ and $\Omega \subset \mathbb{R}^n$ be a bounded domain with boundary of class $C^{1,1}$. For all $u \in \mathbb{X}_\perp^q(\mathbb{T}\times\Omega)$ it holds that*

$$\|u\|_{\mathbb{X}_\perp^q(\mathbb{T}\times\Omega)} \leq C_{16}\big(\|(\partial_t^2-\Delta-\partial_t\Delta)u\|_{L_\perp^q(\mathbb{T}\times\Omega)} + \|\mathrm{Tr}_0\,u\|_{T_{\perp,D_1}^q(\mathbb{T}\times\partial\Omega)}\big),$$
$$(3.2.34)$$

where $C_{16} = C_{16}(n,q,\Omega,\mathcal{T}) > 0$.

Proof. If (3.2.34) does not hold, then we find a sequence $\{u_k\}_{k=1}^\infty \subset \mathbb{X}_\perp^q(\mathbb{T}\times\Omega)$ such that $\|u_k\|_{\mathbb{X}_\perp^q} = 1$ for all $k \in \mathbb{N}$ and

$$\|(\partial_t^2-\Delta-\partial_t\Delta)u_k\|_q + \|\mathrm{Tr}_0\,u_k\|_{T_{\perp,D_1}^q} \to 0 \qquad \text{as} \qquad k \to \infty.$$

Suppressing the notation of subsequences, we thus have the weak convergence $u_k \rightharpoonup u$ in $\mathbb{X}_\perp^q(\mathbb{T} \times \Omega)$, and u solves

$$\begin{cases} \partial_t^2 u - \Delta u - \partial_t \Delta u = 0 & \text{in } \mathbb{T} \times \Omega, \\ \qquad\qquad\qquad u = 0 & \text{on } \mathbb{T} \times \partial\Omega. \end{cases}$$

By Lemma 3.2.8, especially the injectivity property of $\mathcal{K}$, it follows that $u = 0$. Since the domain Ω is bounded, the embedding $\mathbb{X}_\perp^q(\mathbb{T} \times \Omega) \hookrightarrow \mathrm{L}_\perp^q(\mathbb{T} \times \Omega)$ is compact, whence $\|u_k\|_q \to 0$ as $k \to \infty$. This yields the contradiction

$$\begin{aligned} 1 &= \lim_{k \to \infty} \|u_k\|_{\mathbb{X}_\perp^q} \\ &\leq \lim_{k \to \infty} C_{13}\big(\|(\partial_t^2 - \Delta - \partial_t\Delta)u_k\|_q + \|u_k\|_q + \|\mathrm{Tr}_0\, u_k\|_{\mathrm{T}_{\perp,D_1}^q}\big) = 0. \end{aligned}$$

Therefore, (3.2.34) has to hold. $\qquad\qquad\qquad\qquad\qquad\qquad\qquad\square$

With Lemma 3.2.8 and Lemma 3.2.9 at hand we are able to prove existence and uniqueness of a solution to the fully inhomogeneous damped wave equation (3.2.1).

Lemma 3.2.10. *Let $q \in (1, \infty)$, $n \geq 2$ and $\Omega \subset \mathbb{R}^n$ be a bounded domain of class $C^{1,1}$. For any $f \in \mathrm{L}_\perp^q(\mathbb{T} \times \Omega)$ and $g \in \mathrm{T}_{\perp,D_1}^q(\mathbb{T} \times \partial\Omega)$ there exists a unique solution $u \in \mathbb{X}_\perp^q(\mathbb{T} \times \Omega)$ to (3.2.1) which satisfies*

$$\|u\|_{\mathbb{X}_\perp^q(\mathbb{T}\times\Omega)} \leq C_{17}\big(\|f\|_{\mathrm{L}_\perp^q(\mathbb{T}\times\Omega)} + \|g\|_{\mathrm{T}_{\perp,D_1}^q(\mathbb{T}\times\partial\Omega)}\big), \qquad (3.2.35)$$

where $C_{17} = C_{17}(n, q, \Omega, \mathcal{T}) > 0$. If additionally $f \in \mathrm{L}_\perp^s(\mathbb{T} \times \Omega)$ and $g \in \mathrm{T}_{\perp,D_1}^s(\mathbb{T} \times \partial\Omega)$ for some $s \in (1, \infty)$, then also $u \in \mathbb{X}_\perp^s(\mathbb{T} \times \Omega)$.

Proof. The operator $\mathcal{K}$ in Lemma 3.2.8 is injective and has a dense range. By Lemma 3.2.9, the range is also closed. Hence, $\mathcal{K}$ is an isomorphism. The unique solvability of (3.2.1) as well as the estimate (3.2.35) follow. The regularity assertion follows immediately from the unique solvability of (3.2.1) in $\mathbb{X}_\perp^{\min\{s,q\}}(\mathbb{T} \times \Omega)$. $\qquad\qquad\qquad\qquad\square$

After investigating the damped wave equation subject to inhomogeneous Dirichlet boundary values it remains to study the linearized Kuznetsov equation with inhomogeneous Neumann boundary values on a bounded domain. This will be done similarly to the case of Dirichlet boundary conditions. In other words, we first show that the operator

$$\begin{aligned} \mathcal{K} &: \mathbb{X}_\perp^q(\mathbb{T} \times \Omega) \to \mathrm{L}_\perp^q(\mathbb{T} \times \Omega) \times \mathrm{T}_{\perp,N_1}^q(\mathbb{T} \times \partial\Omega), \\ \mathcal{K}(u) &:= \big(\partial_t^2 u - \Delta u - \partial_t \Delta u, \mathrm{Tr}_1\, u\big) \end{aligned} \qquad (3.2.36)$$

is injective and has a dense range. Moreover, we show that for every $u \in \mathbb{X}_\perp^q(\mathbb{T} \times \Omega)$ the estimate

$$\|u\|_{\mathbb{X}_\perp^q} \leq C\big(\|(\partial_t^2 - \Delta - \partial_t \Delta)u\|_q + \|u\|_q + \|\mathrm{Tr}_1\, u\|_{T_{\perp, N_1}^q}\big)$$

holds by utilizing the analogous localization argument as in the proof of Lemma 3.2.8. Therefore, we choose a similar covering of Ω by finitely many balls $B_j \subset \Omega$ that are either of type $\mathbb{R}^n$ or $\mathbb{R}_{\omega_j}^n$. Observe, in the whole space case the L^q estimates are established by following exactly the same argumentation as for Dirichlet boundary values. For balls of type $\mathbb{R}_{\omega_j}^n$ some further calculations are needed due to the additional term occurring in the transformed Neumann boundary condition in the half space. More precisely, let ϕ_{ω_j} be as in (3.2.29) and Φ be the lifting operator $\Phi[u](t, \tilde{x}) := \tilde{u}(t, \tilde{x}) := u(t, \phi_{\omega_j}^{-1}(\tilde{x}))$. Then Φ is a homeomorphism $\Phi \colon \mathrm{T}_{\perp, N_1}^q(\mathbb{T} \times \partial\mathbb{R}_{\omega_j}^n) \to \mathrm{T}_{\perp, N_1}^q(\mathbb{T} \times \partial\mathbb{R}_+^n)$ with

$$
\begin{aligned}
\Phi\left[\mathrm{Tr}_1\, u\right] &= \nabla u \circ \phi_{\omega_j}^{-1} \cdot \nu \circ \phi_{\omega_j}^{-1} \\
&= \nabla\big(u \circ \phi_{\omega_j}^{-1}\big)\nabla\phi_{\omega_j} \cdot \big[|\mathrm{cof}\,\nabla\phi_{\omega_j}\,\nu \circ \phi_{\omega_j}^{-1}|(\mathrm{cof}\,\nabla\phi_{\omega_j})^{-1}\tilde{\nu}\big] \\
&= |\mathrm{cof}\,\nabla\phi_{\omega_j}\,\nu \circ \phi_{\omega_j}^{-1}|\,\nabla\tilde{u}\nabla\phi_{\omega_j} \cdot \nabla\phi_{\omega_j}^\top\tilde{\nu} \\
&= |\mathrm{cof}\,\nabla\phi_{\omega_j}\,\nu \circ \phi_{\omega_j}^{-1}|\big(\nabla\tilde{u} \cdot \tilde{\nu} + \nabla\tilde{u}(\nabla\phi_{\omega_j}\nabla\phi_{\omega_j}^\top - I) \cdot \tilde{\nu}\big) \\
&= |\mathrm{cof}\,\nabla\phi_{\omega_j}\,\nu \circ \phi_{\omega_j}^{-1}|\,\mathrm{Tr}_1\,\tilde{u} + \mathrm{Tr}_0\,\tilde{S}_j\tilde{u},
\end{aligned}
$$

where

$$\tilde{S}_j\tilde{u} := |\mathrm{cof}\,\nabla\phi_{\omega_j}\,\nu \circ \phi_{\omega_j}^{-1}|\,\nabla\tilde{u}\big(\nabla\phi_{\omega_j}\nabla\phi_{\omega_j}^\top - I\big) \cdot \tilde{\nu}. \tag{3.2.37}$$

Here, $\tilde{\nu}$ denotes the outer unit normal vector on $\mathbb{T} \times \mathbb{R}_+^n$ and ν the outer unit normal vector on $\mathbb{T} \times \mathbb{R}_{\omega_j}^n$. It should be understood that $\mathrm{Tr}_1\, u$ denotes the normal trace operator in $\mathbb{T} \times \mathbb{R}_{\omega_j}^n$ and $\mathrm{Tr}_1\,\tilde{u}$ the normal trace operator in $\mathbb{T} \times \mathbb{R}_+^n$. Recalling (3.2.24), we can estimate

$$
\begin{aligned}
\|\tilde{S}_j\tilde{u}\|&_{\mathrm{T}_{\perp, N_1}^q(\mathbb{T} \times \partial\mathbb{R}_+^n)} \\
&\leq c_0\|\mathrm{cof}\,\nabla\phi_{\omega_j}\,\nu \circ \phi_{\omega_j}^{-1}\|_{\mathrm{W}^{1,\infty}(\mathbb{R}^{n-1})} \\
&\qquad\quad \cdot \|\nabla\tilde{u}\big(\nabla\phi_{\omega_j}(\nabla\phi_{\omega_j})^\top - I\big) \cdot \tilde{\nu}\|_{\mathrm{T}_{\perp, N_1}^q(\mathbb{T} \times \mathbb{R}^{n-1})} \\
&\leq c_1\|\nabla\tilde{u}\|_{\mathrm{T}_{\perp, N_1}^q(\mathbb{T} \times \mathbb{R}^{n-1})}\|\big(\nabla\phi_{\omega_j}(\nabla\phi_{\omega_j})^\top - I\big) \cdot \tilde{\nu}\|_{\mathrm{W}^{1,\infty}(\mathbb{R}^{n-1})} \\
&\leq c\|\tilde{u}\|_{\mathbb{X}_\perp^q(\mathbb{T} \times \mathbb{R}_+^n)} \\
&\leq C_{18}\big(\|(\partial_t^2 - \Delta - \partial_t\Delta)\tilde{u}\|_{\mathrm{L}_\perp^q(\mathbb{T} \times \mathbb{R}_+^n)} + \|\mathrm{Tr}_1\,\tilde{u}\|_{\mathrm{T}_{\perp, N_1}^q(\mathbb{T} \times \mathbb{R}_+^n)}\big),
\end{aligned} \tag{3.2.38}
$$

where the third inequality follows from the properties of the trace operator stated in (2.6.14). Note that $c = c_2\delta$ and $C_{18} = c_3\delta$, where $\delta > 0$ is chosen such that $\|\nabla'\omega_j\|_\infty < \delta$. With this observation at hand we finally investigate the Neumann boundary value problem on a bounded domain. Since we are employing the same arguments as for the corresponding Dirichlet boundary value problem collected in Lemma 3.2.8 and 3.2.9, we will not outline these properties in separated lemmas, but prove the following lemma directly.

Lemma 3.2.11. *For $n \geq 2$ let $\Omega \subset \mathbb{R}^n$ be a bounded domain with boundary of class $C^{1,1}$ and let $q \in (1,\infty)$. For any $f \in L^q_\perp(\mathbb{T}\times\Omega)$ and $g \in T^q_{\perp,N_1}(\mathbb{T}\times \partial\Omega)$ there exists a unique solution $u \in \mathbb{X}^q_\perp(\mathbb{T}\times\Omega)$ to*

$$\begin{cases} \partial_t^2 u - \Delta u - \partial_t\Delta u = f & in\ \mathbb{T}\times\Omega, \\ \dfrac{\partial u}{\partial\nu} = g & on\ \mathbb{T}\times\partial\Omega, \end{cases} \qquad (3.2.39)$$

and there is a constant $C_{19} = C_{19}(n,q,\Omega,\mathcal{T}) > 0$ such that

$$\|u\|_{\mathbb{X}^q_\perp(\mathbb{T}\times\Omega)} \leq C_{19}\big(\|f\|_{L^q_\perp(\mathbb{T}\times\Omega)} + \|g\|_{T^q_{\perp,N_1}(\mathbb{T}\times\partial\Omega)}\big) \qquad (3.2.40)$$

holds. If additionally $f \in L^s_\perp(\mathbb{T} \times \Omega)$ and $g \in T^s_{\perp,N_1}(\mathbb{T} \times \partial\Omega)$ for some $s \in (1,\infty)$, then also $u \in \mathbb{X}^s_\perp(\mathbb{T} \times \Omega)$.

Proof. We begin by showing that the operator $\mathcal{K}$, defined in (3.2.36), is injective and has a dense range. For this purpose, we consider for $k \in \frac{2\pi}{\mathcal{T}}\mathbb{Z} \setminus \{0\}$ the Helmholtz equation

$$\begin{cases} -k^2 v - (1 + ik)\,\Delta v = h & in\ \Omega, \\ \dfrac{\partial v}{\partial\nu} = 0 & on\ \partial\Omega, \end{cases} \qquad (3.2.41)$$

and obtain via standard theory for elliptic boundary value problems (see the proof of Lemma 3.2.2) for every $h \in L^q(\Omega)$ a unique solution $v \in W^{2,q}(\Omega)$ to (3.2.41). Observe that since $k \neq 0$ no compatibility condition is required for the data h in the pure Neumann problem (3.2.41). The injectivity of $\mathcal{K}$ and the density of its range follow as in the proof of Lemma 3.2.8.

In order to verify the L^q estimate (3.2.40) we utilize the same localization argument as in the Dirichlet case. As described above, the only case remaining to be considered, is the bent half space case. Therefore, let ψ_j

be as in the proof of Lemma 3.2.8, and observe that for $j \in \{1, \ldots, m\}$ and $v_j := \psi_j u$ we obtain the "bent half space" problem

$$\begin{cases} \partial_t^2 v_j - \Delta v_j - \partial_t \Delta v_j = f_j & \text{in } \mathbb{T} \times (\Omega \cap B_j), \\ \dfrac{\partial v_j}{\partial \nu} = g_j & \text{on } \mathbb{T} \times \partial(\Omega \cap B_j). \end{cases} \tag{3.2.42}$$

where

$$f_j := \psi_j(\partial_t^2 - \Delta - \partial_t \Delta)u - (\Delta \psi_j)(u + \partial_t u) - 2(\nabla \psi_j)(\nabla u + \partial_t \nabla u)$$
$$g_j := \frac{\partial \psi_j}{\partial \nu} u + \psi_j \frac{\partial u}{\partial \nu}.$$

Utilizing the lifting operator Φ to (3.2.42) we find the following (equivalent) half space problem

$$\begin{cases} \partial_t^2 \tilde{v}_j - \Delta \tilde{v}_j - \partial_t \Delta \tilde{v}_j = \tilde{F}_j & \text{in } \mathbb{T} \times \mathbb{R}_+^n, \\ \dfrac{\partial \tilde{v}_j}{\partial \tilde{\nu}} = \tilde{G}_j & \text{on } \mathbb{T} \times \mathbb{R}^{n-1}, \end{cases}$$

with

$$\tilde{F}_j := \tilde{f}_j - \tilde{R}_j \tilde{v}_j, \qquad \text{and} \qquad \tilde{G}_j := \tilde{g}_j - \tilde{S}_j \tilde{v}_j.$$

Recall, $\tilde{R}_j$ and $\tilde{S}_j$ are given as in (3.2.31) and (3.2.37), respectively. Since the operator Φ is a homeomorphism $\Phi \colon \mathrm{T}_{\perp, N_1}^q(\mathbb{T} \times \partial \mathbb{R}_{\omega_j}^n) \to \mathrm{T}_{\perp, N_1}^q(\mathbb{T} \times \partial \mathbb{R}_+^n)$ and $g_j \in \mathrm{T}_{\perp, N_1}^q(\mathbb{T} \times \partial \mathbb{R}_{\omega_j}^n)$, we deduce from Lemma 3.2.7 that

$$\begin{aligned} \|\tilde{v}_j\|_{\mathrm{X}_\perp^q(\mathbb{T} \times \mathbb{R}_+^n)} &\leq c_0 \big(\|\tilde{F}_j\|_{\mathrm{L}_\perp^q(\mathbb{T} \times \mathbb{R}_+^n)} + \|\tilde{G}_j\|_{\mathrm{T}_{\perp, N_1}^q(\mathbb{T} \times \partial \mathbb{R}_+^n)} \big) \\ &\leq c_0 \big(\|\tilde{f}_j\|_{\mathrm{L}_\perp^q(\mathbb{T} \times \mathbb{R}_+^n)} + \|\tilde{R}_j \tilde{v}_j\|_{\mathrm{L}_\perp^q(\mathbb{T} \times \mathbb{R}_+^n)} + \|\tilde{g}_j\|_{\mathrm{T}_{\perp, N_1}^q(\mathbb{T} \times \partial \mathbb{R}_+^n)} \\ &\qquad + \|\tilde{S}_j \tilde{v}_j\|_{\mathrm{T}_{\perp, N_1}^q(\mathbb{T} \times \partial \mathbb{R}_+^n)} \big), \end{aligned}$$

and in view of (3.2.32) and (3.2.38) we deduce

$$\begin{aligned} \|\tilde{v}_j\|_{\mathrm{X}_\perp^q(\mathbb{T} \times \mathbb{R}_+^n)} &\leq c_0 \big(\|\tilde{f}_j\|_{\mathrm{L}_\perp^q(\mathbb{T} \times \mathbb{R}_+^n)} + \|\tilde{g}_j\|_{\mathrm{T}_{\perp, N_1}^q(\mathbb{T} \times \partial \mathbb{R}_+^n)} \\ &\qquad + C_{15}(\mathrm{d}) \|\partial_t \nabla \tilde{v}_j\|_{\mathrm{L}_\perp^q(\mathbb{T} \times \mathbb{R}_+^n)} + C_{14}(\delta) \|\tilde{v}_j\|_{\mathrm{X}_\perp^q(\mathbb{T} \times \mathbb{R}_+^n)} \\ &\qquad + C_{18}(\delta) \big(\|\tilde{v}_j\|_{\mathrm{X}_\perp^q(\mathbb{T} \times \mathbb{R}_+^n)} + \|\mathrm{Tr}_1 \tilde{v}_j\|_{\mathrm{T}_{\perp, N_1}^q(\mathbb{T} \times \partial \mathbb{R}_+^n)} \big) \big). \end{aligned}$$

Again, choosing δ sufficiently small, we find

$$\|\tilde{v}_j\|_{\mathrm{X}_\perp^q(\mathbb{T} \times \mathbb{R}_+^n)} \leq c_1 \big(\|\tilde{f}_j\|_{\mathrm{L}_\perp^q(\mathbb{T} \times \mathbb{R}_+^n)} + \|\tilde{g}_j\|_{\mathrm{T}_{\perp, N_1}^q(\mathbb{T} \times \partial \mathbb{R}_+^n)} + \|\partial_t \nabla \tilde{v}_j\|_{\mathrm{L}_\perp^q(\mathbb{T} \times \mathbb{R}_+^n)} \big).$$

From (3.2.33) we obtain in a similar way as in the proof of Lemma 3.2.8

$$\|v_j\|_{\mathbb{X}_{\perp}^q(\mathbb{T}\times(\Omega\cap B_j))} \leq c_2\big(\|(\partial_t^2 - \Delta - \partial_t\Delta)u\|_{\mathsf{L}_{\perp}^q(\mathbb{T}\times(\Omega\cap B_j))} + \|u\|_{\mathsf{L}_{\perp}^q(\mathbb{T}\times(\Omega\cap B_j))}$$
$$+ \|\nabla u\|_{\mathsf{L}_{\perp}^q(\mathbb{T}\times(\Omega\cap B_j))} + \|\mathrm{Tr}_0\,\psi_j\,\mathrm{Tr}_1\,u\|_{\mathsf{T}_{\perp,N_1}^q(\mathbb{T}\times\partial(\Omega\cap B_j))}$$
$$+ \|\partial_t\nabla u\|_{\mathsf{L}_{\perp}^q(\mathbb{T}\times(\Omega\cap B_j))} + \|\partial_t u\|_{\mathsf{L}_{\perp}^q(\mathbb{T}\times(\Omega\cap B_j))}$$
$$+ \|\mathrm{Tr}_1\,\psi_j\,\mathrm{Tr}_0\,u\|_{\mathsf{T}_{\perp,N_1}^q(\mathbb{T}\times\partial(\Omega\cap B_j))}\big)$$

with $c_2 = c_2(\psi_j,\delta) > 0$. To estimate the final term on the right-hand side, we employ the Dirichlet trace operator from Subsection 2.6.2, and obtain

$$\|\mathrm{Tr}_1\,\psi_j\,\mathrm{Tr}_0\,u\|_{\mathsf{T}_{\perp,N_1}^q(\mathbb{T}\times\partial(\Omega\cap B_j))} \leq \|\mathrm{Tr}_1\,\psi_j\,\mathrm{Tr}_0\,u\|_{\mathsf{T}_{\perp,D_1}^q(\mathbb{T}\times\partial(\Omega\cap B_j))}$$
$$\leq c_3\|\mathrm{Tr}_1\,\psi_j\|_{\mathsf{T}_{\perp,D_1}^\infty(\mathbb{T}\times\partial(\Omega\cap B_j))}\|\mathrm{Tr}_0\,u\|_{\mathsf{T}_{\perp,D_1}^q(\mathbb{T}\times\partial(\Omega\cap B_j))}$$
$$\leq c_4(\psi_j)\|u\|_{\mathbb{X}_{\perp}^q(\mathbb{T}\times(\Omega\cap B_j))},$$

and thus

$$\|v_j\|_{\mathbb{X}_{\perp}^q(\mathbb{T}\times(\Omega\cap B_j))} \leq c_5\big(\|\nabla u\|_{\mathsf{L}_{\perp}^q(\mathbb{T}\times\Omega)} + \|\partial_t u\|_{\mathsf{L}_{\perp}^q(\mathbb{T}\times\Omega)} + \|u\|_{\mathsf{L}_{\perp}^q(\mathbb{T}\times\Omega)}$$
$$+ \|\mathrm{Tr}_1\,u\|_{\mathsf{T}_{\perp,N_1}^q(\mathbb{T}\times\partial\Omega)} + \|(\partial_t^2 - \Delta - \partial_t\Delta)u\|_{\mathsf{L}_{\perp}^q(\mathbb{T}\times\Omega)} + \|\partial_t\nabla u\|_{\mathsf{L}_{\perp}^q(\mathbb{T}\times\Omega)}\big).$$

Again summing up over $j \in \{1, \ldots, m\}$ and using Ehrling's Lemma we finally deduce

$$\|u\|_{\mathbb{X}_{\perp}^q(\mathbb{T}\times\Omega)} \leq c_6\big(\|(\partial_t^2 - \Delta - \partial_t\Delta)u\|_{\mathsf{L}_{\perp}^q(\mathbb{T}\times\Omega)}$$
$$+ \|u\|_{\mathsf{L}_{\perp}^q(\mathbb{T}\times\Omega)} + \|\mathrm{Tr}_1\,u\|_{\mathsf{T}_{\perp,N_1}^q(\mathbb{T}\times\partial\Omega)}\big).$$

Observe, the term $\|u\|_q$ on the right-hand side above can be omitted by the same argument as in Lemma 3.2.9, hence (3.2.40) holds and the operator $\mathcal{K}$, defined in (3.2.36), is an isomorphism. The unique solvability of (3.2.39) as well as (3.2.40) follow. The regularity assertion follows immediately from the unique solvability of (3.2.39) in $\mathbb{X}_{\perp}^{\min\{s,q\}}(\mathbb{T}\times\Omega)$. $\qquad\square$

3.2.5 The Time-Periodic Damped Wave Equation

In this subsection we investigate the $\mathcal{T}$-time-periodic systems (3.2.1) and (3.2.2), where the solution still consists of a steady state and a purely oscillatory part. For this reason, we will utilize the results deduced for the stationary part $u_{\mathrm{s}} = \mathcal{P}u$ and the purely oscillatory part $u_{\mathrm{tp}} = \mathcal{P}_{\perp}u$ in the previous subsections. In case of Dirichlet boundary values the main theorem of this section reads:

Theorem 3.2.12 (Dirichlet problem). *Assume for $n \geq 2$ that either $\Omega = \mathbb{R}^n$, $\Omega = \mathbb{R}^n_+$ or $\Omega \subset \mathbb{R}^n$ is a bounded domain with a $C^{1,1}$-smooth boundary. Let $q \in (1, \infty)$. Then for any $f \in L^q(\mathbb{T}; L^q(\Omega))$ and $g \in T^q_{D_1}(\mathbb{T} \times \partial\Omega)$ the damped wave equation*

$$\begin{cases} \partial_t^2 u - \Delta u - \partial_t \Delta u = f & in\ \mathbb{T} \times \Omega, \\ \qquad\qquad\qquad\ \ u = g & on\ \mathbb{T} \times \partial\Omega, \end{cases} \tag{3.2.43}$$

admits the existence of a solution $u(t, x) = u_s(x) + u_{tp}(t, x)$ with

$$u_s \in \dot{W}^{2,q}(\Omega),$$
$$u_{tp} \in W^{2,q}_\perp\big(\mathbb{T}; L^q(\Omega)\big) \cap W^{1,q}_\perp\big(\mathbb{T}; W^{2,q}(\Omega)\big),$$

satisfying

$$\|\nabla^2 u_s\|_{L^q(\Omega)} \leq C_{20}\big(\|f_s\|_{L^q(\Omega)} + \|g_s\|_{W^{2-\frac{1}{q},q}(\partial\Omega)}\big), \tag{3.2.44}$$

$$\|u_{tp}\|_{\mathbb{X}^q_\perp(\mathbb{T} \times \Omega)} \leq C_{21}\big(\|f_{tp}\|_{L^q(\mathbb{T} \times \Omega)} + \|g_{tp}\|_{T^q_{\perp,D_1}(\mathbb{T} \times \partial\Omega)}\big), \tag{3.2.45}$$

where $C_{20} = C_{20}(n, q, \Omega) > 0$ and $C_{21} = C_{21}(n, q, \Omega, \mathcal{T}) > 0$. If $v = v_s + v_{tp}$ is another solution to (3.2.43) with $v_s \in \dot{W}^{2,q_1}(\Omega)$ and $v_{tp} \in W^{2,q_2}_\perp\big(\mathbb{T}; L^{q_2}(\Omega)\big) \cap W^{1,q_2}_\perp\big(\mathbb{T}; W^{2,q_2}(\Omega)\big)$, $q_1, q_2 \in (1, \infty)$, then $u_s - v_s$ is a polynomial of order 1, if $\Omega = \mathbb{R}^n$ and $\mathbb{R}^n_+$, but $u_s = v_s$ if Ω is a bounded domain, and $u_{tp} = v_{tp}$.

Proof. Consider first the classical Poisson equation (3.2.6), *i.e.*,

$$\begin{cases} -\Delta u_s = f_s & in\ \Omega, \\ \quad\ \ u_s = g_s & on\ \partial\Omega. \end{cases} \tag{3.2.46}$$

Existence of a solution $u_s \in \dot{W}^{2,q}(\Omega)$ to (3.2.46) which satisfies (3.2.44) follows from Lemma 3.2.1 and Lemma 3.2.2. For u_{tp} we employ Lemma 3.2.5 in the case $\Omega = \mathbb{R}^n$, Lemma 3.2.6 in the case $\Omega = \mathbb{R}^n_+$, and Lemma 3.2.10 in the case of a bounded domain to provide a solution $u_{tp} \in \mathbb{X}^q_\perp(\mathbb{T} \times \Omega)$ to

$$\begin{cases} \partial_t^2 u_{tp} - \Delta u_{tp} - \partial_t \Delta u_{tp} = f_{tp} & in\ \mathbb{T} \times \Omega, \\ \qquad\qquad\qquad\qquad\ \ u_{tp} = g_{tp} & on\ \mathbb{T} \times \partial\Omega, \end{cases} \tag{3.2.47}$$

that obeys (3.2.45). Setting $u := u_s + u_{tp}$, we thus obtain the desired solution to the damped wave equation (3.2.43) with inhomogeneous Dirichlet boundary values.

It remains to prove the uniqueness of the solution. Therefore, assume that $v = v_\mathrm{s} + v_\mathrm{tp}$ is another solution to (3.2.43) with $v_\mathrm{s} \in \dot{\mathrm{W}}^{2,q_1}(\Omega)$ and $v_\mathrm{tp} \in \mathrm{W}^{2,q_2}_\perp(\mathbb{T}; \mathrm{L}^{q_2}(\Omega)) \cap \mathrm{W}^{1,q_2}_\perp(\mathbb{T}; \mathrm{W}^{2,q_2}(\Omega))$. Since both u_tp and v_tp solve (3.2.47), the uniqueness statements of the lemmas mentioned above yield $u_\mathrm{tp} = v_\mathrm{tp}$. As solutions of (3.2.46), u_s and v_s differ at most by a polynomial of order 1 when $\Omega = \mathbb{R}^n$ or $\Omega = \mathbb{R}^n_+$, and $u_\mathrm{s} = v_\mathrm{s}$ when $\Omega \subset \mathbb{R}^n$ is a bounded domain, see Lemma 3.2.1 and Lemma 3.2.2. □

In the framework of Neumann boundary conditions we proceed similarly as for Dirichlet boundary values and split the solution into a steady-state and a purely oscillatory part, where the stationary part u_s solves the Neumann boundary value problem (3.2.7), that is

$$\begin{cases} -\Delta u_\mathrm{s} = f_\mathrm{s} & \text{in } \Omega, \\ \dfrac{\partial u_\mathrm{s}}{\partial \nu} = g_\mathrm{s} & \text{on } \partial\Omega, \end{cases} \tag{3.2.48}$$

and u_tp is a solution to (3.2.2) with purely periodic data, that is

$$\begin{cases} \partial_t^2 u_\mathrm{tp} - \Delta u_\mathrm{tp} - \partial_t \Delta u_\mathrm{tp} = f_\mathrm{tp} & \text{in } \mathbb{T} \times \Omega, \\ \dfrac{\partial u_\mathrm{tp}}{\partial \nu} = g_\mathrm{tp} & \text{on } \mathbb{T} \times \partial\Omega. \end{cases} \tag{3.2.49}$$

As mentioned in Subsection 3.2.1, if Ω is a bounded domain it is necessary to further assume that by Gauß's theorem the compatibility condition (3.2.9) holds. With view to the definition of $f_\mathrm{s}(x) = \mathcal{P}f(x) = \frac{1}{\mathcal{T}} \int_0^{\mathcal{T}} f(s,x)\,\mathrm{d}s$ we see that (3.2.9) is equivalent to the compatibility condition

$$\int\limits_0^{\mathcal{T}} \int\limits_\Omega f \,\mathrm{d}x\,\mathrm{d}t + \int\limits_0^{\mathcal{T}} \int\limits_{\partial\Omega} g \,\mathrm{d}S\,\mathrm{d}t = 0. \tag{3.2.50}$$

The main theorem concerning existence of non-resonant solutions to the damped wave equation with Neumann boundary values then reads as follow:

Theorem 3.2.13 (Neumann problem). *Let n, q and Ω be as in Theorem 3.2.12. Furthermore, let $f \in \mathrm{L}^q(\mathbb{T}; \mathrm{L}^q(\Omega))$ and $g \in \mathrm{T}^q_{N_1}(\mathbb{T} \times \partial\Omega)$. If Ω is a bounded domain, assume additionally that f and g satisfy (3.2.50). Then there is a solution $u(t,x) = u_\mathrm{s}(x) + u_\mathrm{tp}(t,x)$ to (3.2.49) with*

$$u_\mathrm{s} \in \dot{\mathrm{W}}^{2,q}(\Omega),$$
$$u_\mathrm{tp} \in \mathrm{W}^{2,q}_\perp(\mathbb{T}; \mathrm{L}^q(\Omega)) \cap \mathrm{W}^{1,q}_\perp(\mathbb{T}; \mathrm{W}^{2,q}(\Omega))$$

satisfying

$$\|\nabla^2 u_{\mathrm{s}}\|_{\mathrm{L}^q(\Omega)} \le C_{22}\big(\|f_{\mathrm{s}}\|_{\mathrm{L}^q(\Omega)} + \|g_{\mathrm{s}}\|_{\mathrm{W}^{1-\frac{1}{q},q}(\partial\Omega)}\big), \tag{3.2.51}$$

$$\|u_{\mathrm{tp}}\|_{\mathrm{X}^q_{\perp}(\mathbb{T}\times\Omega)} \le C_{23}\big(\|f_{\mathrm{tp}}\|_{\mathrm{L}^q_{\perp}(\mathbb{T}\times\Omega)} + \|g_{\mathrm{tp}}\|_{\mathrm{T}^q_{\perp,N_1}(\mathbb{T}\times\partial\Omega)}\big), \tag{3.2.52}$$

where $C_{22} = C_{22}(n, q, \Omega) > 0$ and $C_{23} = C_{23}(n, q, \Omega, \mathcal{T}) > 0$. If $v = v_{\mathrm{s}} + v_{\mathrm{tp}}$ is another solution to (3.2.49) with $v_{\mathrm{s}} \in \dot{\mathrm{W}}^{2,q_1}(\Omega)$ and $v_{\mathrm{tp}} \in \mathrm{W}^{2,q_2}_{\perp}\big(\mathbb{T}; \mathrm{L}^{q_2}(\Omega)\big) \cap \mathrm{W}^{1,q_2}_{\perp}\big(\mathbb{T}; \mathrm{W}^{2,q_2}(\Omega)\big)$, q_1, $q_2 \in (1,\infty)$, then $u_{\mathrm{s}} - v_{\mathrm{s}}$ is a polynomial of order 1, if $\Omega = \mathbb{R}^n$ and $\mathbb{R}^n_+$, but $u_{\mathrm{s}} = v_{\mathrm{s}}$ if Ω is a bounded domain, and $u_{\mathrm{tp}} = v_{\mathrm{tp}}$.

Proof. Let us first consider the steady state Neumann problem (3.2.48). Existence of a solution to the corresponding whole space problem follows from Lemma 3.2.1. In the cases $\Omega = \mathbb{R}^n_+$ or $\Omega \subset \mathbb{R}^n$ a bounded domain, an application of Lemma 3.2.4 yields the existence of a solution to (3.2.48) such that (3.2.51) holds.

For the purely oscillatory Neumann boundary value problem we conclude from Lemma 3.2.5, Lemma 3.2.7 and Lemma 3.2.11 in the same way as in the Dirichlet case the unique solvability of

$$\begin{cases} \partial_t^2 u_{\mathrm{tp}} - \Delta u_{\mathrm{tp}} - \partial_t \Delta u_{\mathrm{tp}} = f_{\mathrm{tp}} & \text{in } \mathbb{T} \times \Omega, \\ \dfrac{\partial u_{\mathrm{tp}}}{\partial \nu} = g_{\mathrm{tp}} & \text{on } \mathbb{T} \times \partial\Omega. \end{cases}$$

Hence, the solution to (3.2.49) is given by $u := u_{\mathrm{s}} + u_{\mathrm{tp}}$. The uniqueness assertion follows as in the proof of Theorem 3.2.12. $\qquad\square$

3.3 The Kuznetsov Equation

In a damped hyperbolic system with periodic forcing, resonance can be avoided if the energy from the external forces accumulated over a period is dissipated via the damping mechanism. The existence of a time-periodic solution would be a manifestation hereof. This section is dedicated to the analysis of the Kuznetsov equation (3.0.1), which is a nonlinear wave equation that describes acoustic wave propagation. As our main result, we show for any periodic forcing term, sufficiently restricted in "size", existence of a time-periodic solution. We shall treat non-homogeneous boundary values of both Dirichlet and Neumann type. We consider spatial domains $\Omega \subset \mathbb{R}^n$ that are either bounded, the half-space or the whole-space.

Well-posedness of the initial-value problem for the Kuznetsov equation has only recently been established [52, 53, 70]. Our result can be viewed as an extension of these results to the corresponding time-periodic problem. Related time-periodic problems have been studied by other authors over the years. In particular, we mention the work of KOKOCKI [56], in which a class of nonlinear wave equations with Kelvin-Voigt damping is investigated. The work of KOKOCKI [56] does not cover the Kuznetsov equation however.

We employ a setting of time-periodic functions, which was introduced in Chapter 2, and therefore take the torus $\mathbb{T}$ as time axis. The Kuznetsov equation with Dirichlet boundary condition then reads

$$\begin{cases} \partial_t^2 u - \Delta u - \dfrac{b}{c^2}\partial_t\Delta u - \partial_t\Big(\dfrac{1}{\rho_0 c^4}\dfrac{B}{2A}(\partial_t u)^2 + |\nabla u|^2\Big) = f & \text{in } \mathbb{T}\times\Omega, \\ \qquad\qquad\qquad\qquad\qquad\qquad\qquad\qquad u = g & \text{on } \mathbb{T}\times\partial\Omega. \end{cases}$$
(3.3.1)

The corresponding Neumann problem reads

$$\begin{cases} \partial_t^2 u - \Delta u - \dfrac{b}{c^2}\partial_t\Delta u - \partial_t\Big(\dfrac{1}{\rho_0 c^4}\dfrac{B}{2A}(\partial_t u)^2 + |\nabla u|^2\Big) = f & \text{in } \mathbb{T}\times\Omega, \\ \qquad\qquad\qquad\qquad\qquad\qquad\qquad\qquad \dfrac{\partial u}{\partial\nu} = g & \text{on } \mathbb{T}\times\partial\Omega. \end{cases}$$
(3.3.2)

Here, ν is the outer unit normal vector at $\partial\Omega$. Taking time derivatives on both sides in (3.0.1), that is (3.3.1)$_1$, one obtains Kuznetsov's equation expressed in terms of the acoustic pressure fluctuation $\partial_t u$. This latter formulation was used in [52, 53, 70].

We will show for time-periodic data f and g of the same period $\mathcal{T} > 0$, and whose norm in appropriate Sobolev spaces are sufficiently small, the existence of a solution u to (3.3.1) and (3.3.2) which is also time-periodic of the same period $\mathcal{T}$. More precisely, we will proof the following theorem:

Theorem 3.3.1. *Let $n \geq 3$ and $\max\{\frac{5}{2}, \frac{5n}{6}\} < q < n$. Assume that either $\Omega = \mathbb{R}^n$, $\Omega = \mathbb{R}_+^n$ or $\Omega \subset \mathbb{R}^n$ is a bounded domain with a $C^{1,1}$-smooth boundary. There is an $\varepsilon > 0$ such that for all $f \in \mathrm{L}^q\big(\mathbb{T}; \mathrm{L}^q(\Omega)\big)$ and $g \in \mathrm{T}_{D_1}^q(\mathbb{T}\times\partial\Omega)$ satisfying*

$$\|f\|_{\mathrm{L}^q(\mathbb{T}\times\Omega)} + \|g\|_{\mathrm{T}_{D_1}^q(\mathbb{T}\times\partial\Omega)} \leq \varepsilon$$

there is a solution u to (3.3.1) with

$$u(t, x) = u_\mathrm{s}(x) + u_\mathrm{tp}(t, x) \in \dot{\mathrm{W}}^{2,q}(\Omega) \oplus \mathbb{X}_\perp^q(\mathbb{T}\times\Omega).$$

Again, for the Neumann boundary value problem a further compatibility condition is necessary. The main theorem then reads as follow.

Theorem 3.3.2. *Let n, q and Ω be as in Theorem 3.3.1. There is an $\varepsilon > 0$ such that for all $f \in L^q(\mathbb{T}; L^q(\Omega))$ and $g \in T^q_{N_1}(\mathbb{T} \times \partial\Omega)$ satisfying*

$$\|f\|_{L^q(\mathbb{T}\times\Omega)} + \|g\|_{T^q_{N_1}(\mathbb{T}\times\partial\Omega)} \leq \varepsilon$$

and

$$\int_0^\mathcal{T} \int_\Omega f \, \mathrm{d}x \, \mathrm{d}t + \int_0^\mathcal{T} \int_{\partial\Omega} g \, \mathrm{d}S \, \mathrm{d}t = 0$$

there is a solution u to (3.3.2) *with*

$$u(t,x) = u_\mathrm{s}(x) + u_\mathrm{tp}(t,x) \in \dot{\mathrm{W}}^{2,q}(\Omega) \oplus \mathbb{X}^q_\perp(\mathbb{T} \times \Omega).$$

The proofs of Theorem 3.3.1 and Theorem 3.3.2 require estimates of the nonlinear terms

$$\partial_t u \partial_t^2 u \qquad \text{and} \qquad \nabla u \cdot \nabla \partial_t u$$

that appear in (3.3.1) and (3.3.2). To this end we utilize the embedding properties of time-periodic Sobolev spaces stated in Theorem 2.6.3. For reasons of simplicity we set without loss of generality each constant occurring in (3.3.1) and (3.3.2) equal to 1. Note, since we are seeking solutions of the form $u = u_\mathrm{s} + u_\mathrm{tp} = \mathcal{P}u + \mathcal{P}_\perp u$, where $u_\mathrm{s} \in \dot{\mathrm{W}}^{2,q}(\Omega)$ is the solution to the linear elliptic problem

$$\begin{cases} -\Delta u_\mathrm{s} = f_\mathrm{s} & \text{in } \Omega, \\ \quad\ u_\mathrm{s} = g_\mathrm{s} & \text{on } \partial\Omega, \end{cases} \tag{3.3.3}$$

and u_tp solves the nonlinear problem

$$\begin{cases} \partial_t^2 u_\mathrm{tp} - \Delta u_\mathrm{tp} - \partial_t \Delta u_\mathrm{tp} - \partial_t\big((\partial_t u_\mathrm{tp})^2 + |\nabla u_\mathrm{tp}|^2\big) \\ \qquad\qquad\qquad\qquad -2\nabla u_\mathrm{s} \cdot \nabla \partial_t u_\mathrm{tp} = f_\mathrm{tp} & \text{in } \mathbb{T} \times \Omega, \\ \qquad\qquad\qquad\qquad\qquad\qquad\qquad\quad u_\mathrm{tp} = g_\mathrm{tp} & \text{on } \mathbb{T} \times \partial\Omega, \end{cases} \tag{3.3.4}$$

the nonlinear terms only appear in (3.3.4) and become

$$\partial_t u \partial_t^2 u = \partial_t (u_\mathrm{s} + u_\mathrm{tp}) \partial_t^2 (u_\mathrm{s} + u_\mathrm{tp}) = \partial_t u_\mathrm{tp} \partial_t^2 u_\mathrm{tp},$$
$$\nabla u \cdot \nabla \partial_t u = \nabla(u_\mathrm{s} + u_\mathrm{tp}) \cdot \nabla \partial_t (u_\mathrm{s} + u_\mathrm{tp}) = \nabla u_\mathrm{tp} \cdot \nabla \partial_t u_\mathrm{tp} + \nabla u_\mathrm{s} \cdot \nabla \partial_t u_\mathrm{tp}.$$

For $\partial_t u_\mathrm{tp} \partial_t^2 u_\mathrm{tp}$ and $\nabla u_\mathrm{tp} \cdot \nabla \partial_t u_\mathrm{tp}$ we deduce the following estimates.

Lemma 3.3.3. *Let n, q and Ω be as in Theorem 3.3.1. Then*

$$\|\partial_t v_{\mathrm{tp}} \partial_t^2 u_{\mathrm{tp}}\|_{\mathrm{L}^q_\perp(\mathbb{T}\times\Omega)} + \|\nabla v_{\mathrm{tp}} \cdot \nabla\partial_t u_{\mathrm{tp}}\|_{\mathrm{L}^q_\perp(\mathbb{T}\times\Omega)}$$
$$\leq C_{24}\|v_{\mathrm{tp}}\|_{\mathbb{X}^q_\perp(\mathbb{T}\times\Omega)}\|u_{\mathrm{tp}}\|_{\mathbb{X}^q_\perp(\mathbb{T}\times\Omega)}$$

holds for any $u_{\mathrm{tp}}, v_{\mathrm{tp}} \in \mathbb{X}^q_\perp(\mathbb{T}\times\Omega)$.

Proof. Clearly, for any $v_{\mathrm{tp}}, u_{\mathrm{tp}} \in \mathbb{X}^q_\perp(\mathbb{T}\times\Omega)$ we have that $\nabla\partial_t u_{\mathrm{tp}} \in \mathrm{L}^q_\perp(\mathbb{T}; \mathrm{W}^{1,q}(\Omega))$ and $\partial_t v_{\mathrm{tp}} \in \mathrm{W}^{1,q}_\perp(\mathbb{T}; \mathrm{L}^q(\Omega)) \cap \mathrm{L}^q_\perp(\mathbb{T}; \mathrm{W}^{2,q}(\Omega))$, *i.e.*, Theorem 2.6.3 $(m=1)$ yields for $\max\{\frac{5}{2}, \frac{5n}{6}\} < q < n$

$$\alpha = \frac{4}{5}: \quad \|\partial_t v_{\mathrm{tp}}\|_\infty \leq c_0\|\partial_t v_{\mathrm{tp}}\|_{\mathrm{W}^{1,2,q}_\perp} \leq c_0\|v_{\mathrm{tp}}\|_{\mathbb{X}^q_\perp},$$

$$\alpha = 0: \quad \|\nabla\partial_t u_{\mathrm{tp}}\|_{\mathrm{L}^q_\perp(\mathbb{T}; \mathrm{L}^{\frac{nq}{n-q}}(\Omega))} \leq c_1\|\partial_t u_{\mathrm{tp}}\|_{\mathrm{W}^{1,2,q}_\perp} \leq c_1\|u_{\mathrm{tp}}\|_{\mathbb{X}^q_\perp}.$$

Note that $\nabla v_{\mathrm{tp}} \in \mathrm{L}^q_\perp(\mathbb{T}; \mathrm{W}^{1,q}(\Omega))$ for any $v_{\mathrm{tp}} \in \mathbb{X}^q_\perp(\mathbb{T}\times\Omega) \subset \mathrm{W}^{1,2,q}_\perp(\mathbb{T}\times\Omega)$. Hence, it is allowed to employ Theorem 2.6.3 with $m=1$ and $\alpha \in [0,1]$, which implies for $q > \frac{5n}{6}$ that

$$\alpha = \frac{4}{5}: \quad \|\nabla v_{\mathrm{tp}}\|_{\mathrm{L}^\infty_\perp(\mathbb{T}; \mathrm{L}^n(\Omega))} \leq c_2\|v_{\mathrm{tp}}\|_{\mathrm{W}^{1,2,q}_\perp(\mathbb{T}\times\Omega)} \leq c_2\|v_{\mathrm{tp}}\|_{\mathbb{X}^q_\perp(\mathbb{T}\times\Omega)}.$$

Utilizing Hölder's inequality, we can therefore deduce

$$\|\partial_t v_{\mathrm{tp}} \partial_t^2 u_{\mathrm{tp}}\|_{\mathrm{L}^q_\perp(\mathbb{T}\times\Omega)} + \|\nabla v_{\mathrm{tp}} \cdot \nabla\partial_t u_{\mathrm{tp}}\|_{\mathrm{L}^q_\perp(\mathbb{T}\times\Omega)}$$
$$\leq \|\partial_t v_{\mathrm{tp}}\|_{\mathrm{L}^\infty_\perp(\mathbb{T}\times\Omega)}\|\partial_t^2 u_{\mathrm{tp}}\|_{\mathrm{L}^q_\perp(\mathbb{T}\times\Omega)}$$
$$+ \|\nabla v_{\mathrm{tp}}\|_{\mathrm{L}^\infty_\perp(\mathbb{T}; \mathrm{L}^n(\Omega))}\|\nabla\partial_t u_{\mathrm{tp}}\|_{\mathrm{L}^q_\perp(\mathbb{T}; \mathrm{L}^{\frac{nq}{n-q}}(\Omega))}$$
$$\leq C_{24}\|v_{\mathrm{tp}}\|_{\mathbb{X}^q_\perp(\mathbb{T}\times\Omega)}\|u_{\mathrm{tp}}\|_{\mathbb{X}^q_\perp(\mathbb{T}\times\Omega)},$$

with $C_{24} > 0$. $\qquad\square$

Observe that the restriction $q > \max\{\frac{5}{2}, \frac{5n}{6}\}$ in Lemma 3.3.3 is required to obtain the estimates for $\partial_t u_{\mathrm{tp}}$ and ∇v_{tp} whereas $1 < q < n$ is needed for the remaining one.

Proof of Theorem 3.3.1. We shall establish existence of a solution u to (3.3.1) of the form $u = u_{\mathrm{s}} + u_{\mathrm{tp}}$, where $u_{\mathrm{s}} \in \dot{\mathrm{W}}^{2,q}(\Omega)$ is a solution to the steady-state problem (3.3.3) and $u_{\mathrm{tp}} \in \mathbb{X}^q_\perp(\mathbb{T}\times\Omega)$ is a solution to the purely oscillatory problem (3.3.4). Standard theory for elliptic boundary value problems (see Lemma 3.2.1 and Lemma 3.2.2) yields for every $q \in$

$(1, n)$, $f_\mathrm{s} \in \mathrm{L}^q(\Omega)$ and $g_\mathrm{s} \in \mathrm{W}^{2-\frac{1}{q}, q}(\partial\Omega)$ a solution $u_\mathrm{s} \in \dot{\mathrm{W}}^{2,q}(\Omega)$ to (3.3.3) with

$$\|\nabla u_\mathrm{s}\|_{\frac{nq}{n-q}} \leq \|\nabla^2 u_\mathrm{s}\|_q \leq c_0 \left(\|f_\mathrm{s}\|_q + \|g_\mathrm{s}\|_{2-\frac{1}{q}, q} \right). \tag{3.3.5}$$

The solution to (3.3.4) shall be obtained as a fixed point of the mapping

$$\mathcal{N} \colon \mathbb{X}_\perp^q(\mathbb{T} \times \Omega) \to \mathbb{X}_\perp^q(\mathbb{T} \times \Omega),$$
$$\mathcal{N}(u_\mathrm{tp}) := \mathrm{A_D}^{-1} \left(\partial_t \left((\partial_t u_\mathrm{tp})^2 + |\nabla u_\mathrm{tp}|^2 \right) + 2\nabla u_\mathrm{s} \cdot \nabla \partial_t u_\mathrm{tp} + f_\mathrm{tp}, g_\mathrm{tp} \right)$$

with $\mathrm{A_D}$ defined as

$$\begin{aligned} \mathrm{A_D} \colon & \mathbb{X}_\perp^q(\mathbb{T} \times \Omega) \to \mathrm{L}_\perp^q\left(\mathbb{T}; \mathrm{L}^q(\Omega)\right) \times \mathrm{T}_{\perp, D_1}^q(\mathbb{T} \times \partial\Omega), \\ \mathrm{A_D}(u_\mathrm{tp}) & := \left(\partial_t^2 u_\mathrm{tp} - \Delta u_\mathrm{tp} - \partial_t \Delta u_\mathrm{tp}, \mathrm{Tr}_0\, u_\mathrm{tp} \right). \end{aligned} \tag{3.3.6}$$

We shall verify that $\mathcal{N}$ is a contracting self-mapping on a closed ball of sufficiently small radius. For this purpose, let $\rho > 0$ and consider some $u_\mathrm{tp} \in \mathbb{X}_\perp^q(\mathbb{T} \times \Omega) \cap B_\rho$. Since $\mathrm{A_D}$ is an isomorphism, we obtain

$$\begin{aligned} \|\mathcal{N}(u_\mathrm{tp})\|_{\mathbb{X}_\perp^q} \leq c_1 \|\mathrm{A_D}^{-1}\| & \left(\|\partial_t u_\mathrm{tp} \partial_t^2 u_\mathrm{tp}\|_q + \|\nabla u_\mathrm{tp} \cdot \partial_t \nabla u_\mathrm{tp}\|_q \right. \\ & \left. + \|\nabla u_\mathrm{s} \cdot \nabla \partial_t u_\mathrm{tp}\|_q + \|f_\mathrm{tp}\|_q + \|g_\mathrm{tp}\|_{\mathrm{T}_{\perp, D_1}^q} \right). \end{aligned}$$

Utilizing Lemma 3.3.3, we find that

$$\|\partial_t u_\mathrm{tp} \partial_t^2 u_\mathrm{tp}\|_q + \|\nabla u_\mathrm{tp} \cdot \partial_t \nabla u_\mathrm{tp}\|_q \leq c_2 \|u_\mathrm{tp}\|_{\mathbb{X}_\perp^q}^2.$$

Employing (3.3.5) and Theorem 2.6.3 with $m = 1$ and $\alpha = 0$, we further obtain

$$\begin{aligned} \|\nabla u_\mathrm{s} \cdot \nabla \partial_t u_\mathrm{tp}\|_{\mathrm{L}_\perp^q(\mathbb{T} \times \Omega)} & \leq \|\nabla u_\mathrm{s}\|_{\mathrm{L}^{\frac{nq}{n-q}}(\Omega)} \|\nabla \partial_t u_\mathrm{tp}\|_{\mathrm{L}_\perp^q(\mathbb{T}; \mathrm{L}^n(\Omega))} \\ & \leq c_3 \|\nabla^2 u_\mathrm{s}\|_{\mathrm{L}^q(\Omega)} \|u_\mathrm{tp}\|_{\mathbb{X}_\perp^q(\mathbb{T} \times \Omega)}. \end{aligned}$$

Consequently,

$$\|\mathcal{N}(u_\mathrm{tp})\|_{\mathbb{X}_\perp^q(\mathbb{T} \times \Omega)} \leq c_4 \left(\rho^2 + \varepsilon\rho + \varepsilon \right).$$

Choosing $\varepsilon = \rho^2$ and ρ sufficiently small, we have $c_4(\rho^2 + \varepsilon\rho + \varepsilon) \leq \rho$, *i.e.*, $\mathcal{N}$ is a self-mapping on B_ρ. Moreover, we have

$$\begin{aligned} \|\mathcal{N}(u_\mathrm{tp}) - \mathcal{N}(v_\mathrm{tp})\|_{\mathbb{X}_\perp^q(\mathbb{T} \times \Omega)} \leq c_5 \Big(& \|\partial_t u_\mathrm{tp} \partial_t^2 u_\mathrm{tp} - \partial_t v_\mathrm{tp} \partial_t^2 v_\mathrm{tp}\|_{\mathrm{L}_\perp^q(\mathbb{T} \times \Omega)} \\ & + \|\nabla u_\mathrm{tp} \cdot \partial_t \nabla u_\mathrm{tp} - \nabla v_\mathrm{tp} \cdot \partial_t \nabla v_\mathrm{tp}\|_{\mathrm{L}_\perp^q(\mathbb{T} \times \Omega)} \\ & + \|\nabla u_\mathrm{s} \cdot \partial_t \nabla u_\mathrm{tp} - \nabla u_\mathrm{s} \cdot \partial_t \nabla v_\mathrm{tp}\|_{\mathrm{L}_\perp^q(\mathbb{T} \times \Omega)} \Big), \end{aligned}$$

and by adding some zeros and utilizing the triangle inequality we further estimate

$$\|\mathcal{N}(u_{\mathrm{tp}}) - \mathcal{N}(v_{\mathrm{tp}})\|_{\mathbb{X}_{\perp}^q(\mathbb{T}\times\Omega)}$$
$$\leq c_5\Big(\|\nabla u_{\mathrm{s}} \cdot \partial_t \nabla (u_{\mathrm{tp}} - v_{\mathrm{tp}})\|_{\mathrm{L}_{\perp}^q(\mathbb{T}\times\Omega)} + \|\partial_t^2 v_{\mathrm{tp}} \partial_t (u_{\mathrm{tp}} - v_{\mathrm{tp}})\|_{\mathrm{L}_{\perp}^q(\mathbb{T}\times\Omega)}$$
$$+ \|\partial_t u_{\mathrm{tp}} \partial_t^2 (u_{\mathrm{tp}} - v_{\mathrm{tp}})\|_{\mathrm{L}_{\perp}^q(\mathbb{T}\times\Omega)} + \|\nabla u_{\mathrm{tp}} \cdot \partial_t \nabla (u_{\mathrm{tp}} - v_{\mathrm{tp}})\|_{\mathrm{L}_{\perp}^q(\mathbb{T}\times\Omega)}$$
$$+ \|\partial_t \nabla v_{\mathrm{tp}} \cdot \nabla (u_{\mathrm{tp}} - v_{\mathrm{tp}})\|_{\mathrm{L}_{\perp}^q(\mathbb{T}\times\Omega)}\Big).$$

In view of Lemma 3.3.3 and the choice of ε and ρ we deduce

$$\|\mathcal{N}(u_{\mathrm{tp}}) - \mathcal{N}(v_{\mathrm{tp}})\|_{\mathbb{X}_{\perp}^q(\mathbb{T}\times\Omega)} \leq c_6(4\rho + \varepsilon)\|u_{\mathrm{tp}} - v_{\mathrm{tp}}\|_{\mathbb{X}_{\perp}^q(\mathbb{T}\times\Omega)}$$
$$\leq c_7(4\rho + \rho^2)\|u_{\mathrm{tp}} - v_{\mathrm{tp}}\|_{\mathbb{X}_{\perp}^q(\mathbb{T}\times\Omega)}.$$

Therefore, if ρ is sufficiently small, $\mathcal{N}$ becomes a contracting self-mapping. By the contraction mapping principle, existence of a fixed point of $\mathcal{N}$ follows. This concludes the proof. $\square$

Proof of Theorem 3.3.2. The assertion follows similarly to the proof of Theorem 3.3.1, by considering

$$A_{\mathrm{N}}\colon \mathbb{X}_{\perp}^q(\mathbb{T}\times\Omega) \to \mathrm{L}_{\perp}^q(\mathbb{T};\mathrm{L}^q(\Omega)) \times \mathrm{T}_{\perp,N_1}^q(\mathbb{T}\times\partial\Omega),$$
$$A_{\mathrm{N}}(u_{\mathrm{tp}}) := (\partial_t^2 u_{\mathrm{tp}} - \Delta u_{\mathrm{tp}} - \partial_t \Delta u_{\mathrm{tp}}, \mathrm{Tr}_1 u_{\mathrm{tp}})$$

instead of A_{D}. $\square$

3.4 The Blackstock-Crighton Equation

In the present section we are going to analyze the Blackstock-Crighton-Kuznetsov equation (3.0.2) and the Blackstock-Crighton-Westervelt equation (3.0.3). That is

$$(a\Delta - \partial_t)\left(\partial_t^2 u - c^2 \Delta u - b\partial_t \Delta u\right) - \partial_t^2\left(\lambda(\partial_t u)^2 + s|\nabla u|^2\right) = f \quad (3.4.1)$$

where we have used the same notation $\lambda := \frac{1}{c^2}\left((1-s)+\frac{B}{2A}\right)$ and $s \in \{0,1\}$ as in [9]. Note, that (3.4.1) is the Blackstock-Crighton-Kuznetsov equation (3.0.2) if $s = 1$, and for $s = 0$ the Blackstock-Crighton-Westervelt equation (3.0.3). Instead of considering these two boundary value problems separately, we will study (3.4.1). The initial-value problem corresponding to the boundary value problems (3.4.1) has been investigated recently in the articles by BRUNNHUBER and KALTENBACHER [10], BRUNNHUBER [9], and BRUNNHUBER and MEYER [11]. Finally, we would like to draw the readers attention to the recent work of TANI [86], in which a new model for nonlinear wave propagation that addresses some shortcomings in the Blackstock-Crighton model concerning extreme propagation distances is derived and analyzed.

In order to show the absence of resonance in (3.4.1), we will proceed similarly to the previous section and employ a fixed-point argument based on the L^q estimates satisfied by the solution to the corresponding linearized system (3.0.5), that is

$$(a\Delta - \partial_t)\left(\partial_t^2 u - c^2 \Delta u - b\partial_t \Delta u\right) = f. \tag{3.4.2}$$

Compared to the damped wave (3.0.4), we will study (3.4.2) only on bounded domains. The reason hereof is that u_s as a solution to the stationary part of (3.4.2), namely the Bi-Laplacian, in an unbounded domain does not have enough regularity to carry out a fixed-point argument, when the Blackstock-Crighton equation (3.4.1) is examined. Note that (3.4.2) is in particular the damped wave equation coupled with the heat equation. Based on this observation we decompose (3.4.2) into the time-periodic heat equation investigated in [61] and the time-periodic damped wave equation, which has been subject of Section 3.2.

We shall treat (3.4.2) with inhomogeneous boundary values of both Dirichlet and Neumann type. Furthermore, we let $n \geq 2$ and denote by $\Omega \subset \mathbb{R}^n$ a bounded (spatial) domain of class $C^{3,1}$. The Blackstock-Crighton equation (in generalized form) with Dirichlet boundary condition then reads

$$\begin{cases} (a\Delta - \partial_t)\left(\partial_t^2 u - c^2 \Delta u - b\partial_t \Delta u\right) \\ \qquad\qquad -\partial_t^2\left(\lambda(\partial_t u)^2 + s|\nabla u|^2\right) = f & \text{in } \mathbb{T} \times \Omega, \\ \qquad\qquad\qquad (u, \Delta u) = (g, h) & \text{on } \mathbb{T} \times \partial\Omega, \end{cases} \tag{3.4.3}$$

with λ and s as above. The corresponding Neumann problem reads

$$\begin{cases} (a\Delta - \partial_t)\left(\partial_t^2 u - c^2 \Delta u - b\partial_t \Delta u\right) \\ \qquad\qquad -\partial_t^2\left(\lambda(\partial_t u)^2 + s|\nabla u|^2\right) = f & \text{in } \mathbb{T} \times \Omega, \\ \qquad\qquad\qquad (\partial_\nu u, \partial_\nu \Delta u) = (g, h) & \text{on } \mathbb{T} \times \partial\Omega. \end{cases} \tag{3.4.4}$$

As the main result in this section, we show for given $\mathcal{T}$-time-periodic data f, g and h in appropriate function spaces, and sufficiently restricted in size, the existence of a time-periodic solution u to (3.4.3) and (3.4.4). In the case of Dirichlet boundary conditions, the result can be stated as follows:

Theorem 3.4.1. *Let $\Omega \subset \mathbb{R}^n$, $n \geq 2$, be a bounded domain with a $C^{3,1}$-smooth boundary, and $\max\{2, \frac{n}{2}\} < q < \infty$. There is an $\varepsilon > 0$ such that for all $f \in \mathrm{L}^q(\mathbb{T}; \mathrm{L}^q(\Omega))$, $g \in \mathrm{T}^q_{D_2}(\mathbb{T} \times \partial\Omega)$ and $h \in \mathrm{T}^q_{D_1}(\mathbb{T} \times \partial\Omega)$ satisfying*

$$\|f\|_{\mathrm{L}^q(\mathbb{T} \times \Omega)} + \|g\|_{\mathrm{T}^q_{D_2}(\mathbb{T} \times \partial\Omega)} + \|h\|_{\mathrm{T}^q_{D_1}(\mathbb{T} \times \partial\Omega)} \leq \varepsilon$$

there is a solution $u \in \mathbb{Y}^q(\mathbb{T} \times \Omega)$ to (3.4.3).

A solution to the Neumann problem (3.4.4) only exists when the data satisfies certain compatibility conditions. More precisely, we obtain:

Theorem 3.4.2. *Let n, Ω and q be as in Theorem 3.4.1. There is an $\varepsilon > 0$ such that for all $f \in \mathrm{L}^q(\mathbb{T}; \mathrm{L}^q(\Omega))$, $g \in \mathrm{T}^q_{N_2}(\mathbb{T} \times \partial\Omega)$ and $h \in \mathrm{T}^q_{N_1}(\mathbb{T} \times \partial\Omega)$ satisfying*

$$\|f\|_{\mathrm{L}^q(\mathbb{T} \times \Omega)} + \|g\|_{\mathrm{T}^q_{N_2}(\mathbb{T} \times \partial\Omega)} + \|h\|_{\mathrm{T}^q_{N_1}(\mathbb{T} \times \partial\Omega)} \leq \varepsilon$$

and

$$\int\limits_0^{\mathcal{T}} \int\limits_\Omega f \, \mathrm{d}x \, \mathrm{d}t + ac^2 \int\limits_0^{\mathcal{T}} \int\limits_{\partial\Omega} h \, \mathrm{d}S \, \mathrm{d}t = 0$$

there is a solution $u \in \mathbb{Y}^q(\mathbb{T} \times \Omega)$ to (3.4.4).

We proceed similarly to the analysis of the Kuznetsov equation and consider first the linearized Blackstock-Crighton equation. Existence of a solution will be proven. Moreover, we shall establish L^q estimates for the solution to the linearized problem. This is outlined in the following subsection. But before we consider the linearized systems corresponding to (3.4.3) and (3.4.4) we again recall some notation for the function spaces occurring in this section. As introduced in Subsection 2.6 we denote by

$$\mathbb{Y}^q_\perp(\mathbb{T} \times \Omega) := \mathcal{P}_\perp \mathrm{W}^{3,q}(\mathbb{T}; \mathrm{L}^q(\Omega)) \cap \mathcal{P}_\perp \mathrm{W}^{1,q}(\mathbb{T}; \mathrm{W}^{4,q}(\Omega))$$
$$= \mathrm{W}^{3,q}_\perp(\mathbb{T}; \mathrm{L}^q(\Omega)) \cap \mathrm{W}^{1,q}_\perp(\mathbb{T}; \mathrm{W}^{4,q}(\Omega)),$$

the anisotropic Sobolev space for the solution to the systems (3.4.3) and (3.4.4), equipped with the canonical Sobolev norm

$$\|f\|_{\mathbb{Y}_\perp^q(\mathbb{T}\times\Omega)} := \left(\|f\|_{\mathrm{W}_\perp^{3,q}(\mathbb{T};\mathrm{L}^q(\Omega))}^q + \|f\|_{\mathrm{W}_\perp^{1,q}(\mathbb{T};\mathrm{W}^{4,q}(\Omega))}^q \right)^{\frac{1}{q}}.$$

Observe, since the Blackstock-Crighton equation is a partial differential equation of order four, a second boundary condition in (3.4.3) and (3.4.4) is needed, and the trace operators associated to the Dirichlet and Neumann trace are defined as in Section 2.6, *i.e.*,

$$\mathrm{Tr}_D \colon \mathbb{Y}_\perp^q(\mathbb{T}\times\Omega) \to \mathrm{T}_{\perp,D_2}^q(\mathbb{T}\times\partial\Omega) \times \mathrm{T}_{\perp,D_1}^q(\mathbb{T}\times\partial\Omega),$$
$$\mathrm{Tr}_D(u) := (u, \Delta u)_{|\mathbb{T}\times\partial\Omega},$$

and

$$\mathrm{Tr}_N \colon \mathbb{Y}_\perp^q(\mathbb{T}\times\Omega) \to \mathrm{T}_{\perp,N_2}^q(\mathbb{T}\times\partial\Omega) \times \mathrm{T}_{\perp,N_1}^q(\mathbb{T}\times\partial\Omega),$$
$$\mathrm{Tr}_N(u) := (\partial_\nu u, \partial_\nu \Delta u)_{|\mathbb{T}\times\partial\Omega}.$$

Here, the trace spaces $\mathrm{T}_{\perp,D_1}^q(\mathbb{T}\times\partial\Omega)$ and $\mathrm{T}_{\perp,N_1}^q(\mathbb{T}\times\partial\Omega)$ are defined as in Section 3.2 and the other ones are given by

$$\mathrm{T}_{\perp,D_2}^q(\mathbb{T}\times\partial\Omega) := \mathrm{W}_\perp^{3-\frac{1}{2q},q}\big(\mathbb{T};\mathrm{L}^q(\partial\Omega)\big) \cap \mathrm{W}_\perp^{1,q}\big(\mathbb{T};\mathrm{W}^{4-\frac{1}{q},q}(\partial\Omega)\big),$$
$$\mathrm{T}_{\perp,N_2}^q(\mathbb{T}\times\partial\Omega) := \mathrm{W}_\perp^{\frac{5}{2}-\frac{1}{2q},q}\big(\mathbb{T};\mathrm{L}^q(\partial\Omega)\big) \cap \mathrm{W}_\perp^{1,q}\big(\mathbb{T};\mathrm{W}^{3-\frac{1}{q},q}(\partial\Omega)\big),$$

3.4.1 The Linearized Blackstock-Crighton Equation

Finally we are going to investigate the linearized Blackstock-Crighton equation (3.4.2), which is a slightly different version of the damped wave equation (3.0.4). To be more specific, we will study

$$(a\Delta - \partial_t)\left(\partial_t^2 v - c^2\Delta v - b\partial_t \Delta v\right) = f,$$

which is a combination of the damped wave equation (3.0.4) and the heat equation. Based on this observation, we decompose (3.4.2) into a time-periodic damped wave equation

$$\partial_t^2 v - c^2\Delta v - b\partial_t\Delta v = u, \tag{3.4.5}$$

studied in the previous section, coupled with a time-periodic heat equation

$$-\partial_t u + a\Delta u = f. \tag{3.4.6}$$

This allows us to employ the results from the previous section and some known results for the time-periodic heat equation, see [61]. Here, a, b and c are positive constants as described in the introduction of this chapter. As for the damped wave equation, we will study (3.4.2) with both, Dirichlet and Neumann boundary values. In the Dirichlet case we will investigate

$$\begin{cases} \left(a\Delta - \partial_t\right)\left(\partial_t^2 v - c^2\Delta v - b\partial_t\Delta v\right) = f & \text{in } \mathbb{T} \times \Omega, \\ \left(v, \Delta v\right) = \left(g, h\right) & \text{on } \mathbb{T} \times \partial\Omega, \end{cases} \tag{3.4.7}$$

whereas the corresponding Neumann boundary value problem reads

$$\begin{cases} \left(a\Delta - \partial_t\right)\left(\partial_t^2 v - c^2\Delta v - b\partial_t\Delta v\right) = f & \text{in } \mathbb{T} \times \Omega, \\ \left(\dfrac{\partial v}{\partial \nu}, \dfrac{\partial \Delta v}{\partial \nu}\right) = \left(g, h\right) & \text{on } \mathbb{T} \times \partial\Omega. \end{cases} \tag{3.4.8}$$

Problem (3.4.7) and (3.4.8) will be investigated in a bounded domain $\Omega \subset \mathbb{R}^n$, where $n \geq 2$. Without loss of generality, we set $a = b = c = 1$.

Employing $\mathcal{P}$ and $\mathcal{P}_\perp$, we decompose (3.4.7) into the corresponding stationary problem

$$\begin{cases} -\Delta^2 v_{\mathrm{s}} = f_{\mathrm{s}} & \text{in } \Omega, \\ \left(v_{\mathrm{s}}, \Delta v_{\mathrm{s}}\right) = \left(g_{\mathrm{s}}, h_{\mathrm{s}}\right) & \text{on } \partial\Omega, \end{cases} \tag{3.4.9}$$

and the purely oscillatory problem

$$\begin{cases} \left(\Delta - \partial_t\right)\left(\partial_t^2 v_{\mathrm{tp}} - \Delta v_{\mathrm{tp}} - \partial_t\Delta v_{\mathrm{tp}}\right) = f_{\mathrm{tp}} & \text{in } \mathbb{T} \times \Omega, \\ \left(v_{\mathrm{tp}}, \Delta v_{\mathrm{tp}}\right) = \left(g_{\mathrm{tp}}, h_{\mathrm{tp}}\right) & \text{on } \mathbb{T} \times \partial\Omega. \end{cases} \tag{3.4.10}$$

The corresponding Neumann boundary value problem splits up into

$$\begin{cases} -\Delta^2 v_{\mathrm{s}} = f_{\mathrm{s}} & \text{in } \Omega, \\ \left(\dfrac{\partial v_{\mathrm{s}}}{\partial \nu}, \dfrac{\partial \Delta v_{\mathrm{s}}}{\partial \nu}\right) = \left(g_{\mathrm{s}}, h_{\mathrm{s}}\right) & \text{on } \partial\Omega, \end{cases} \tag{3.4.11}$$

and

$$\begin{cases} \left(\Delta - \partial_t\right)\left(\partial_t^2 v_{\mathrm{tp}} - \Delta v_{\mathrm{tp}} - \partial_t\Delta v_{\mathrm{tp}}\right) = f_{\mathrm{tp}} & \text{in } \mathbb{T} \times \Omega, \\ \left(\dfrac{\partial v_{\mathrm{tp}}}{\partial \nu}, \dfrac{\partial \Delta v_{\mathrm{tp}}}{\partial \nu}\right) = \left(g_{\mathrm{tp}}, h_{\mathrm{tp}}\right) & \text{on } \mathbb{T} \times \partial\Omega. \end{cases} \tag{3.4.12}$$

We begin by investigating the purely oscillatory problem (3.4.10) and (3.4.12) via the approach described above. That is, we decompose (3.4.10) into the time-periodic heat equation (3.4.6) and the time-periodic damped wave equation (3.4.5) with inhomogeneous Dirichlet boundary values and study those problems separately. Similarly, (3.4.12) is split into the same systems, but instead of Dirichlet we have Neumann boundary conditions. Consequently, we can employ the results from Theorem 3.2.12, Theorem 3.2.13 and [61], and prove existence of nonresonant solutions to (3.4.10) and (3.4.12). Since the proof is highly based on this decomposition and the results for (3.4.5) and (3.4.6), we will treat (3.4.10) and (3.4.12) together in the following Theorem.

Theorem 3.4.3. *Assume that $\Omega \subset \mathbb{R}^n$, $n \geq 2$, is a bounded domain with a $C^{3,1}$-smooth boundary. Let $q \in (1, \infty)$. Then*

$$\mathrm{A_D} \colon \mathbb{Y}_\perp^q(\mathbb{T} \times \Omega) \to \mathrm{L}_\perp^q\big(\mathbb{T}; \mathrm{L}^q(\Omega)\big) \times \mathrm{T}_{\perp, D_2}^q(\mathbb{T} \times \partial\Omega) \times \mathrm{T}_{\perp, D_1}^q(\mathbb{T} \times \partial\Omega),$$
$$\mathrm{A_D}(v) := \big((\Delta - \partial_t)\big(\partial_t^2 v - \Delta v - \partial_t \Delta v\big), \mathrm{Tr}_D \, v\big)$$

and

$$\mathrm{A_N} \colon \mathbb{Y}_\perp^q(\mathbb{T} \times \Omega) \to \mathrm{L}_\perp^q\big(\mathbb{T}; \mathrm{L}^q(\Omega)\big) \times \mathrm{T}_{\perp, N_2}^q(\mathbb{T} \times \partial\Omega) \times \mathrm{T}_{\perp, N_1}^q(\mathbb{T} \times \partial\Omega),$$
$$\mathrm{A_N}(v) := \big((\Delta - \partial_t)\big(\partial_t^2 v - \Delta v - \partial_t \Delta v\big), \mathrm{Tr}_N \, v\big),$$

are isomorphisms.

Proof. On the strength of the embedding (2.6.1), we observe that Δ is a bounded operator

$$\Delta \colon \mathbb{Y}_\perp^q(\mathbb{T} \times \Omega) \to \mathbb{X}_\perp^q(\mathbb{T} \times \Omega).$$

Together with the continuity of the trace operators established in Lemma 2.6.7, this implies that the operators $\mathrm{A_D}$ and $\mathrm{A_N}$ are well-defined as bounded operators in the given setting. We start by showing that the operators are surjective. We concentrate on $\mathrm{A_D}$, as the operator $\mathrm{A_N}$ can be treated in a completely similar manner. To this end, let

$$(f, g, h) \in \mathrm{L}_\perp^q\big(\mathbb{T} \times \Omega\big) \times \mathrm{T}_{\perp, D_2}^q(\mathbb{T} \times \partial\Omega) \times \mathrm{T}_{\perp, D_1}^q(\mathbb{T} \times \partial\Omega).$$

Consider the coupled systems

$$\begin{cases} \partial_t^2 w - \Delta w - \partial_t \Delta w = f & \text{in } \mathbb{T} \times \Omega, \\ \qquad\qquad\qquad\quad w = h - \partial_t g & \text{on } \mathbb{T} \times \partial\Omega, \end{cases} \tag{3.4.13}$$

and

$$\begin{cases} \Delta v - \partial_t v = w & \text{in } \mathbb{T} \times \Omega, \\ \qquad\quad v = g & \text{on } \mathbb{T} \times \partial\Omega. \end{cases} \tag{3.4.14}$$

We recognize (3.4.13) as the time-periodic strongly damped wave equation, which was studied in Section 3.2, and (3.4.14) as the time-periodic heat equation, which was subject of the investigation in [61]. Recalling the embedding (2.6.19), we see that ∂_t is bounded as an operator

$$\partial_t : \mathrm{T}^q_{\perp,D_2}(\mathbb{T} \times \partial\Omega) \to \mathrm{T}^q_{\perp,D_1}(\mathbb{T} \times \partial\Omega), \tag{3.4.15}$$

whence $h - \partial_t g \in \mathrm{T}^q_{\perp,D_1}(\mathbb{T} \times \partial\Omega)$. Consequently, we obtain directly from Theorem 3.2.12 existence of a unique solution

$$w \in \mathbb{X}^q_\perp(\mathbb{T} \times \Omega)$$

to (3.4.13).

We now turn to (3.4.14). From [61, Theorem 2.1] and a standard regularity and lifting argument, based on the mapping property of the trace operator Tr_0 established in Lemma 2.6.7, we obtain a unique solution $v \in \mathbb{Y}^q_\perp(\mathbb{T} \times \Omega)$ to (3.4.14). Recalling the embedding (2.6.2), we see that ∂_t is bounded as an operator

$$\partial_t : \mathbb{Y}^q_\perp(\mathbb{T} \times \Omega) \to \mathbb{X}^q_\perp(\mathbb{T} \times \Omega). \tag{3.4.16}$$

Since the operators ∂_t and Tr_0 commute on spaces of smooth functions, it follows from (3.4.15), (3.4.16) and the mapping property of the trace operator Tr_0 asserted in Lemma 2.6.7 that they commute in the setting

$$\partial_t \circ \mathrm{Tr}_0 = \mathrm{Tr}_0 \circ \partial_t : \mathbb{Y}^q_\perp(\mathbb{T} \times \Omega) \to \mathrm{T}^q_{\perp,D_2}(\mathbb{T} \times \partial\Omega).$$

We thus deduce from (3.4.14) and the boundary condition in (3.4.13) that

$$\begin{aligned} \mathrm{Tr}_D\, v &= \big(g, \mathrm{Tr}_0(\Delta v)\big) = \big(g, \mathrm{Tr}_0\,(\partial_t v + w)\big) \\ &= \big(g, \partial_t g + w\big) = \big(g, h\big). \end{aligned}$$

It follows that $\mathrm{A}_D(v) = (f, g, h)$, and we conclude that A_D is surjective.

To show that A_D is injective, consider $v \in \mathbb{Y}^q_\perp(\mathbb{T} \times \Omega)$ with $\mathrm{A}_D(v) = (0, 0, 0)$. Unique solvability of the time-periodic strongly damped wave equation implies $\Delta v - \partial_t v = 0$, see Theorem 3.2.12. In turn, unique solvability of the time-periodic heat equation [61, Theorem 2.1] implies $v = 0$. Consequently, A_D is injective. By the open mapping theorem, A_D is an isomorphism. □

Observe, the $C^{3,1}$-smooth boundary was required to carry out the regularity and lifting argument to deduce higher regularity for the solution $v \in \mathbb{Y}_\perp^q(\mathbb{T} \times \Omega)$ to (3.4.14).

Let us now consider the elliptic problems (3.4.9) and (3.4.11), and prove existence of a unique solution to these systems. For the stationary Dirichlet problem the main results reads as follow:

Lemma 3.4.4. *Let $\Omega \subset \mathbb{R}^n$, $n \geq 2$, be a bounded domain with a $C^{3,1}$-smooth boundary and $q \in (1, \infty)$. For any $f \in L^q(\Omega)$, $g \in W^{4-\frac{1}{q},q}(\partial\Omega)$ and $h \in W^{2-\frac{1}{q},q}(\partial\Omega)$ there exists a unique solution $v \in W^{4,q}(\Omega)$ to (3.4.9) satisfying*

$$\|v\|_{W^{4,q}(\Omega)} \leq C_{25}\big(\|f\|_{L^q(\Omega)} + \|g\|_{W^{4-\frac{1}{q},q}(\partial\Omega)} + \|h\|_{W^{2-\frac{1}{q},q}(\partial\Omega)}\big), \quad (3.4.17)$$

with $C_{25} = C_{25}(n, q, \Omega) > 0$.

Proof. In the case $q = 2$, it can be verified directly that the operator

$$A : W^{4,q}(\Omega) \to L^q(\Omega) \times W^{4-\frac{1}{q},q}(\partial\Omega) \times W^{2-\frac{1}{q},q}(\partial\Omega),$$
$$A v := \big(-\Delta^2 v, v, \Delta v\big),$$

is Fredholm of index 0; see [89, Example 16.6]. Moreover, the kernel is trivial. The operator is therefore an isomorphism in this case. By the celebrated result of GEYMONAT [43] (see also [79]), both the Fredholm index and the kernel are independent on $q \in (1, \infty)$, whence the operator is in fact an isomorphism for all such q. $\square$

For the corresponding Neumann problem a necessary compatibility condition on the data f and h is needed. Similarly to the compatibility condition for the Poisson-Neumann problem (3.2.7), this can be seen by integrating both sides of (3.4.11)$_1$. This implies the following result on existence of stationary solutions to (3.4.11).

Lemma 3.4.5. *Let n, Ω and q be as in Lemma 3.4.4. For any $f \in L^q(\Omega)$, $g \in W^{3-\frac{1}{q},q}(\partial\Omega)$ and $h \in W^{1-\frac{1}{q},q}(\partial\Omega)$ satisfying*

$$\int_\Omega f \, dx + \int_{\partial\Omega} h \, dS = 0 \qquad (3.4.18)$$

there exists a unique solution $v \in W^{4,q}(\Omega)$ to (3.4.11) satisfying

$$\|v\|_{W^{4,q}(\Omega)} \leq C_{26}\big(\|f\|_{L^q(\Omega)} + \|g\|_{W^{3-\frac{1}{q},q}(\partial\Omega)} + \|h\|_{W^{1-\frac{1}{q},q}(\partial\Omega)}\big), \quad (3.4.19)$$

with $C_{26} = C_{26}(n, q, \Omega) > 0$.

Proof. In the case $q = 2$, it can be verified directly that also the operator

$$A : W^{4,q}(\Omega) \to L^q(\Omega) \times W^{3-\frac{1}{q},q}(\partial\Omega) \times W^{1-\frac{1}{q},q}(\partial\Omega),$$
$$A v := \left(-\Delta^2 v, \partial_\nu v, \partial_\nu \Delta v \right),$$

is Fredholm of index 0; see again [89, Example 16.6]. It is easy to see that its kernel consists of the constants, and is therefore one-dimensional. It follows that also the kernel of the adjoint operator is one-dimensional and is given by the span of the constant $(1, 0, 1)$. Since $\mathscr{R}(A) = \mathscr{N}(A^*)^\perp$, data (f, g, h) satisfying (3.4.18) lie in the range of A. Since the range of a Fredholm operator is closed, existence of a solution $v \in W^{4,q}(\Omega)$ to (3.4.11) satisfying (3.4.19) follows. $\qquad\square$

Summarizing the results established in this subsection, we have shown the existence of a $\mathcal{T}$-time-periodic solution to (3.4.7) and (3.4.8) under periodic forcing. Based on these results, we will investigate the fully inhomogeneous Blackstock-Crighton equation with both Dirichlet and Neumann boundary values.

3.4.2 The Nonlinear Problem

In order to prove Theorem 3.4.1 and Theorem 3.4.2 we shall proceed similarly to the investigation of the Kuznetsov equation (3.0.1) in Section 3.3, that is, we utilize the contraction mapping principle and the embedding Theorem 2.6.3, which are based on the *a priori* L^q estimates for the corresponding linearizations (3.4.7) and (3.4.8) of (3.4.3) and (3.4.4), respectively. Therefore, we have to find appropriate L^q estimates for the nonlinear terms occurring in

$$(a\Delta - \partial_t)\left(\partial_t^2 u - c^2\Delta u - b\partial_t\Delta u\right) - \partial_t^2\left(k(\partial_t u)^2 + s|\nabla u|^2\right) = f \quad \text{in } \mathbb{T} \times \Omega. \tag{3.4.20}$$

Since we are again seeking solutions of the form $u = u_\text{s} + u_\text{tp}$, where u_s is independent of time and u_tp satisfies a mean-zero condition in time, the nonlinear terms are given by

$$\partial_t^2\left((\partial_t u)^2\right) = 2\left((\partial_t^2 u_\text{tp})^2 + \partial_t u_\text{tp}\partial_t^3 u_\text{tp}\right),$$
$$\partial_t^2\left(|\nabla u|^2\right) = 2\left(|\nabla\partial_t u_\text{tp}|^2 + \nabla u_\text{s} \cdot \nabla\partial_t^2 u_\text{tp} + \nabla u_\text{tp} \cdot \nabla\partial_t^2 u_\text{tp}\right).$$

These terms can be estimated as follows:

Lemma 3.4.6. *For $n \geq 2$ let $\Omega \subset \mathbb{R}^n$ be a bounded domain with a $C^{3,1}$ boundary and $\max\{2, \frac{n}{2}\} < q < \infty$. Then*

$$\|\partial_t v_{\mathrm{tp}} \, \partial_t^3 u_{\mathrm{tp}}\|_{\mathrm{L}_\perp^q(\mathbb{T}\times\Omega)} + \|\partial_t^2 v_{\mathrm{tp}} \, \partial_t^2 u_{\mathrm{tp}}\|_{\mathrm{L}_\perp^q(\mathbb{T}\times\Omega)}$$
$$+ \|\nabla \partial_t v_{\mathrm{tp}} \cdot \nabla \partial_t u_{\mathrm{tp}}\|_{\mathrm{L}_\perp^q(\mathbb{T}\times\Omega)} + \|\nabla v_{\mathrm{tp}} \cdot \nabla \partial_t^2 u_{\mathrm{tp}}\|_{\mathrm{L}_\perp^q(\mathbb{T}\times\Omega)}$$
$$\leq C_{27} \|v_{\mathrm{tp}}\|_{\mathbb{Y}_\perp^q(\mathbb{T}\times\Omega)} \|u_{\mathrm{tp}}\|_{\mathbb{Y}_\perp^q(\mathbb{T}\times\Omega)}$$

holds for any $u_{\mathrm{tp}}, v_{\mathrm{tp}} \in \mathbb{Y}_{\mathrm{per},\perp}^q(\mathbb{T}\times\Omega)$.

Proof. Observe that $\partial_t v_{\mathrm{tp}} \in \mathrm{W}_\perp^{2,4,q}(\mathbb{T}\times\Omega)$ for all $v_{\mathrm{tp}} \in \mathbb{Y}_\perp^q(\mathbb{T}\times\Omega)$. For any $q \in (\max\{2, \frac{n}{2}\}, \infty)$, Theorem 2.6.3 with $m = 2$ yields

$$\alpha = 1: \quad \|\partial_t v_{\mathrm{tp}}\|_\infty \leq c_0 \|v_{\mathrm{tp}}\|_{\mathbb{Y}_\perp^q(\mathbb{T}\times\Omega)},$$
$$\alpha = 0: \quad \|\nabla \partial_t v_{\mathrm{tp}}\|_{\mathrm{L}_\perp^q(\mathbb{T};\mathrm{L}^\infty(\Omega))} \leq c_1 \|v_{\mathrm{tp}}\|_{\mathbb{Y}_\perp^q(\mathbb{T}\times\Omega)},$$
$$\alpha = 3: \quad \|\nabla \partial_t u_{\mathrm{tp}}\|_{\mathrm{L}_\perp^\infty(\mathbb{T};\mathrm{L}^q(\Omega))} \leq c_2 \|u_{\mathrm{tp}}\|_{\mathbb{Y}_\perp^q(\mathbb{T}\times\Omega)}.$$

For the remaining terms Theorem 2.6.3 implies $(m = 1)$

$$\alpha = 2: \quad \|\partial_t^2 v_{\mathrm{tp}}\|_{\mathrm{L}_\perp^\infty(\mathbb{T};\mathrm{L}^q(\Omega))} \leq c_3 \|v_{\mathrm{tp}}\|_{\mathbb{Y}_\perp^q(\mathbb{T}\times\Omega)},$$
$$\alpha = 0: \quad \|\partial_t^2 u_{\mathrm{tp}}\|_{\mathrm{L}_\perp^q(\mathbb{T};\mathrm{L}^\infty(\Omega))} + \|\nabla u_{\mathrm{tp}}\|_{\mathrm{L}_\perp^q(\mathbb{T};\mathrm{L}^\infty(\Omega))}$$
$$\leq c_4 \|u_{\mathrm{tp}}\|_{\mathbb{Y}_\perp^q(\mathbb{T}\times\Omega)},$$
$$\alpha = 1: \quad \|\nabla \partial_t^2 u_{\mathrm{tp}}\|_{\mathrm{L}_\perp^\infty(\mathbb{T};\mathrm{L}^q(\Omega))} \leq c_5 \|u_{\mathrm{tp}}\|_{\mathbb{Y}_\perp^q(\mathbb{T}\times\Omega)}.$$

The restriction $q > 2$ is required to obtain the last estimate above whereas $q > \frac{n}{2}$ is necessary for the remaining ones. Utilizing Hölder's inequality, we can therefore deduce

$$\|\nabla v_{\mathrm{tp}} \cdot \nabla \partial_t^2 u_{\mathrm{tp}}\|_q + \|\partial_t v_{\mathrm{tp}} \, \partial_t^3 u_{\mathrm{tp}}\|_q + \|\partial_t^2 v_{\mathrm{tp}} \, \partial_t^2 u_{\mathrm{tp}}\|_q + \|\nabla \partial_t v_{\mathrm{tp}} \cdot \nabla \partial_t u_{\mathrm{tp}}\|_q$$
$$\leq \|\nabla v_{\mathrm{tp}}\|_{\mathrm{L}_\perp^q(\mathbb{T};\mathrm{L}^\infty(\Omega))} \|\nabla \partial_t^2 u_{\mathrm{tp}}\|_{\mathrm{L}_\perp^\infty(\mathbb{T};\mathrm{L}^q(\Omega))} + \|\partial_t v_{\mathrm{tp}}\|_\infty \|\partial_t^3 u_{\mathrm{tp}}\|_q$$
$$+ \|\partial_t^2 v_{\mathrm{tp}}\|_{\mathrm{L}_\perp^\infty(\mathbb{T};\mathrm{L}^q(\Omega))} \|\partial_t^2 u_{\mathrm{tp}}\|_{\mathrm{L}_\perp^q(\mathbb{T};\mathrm{L}^\infty(\Omega))}$$
$$+ \|\nabla \partial_t v_{\mathrm{tp}}\|_{\mathrm{L}_\perp^q(\mathbb{T};\mathrm{L}^\infty(\Omega))} \|\nabla \partial_t u_{\mathrm{tp}}\|_{\mathrm{L}_\perp^\infty(\mathbb{T};\mathrm{L}^q(\Omega))}$$
$$\leq C_{27} \|v_{\mathrm{tp}}\|_{\mathbb{Y}_\perp^q(\mathbb{T}\times\Omega)} \|u_{\mathrm{tp}}\|_{\mathbb{Y}_\perp^q(\mathbb{T}\times\Omega)},$$

with $C_{27} = c_5 c_1 + c_4 c_3 + c_1 c_2 + c_0$. $\qquad\square$

Observe that since we consider the Blackstock-Crighton equation only in a bounded domain u_{s} and u_{tp} have better regularity than the solutions we deduced in the previous section, and therefore u_{s} and u_{tp} satisfy better embedding properties. To be more precisely, we obtain a larger range for the Lebesgue parameter q than for the solution to the Kuznetsov equation.

With this estimates at hand we are able to prove Theorem 3.4.1.

Proof of Theorem 3.4.1. Let us consider functions $f \in \mathrm{L}^q_{\mathrm{per}}\big(\mathbb{T}; \mathrm{L}^q(\Omega)\big)$, $g \in \mathrm{T}^q_{D_2}(\mathbb{T} \times \partial\Omega)$ and $h \in \mathrm{T}^q_{D_1}(\mathbb{T} \times \partial\Omega)$ with $\|f\|_q + \|g\|_{\mathrm{T}^q_{D_2}} + \|h\|_{\mathrm{T}^q_{D_1}} \le \varepsilon$, where $\varepsilon > 0$ has to be specified later. We shall establish existence of a solution u to (3.4.3) of the form $u = u_\mathrm{s} + u_\mathrm{tp}$, where $u_\mathrm{tp} \in \mathbb{Y}^q_\perp(\mathbb{T} \times \Omega)$ is a solution to the purely oscillatory problem

$$
\begin{cases}
\big(a\Delta - \partial_t\big)\big(\partial_t^2 u_\mathrm{tp} - c^2 \Delta u_\mathrm{tp} - b\partial_t \Delta u_\mathrm{tp}\big) \\
\qquad -2\nabla u_\mathrm{s} \cdot \nabla \partial_t^2 u_\mathrm{tp} \\
\qquad +\partial_t^2 \big(\dfrac{1}{c^2}\dfrac{B}{2A}(\partial_t u_\mathrm{tp})^2 + |\nabla u_\mathrm{tp}|^2\big) = f_\mathrm{tp} & \text{in } \mathbb{T} \times \Omega, \\
\qquad\qquad\qquad (u_\mathrm{tp}, \Delta u_\mathrm{tp}) = (g_\mathrm{tp}, h_\mathrm{tp}) & \text{on } \mathbb{T} \times \partial\Omega,
\end{cases}
\tag{3.4.21}
$$

and $u_\mathrm{s} \in \mathrm{W}^{4,q}(\Omega)$ is a solution to the elliptic problem

$$
\begin{cases}
-ac^2 \Delta^2 u_\mathrm{s} = f_\mathrm{s} & \text{in } \Omega, \\
(u_\mathrm{s}, \Delta u_\mathrm{s}) = (g_\mathrm{s}, h_\mathrm{s}) & \text{on } \partial\Omega.
\end{cases}
\tag{3.4.22}
$$

Lemma 3.4.4 yields a solution $u_\mathrm{s} \in \mathrm{W}^{4,p}(\Omega)$ to (3.4.22) satisfying

$$
\|u_\mathrm{s}\|_{4,q} \le c_0 \big(\|f_\mathrm{s}\|_q + \|g_\mathrm{s}\|_{4-\frac{1}{q},q} + \|h_\mathrm{s}\|_{2-\frac{1}{q},q}\big).
$$

Sobolev's embedding theorem implies

$$
\|\nabla u_\mathrm{s}\|_\infty \le c_1 \|\nabla u_\mathrm{s}\|_{3,q} \le c_1 \|u_\mathrm{s}\|_{4,q} \le c_2 \varepsilon,
\tag{3.4.23}
$$

see [2, 4.12 Theorem]. The solution to (3.4.21) shall be obtained as a fixed point of the mapping

$$
\mathcal{N} \colon \mathbb{Y}^q_\perp(\mathbb{T} \times \Omega) \to \mathbb{Y}^q_\perp(\mathbb{T} \times \Omega),
$$

$$
\mathcal{N}(u_\mathrm{tp}) := \mathrm{A_D}^{-1} \bigg(\partial_t^2 \big(\frac{1}{c^2}\frac{B}{2A}(\partial_t u_\mathrm{tp})^2 + |\nabla u_\mathrm{tp}|^2\big)
$$

$$
+ 2\nabla u_\mathrm{s} \cdot \partial_t^2 \nabla u_\mathrm{tp} + f_\mathrm{tp}, g_\mathrm{tp}, h_\mathrm{tp}\bigg)
$$

with $\mathrm{A_D}$ as in Theorem 3.4.3. For this purpose, let $\rho > 0$ and consider some $u_\mathrm{tp} \in \mathbb{Y}^q_\perp(\mathbb{T} \times \Omega) \cap B_\rho$. Since $\mathrm{A_D}$ is an isomorphism, we conclude from Lemma 3.4.6 and (3.4.23) the estimate

$$
\begin{aligned}
\|\mathcal{N}(u_{\mathrm{tp}})\|_{\mathbb{Y}_\perp^q} &\leq \|\mathrm{A_D}^{-1}\| \Big(\|\partial_t u_{\mathrm{tp}} \partial_t^3 u_{\mathrm{tp}}\|_q + \|\partial_t^2 u_{\mathrm{tp}} \partial_t^2 u_{\mathrm{tp}}\|_q + \|f_{\mathrm{tp}}\|_q \\
&\quad + \|\partial_t \nabla u_{\mathrm{tp}} \cdot \partial_t \nabla u_{\mathrm{tp}}\|_q + \|\nabla u_{\mathrm{tp}} \cdot \partial_t^2 \nabla u_{\mathrm{tp}}\|_q + \|g_{\mathrm{tp}}\|_{\mathrm{T}_{\perp,D_2}^q} \\
&\quad + \|\nabla u_{\mathrm{s}} \cdot \nabla \partial_t^2 u_{\mathrm{tp}}\|_q + \|h_{\mathrm{tp}}\|_{\mathrm{T}_{\perp,D_1}^q} \Big) \\
&\leq c_3 (\|u_{\mathrm{tp}}\|_{\mathbb{Y}_\perp^q}^2 + \|\nabla u_{\mathrm{s}}\|_{\mathrm{L}^\infty(\Omega)} \|\nabla \partial_t^2 u_{\mathrm{tp}}\|_{\mathrm{L}^q(\mathrm{T};\mathrm{L}^q(\Omega))} + \varepsilon) \\
&\leq c_4 (\rho^2 + \varepsilon \rho + \varepsilon).
\end{aligned}
$$

Choosing $\rho := \sqrt{\varepsilon}$ and ε sufficiently small, we have $c_4 (\rho^2 + \varepsilon\rho + \varepsilon) \leq \rho$, *i.e.*, $\mathcal{N}$ becomes a self-mapping on B_ρ. Furthermore, observe that

$$
\begin{aligned}
\|\mathcal{N}(u_{\mathrm{tp}}) - \mathcal{N}(v_{\mathrm{tp}})\|_{\mathbb{Y}_\perp^q} &\leq c_5 \|\mathrm{A_D}^{-1}\| \Big(\|\partial_t u_{\mathrm{tp}} \partial_t^3 u_{\mathrm{tp}} - \partial_t v_{\mathrm{tp}} \partial_t^3 v_{\mathrm{tp}}\|_q \\
&\quad + \||\nabla \partial_t u_{\mathrm{tp}}|^2 - |\nabla \partial_t v_{\mathrm{tp}}|^2\|_q + \|\nabla u_{\mathrm{tp}} \cdot \partial_t^2 \nabla u_{\mathrm{tp}} - \nabla v_{\mathrm{tp}} \cdot \partial_t^2 \nabla v_{\mathrm{tp}}\|_q \\
&\quad + \|(\partial_t^2 u_{\mathrm{tp}})^2 - (\partial_t^2 v_{\mathrm{tp}})^2\|_q + \|\nabla u_{\mathrm{s}} \cdot \nabla \partial_t^2 u_{\mathrm{tp}} - \nabla u_{\mathrm{s}} \cdot \nabla \partial_t^2 v_{\mathrm{tp}}\|_q \Big) \\
&\leq c_6 \Big(\|\partial_t u_{\mathrm{tp}} \partial_t^3 (u_{\mathrm{tp}} - v_{\mathrm{tp}})\|_q + \|\partial_t^3 v_{\mathrm{tp}} \partial_t (u_{\mathrm{tp}} - v_{\mathrm{tp}})\|_q \\
&\quad + \|\nabla \partial_t v_{\mathrm{tp}} \cdot \nabla \partial_t (u_{\mathrm{tp}} - v_{\mathrm{tp}})\|_q + \|\nabla u_{\mathrm{tp}} \cdot \nabla \partial_t^2 (u_{\mathrm{tp}} - v_{\mathrm{tp}})\|_q \\
&\quad + \|\nabla \partial_t^2 v_{\mathrm{tp}} \cdot \nabla (u_{\mathrm{tp}} - v_{\mathrm{tp}})\|_q + \|\partial_t^2 u_{\mathrm{tp}} \partial_t^2 (u_{\mathrm{tp}} - v_{\mathrm{tp}})\|_q \\
&\quad + \|\partial_t^2 v_{\mathrm{tp}} \partial_t^2 (u_{\mathrm{tp}} - v_{\mathrm{tp}})\|_q + \|\nabla u_{\mathrm{s}} \cdot \nabla \partial_t^2 (u_{\mathrm{tp}} - v_{\mathrm{tp}})\|_q \\
&\quad + \|\nabla \partial_t u_{\mathrm{tp}} \cdot \nabla \partial_t (u_{\mathrm{tp}} - v_{\mathrm{tp}})\|_q \Big),
\end{aligned}
$$

and by utilizing Lemma 3.4.6, the estimate

$$
\begin{aligned}
\|\mathcal{N}(u_{\mathrm{tp}}) - \mathcal{N}(v_{\mathrm{tp}})\|_{\mathbb{Y}_\perp^q} &\leq c_7 \left(8\rho \|u_{\mathrm{tp}} - v_{\mathrm{tp}}\|_{\mathbb{Y}_\perp^q} + \varepsilon \|u_{\mathrm{tp}} - v_{\mathrm{tp}}\|_{\mathbb{Y}_\perp^q} \right) \\
&= c_7 (8\rho + \rho^2) \|u_{\mathrm{tp}} - v_{\mathrm{tp}}\|_{\mathbb{Y}_\perp^q},
\end{aligned}
$$

holds. Therefore, choosing ε sufficiently small $\mathcal{N}$ also becomes a contracting self-mapping. By the contraction mapping principle, existence of a fixed point for $\mathcal{N}$ follows. This concludes the proof. $\square$

Proof of Theorem 3.4.2. The assertion of Theorem 3.4.2 follow with similar arguments as in the proof Theorem 3.4.1 by employing the Neumann operator $\mathrm{A_N}$ (see Theorem 3.4.3) instead of $\mathrm{A_D}$, and utilizing Lemma 3.4.5 instead of Lemma 3.4.4. $\square$

4 Fluid-Structure Interaction

In many physical and medical applications the interaction of a fluid with a moving or deformable structure plays a fundamental role. For instance, such fluid-structure-interaction problems appear in aeroelasticity, biomechanics or hydroelasticity. To be more specific, in aeroelasticity fluid-structure interaction occurs in the stability analysis of airplane wings (see [54]), and an example from biomechanics is the flow of blood in arteries (see [78]). In the case of blood flow, the blood corpuscle is an elastic structure within the blood which is contained in a cylindrical domain with deformable elastic boundary. In these cases, we cannot assume anymore that the fluid domain is fixed and independent of the displacement of the structure.

Fluid-structure-interaction problems appear in different forms. One example is to consider an elastic structure that is immersed in a viscous liquid inside a domain (for example the whole space). These types of problems can be treated as systems in an exterior domain with a deformable and therefore time-depending boundary. Another kind of problem occurring in the study of fluid-structure interaction is to consider a domain $\Omega(t)$ filled with liquid, where one part of the boundary $\Gamma(t) \subset \Omega(t)$ (or even the whole boundary) is modeled as an elastic structure and is therefore deformable. However, in both these cases the boundary of the fluid domain depends on time and the underlying problem is called a *free boundary problem*. Here, we use the term "free boundary" to denote the part of the boundary of $\Omega(t)$ that is unknown at the outset of our investigation. Just to mention some references, we draw the reader's attention to [25, 47, 26, 33] and the references within for a brief overview of some problems in fluid-structure interaction. Indeed, in all those studies the fluid is described by the Navier-Stokes equations and the elastic structure lies inside the fluid. For the latter case of a free boundary problem, we refer to [16, 20, 24]. Here, the fluid is contained in a domain (for example the half space) and interacts with an elastic structure which is located at one part of the boundary.

Subject of this chapter is the investigation of a free boundary problem similar to the one studied in [16]. We consider a viscous fluid contained in a three-dimensional fluid domain

$$\Omega_\eta(t) := \omega \times (-\eta(t, x'), 1)$$

that interacts with a thin elastic (and thus deformable) plate located at the bottom $\Gamma_\eta(t) := \omega \times \{-\eta(t, x')\}$. Here, η denotes the evolution of the transversal displacement of the interface $\Gamma_\eta(t)$, and ω denotes the two-dimensional fluid-solir interface. In tangential directions ($x_1 x_2$-direction) we assume periodic boundary conditions. The upper part of the boundary is rigid. The time-space domain, in which the interaction takes part, is defined by

$$\Omega_\eta^\mathbb{T} := \bigcup_{t \in \mathbb{T}} \{t\} \times \Omega_\eta(t)$$

with $\mathbb{T} = \mathbb{R}/\mathcal{T}\mathbb{Z}$ and $\mathcal{T} > 0$. We are interested in the occurrence of resonance under periodic forcing, that is, we address the question whether the damping mechanism induced by the fluid is strong enough to avoid resonance in the elastic structure, and therefore allows the existence of a time-periodic solution, with the same period in time as the data. Throughout this chapter we will consider a (three-dimensional) incompressible viscous flow, which is governed by the Navier-Stokes equations

$$\begin{cases} \partial_t u - \mu_f \Delta u + (u \cdot \nabla)u + \nabla p = f_u & \text{in } \Omega_\eta^\mathbb{T}, \\ \operatorname{div} u = 0 & \text{in } \Omega_\eta^\mathbb{T}, \end{cases} \tag{4.0.1}$$

and interacts with a (two-dimensional) thin elastic plate, the motion of which is governed by the (hyperbolic) plate equation

$$\partial_t^2 \eta + \Delta'^2 \eta - \mu_s \Delta' \partial_t \eta = f_\eta - \mathrm{T}_\eta \qquad \text{in } \mathbb{T} \times \omega, \tag{4.0.2}$$

with $\mu_s, \mu_f > 0$. Here, $u \colon \Omega_\eta^\mathbb{T} \to \mathbb{R}^3$ denotes the fluid velocity, and $p \colon \Omega_\eta^\mathbb{T} \to \mathbb{R}$ the associated pressure field. As mentioned above, $\eta \colon \mathbb{T} \times \omega \to \mathbb{R}$ describes the evolution of the transversal displacement of $\Gamma_\eta(t)$. The body forces $f_u \colon \Omega_\eta^\mathbb{T} \to \mathbb{R}^3$ and $f_\eta \colon \mathbb{T} \times \omega \to \mathbb{R}$ are given, whereas

$$\mathrm{T}_\eta := e_3 \cdot \big((\mathrm{T}(u, p)\nu_t) \circ \phi_\eta\big)_{|x_3 = 0}$$

denotes the surface force exerted by the fluid on the structure. Here, ϕ_η is a coordinate transform which allows us to reformulate the current problem with respect to a so-called *reference configuration* and will be given in the next section. Moreover, at the fluid–solid interface we demand that the velocity of the elastic structure equals the fluid velocity, that is,

$$u = -(\partial_t \eta \circ \phi_\eta^{-1})e_3 \qquad \text{on } \Gamma_\eta^\mathbb{T}$$

with

$$\Gamma_\eta^{\mathbb{T}} := \bigcup_{t \in \mathbb{T}} \{t\} \times \Gamma_\eta(t),$$

and at the rigid part of the boundary, *i.e.*, on $\omega \times \{1\}$ we have no-slip boundary conditions. Recall that the Laplace operator Δ' and the Bi-Laplacian Δ'^2 are defined as in (2.1.1). For more details on the coupled hyperbolic-parabolic system, especially on the coupling and the domains, see Section 4.1. For a similar two-dimensional (initial value) problem, existence of strong solutions locally in time for small data can be found in [20].

A classical approach to free boundary problems in fluid-structure interaction relies heavily on maximal-regularity frameworks that include inhomogeneous boundary data. When time-periodic driving forces are studied in such settings, time-periodic Stokes equations appear in the linearization, see Section 4.2 below. A particular example is the investigation of resonance in thin elastic structures that interact with liquids (see for example [38, 20, 16, 17, 46]). Mathematically, this problem can be studied as a coupled fluid-structure system where the fluid occupies a three-dimensional cavity $\Omega \subset \mathbb{R}^3$ and interacts with a thin elastic structure located at one part of the boundary. The remaining part is considered to be rigid. If we assume an incompressible flow, the viscous fluid is governed by the Navier-Stokes equations, and resonance under time-periodic forcing in the hyperbolic elastic structure may be avoided due to energy dissipation in the fluid. The mathematical investigation hereof requires a functional analytic framework for the linearized equations of motion, which leads us to the time-periodic Stokes system

$$\begin{cases} \partial_t u - \Delta u + \nabla p = f & \text{in } \mathbb{T} \times \Omega, \\ \operatorname{div} u = g & \text{in } \mathbb{T} \times \Omega, \\ u = h & \text{on } \mathbb{T} \times \partial\Omega. \end{cases} \tag{4.0.3}$$

In Section 4.4 the Stokes problem (4.0.3) will be studied in a half space $\Omega = \mathbb{R}^n_+$, $n \geq 2$, and in the periodic half space $\Omega = \omega \times \mathbb{R}_+$. Here, $u \colon \mathbb{T} \times \Omega \to \mathbb{R}^n$ denotes the velocity field and $p \colon \mathbb{T} \times \Omega \to \mathbb{R}$ the pressure term. As customary in the formulation of time-periodic problems, the time axis is taken to be $\mathbb{T}$. Data $f \colon \mathbb{T} \times \Omega \to \mathbb{R}^n$, $g \colon \mathbb{T} \times \Omega \to \mathbb{R}$ and $h \colon \mathbb{T} \times \partial\Omega \to \mathbb{R}^n$ that are also $\mathcal{T}$-time-periodic are considered.

This chapter is structured as follows: We begin by presenting the coupled free boundary problem of interest and introduce some necessary assumptions on the displacement η of the elastic structure, see Section 4.1.

The subsequent section (Section 4.2) is dedicated to the introduction of an outline of our approach. The mathematical analysis of the hyperbolic-parabolic coupled system starts in Section 4.3. Here, we show that a solution to the corresponding linearization of the free boundary problem is unique. Before addressing the question of existence and regularity of a solution to the free boundary problem, we have to introduce some results on the Stokes system and the resolvent problem corresponding to the coupled system. This will be done in Section 4.4 and Section 4.5, respectively. Based on this results for the Stokes equations, in Section 4.6 we derive L^q estimates. Finally in Section 4.8 we merge the results from Section 4.3 to Section 4.6 to prove existence and regularity of a unique solution. Furthermore, we establish L^q estimates which will be used in the final section of this doctoral thesis to show existence of a solution to the (nonlinear) free boundary problem.

4.1 Viscous Fluid Flow on an Elastic Plate

Throughout this chapter we will study the interaction of a viscous fluid which is contained in a three-dimensional cell the bottom of which is a thin elastic plate. The top of the container is rigid, whereas for tangential direction (x_1x_2- direction) we assume periodic boundary conditions. All in all we assume to have a periodic layer-like domain where the interaction of the fluid with the elastic structure takes place. To be more precise, we and define the three-dimensional fluid domain as

$$\Omega_\eta(t) := \{x = (x', x_3) \in \mathbb{R}^3 \mid x' \in \omega, \ -\eta(t, x') < x_3 < 1\}.$$

Hence, the moving part of the boundary $\partial\Omega_\eta(t)$ is given by $\Gamma_\eta(t) := \omega \times \{-\eta(t, x')\}$. Note, for $\Omega_\eta(t)$ to be a well-defined domain, we have that

$$-\eta(t, x') < 1 \qquad \text{for all } (t, x') \in \mathbb{T} \times \omega. \tag{4.1.1}$$

To simplify notation, we will omit the time dependency of the underlying domains and write Ω_η and Γ_η instead of $\Omega_\eta(t)$ and $\Gamma_\eta(t)$, respectively. We consider an incompressible viscous fluid, which is governed by the Navier-Stokes equations (4.0.1), and the elastic structure, which in our case is a thin elastic plate, is governed by the plate equation (4.0.2). Moreover, observe that (4.0.1) and (4.0.2) are defined on different domains. The Navier-Stokes equations are formulated on a moving domain, whereas the plate equation is described on the boundary of a reference configuration.

A canonical choice of reference configuration is the rest state

$$\Omega := \omega \times (0, 1)$$

of the system, that is, the physical configuration when no external forces act on the system. In order to couple these two problems in an appropriate way, we have to introduce the coordinate transform

$$\phi_\eta \colon \mathbb{T} \times \Omega \to \Omega_\eta^{\mathbb{T}}, \qquad (t, \tilde{x}) \mapsto (t, x', x_3 - (1 - x_3)\eta(t, x')), \qquad (4.1.2)$$

which allows us to reformulate the fluid-structure-interaction problem on the reference configuration and to investigate the resulting nonlinear problem. Observe that due to (4.1.1) on η, the transform ϕ_η is a bijection with inverse given by

$$\phi_\eta^{-1} \colon \Omega_\eta^{\mathbb{T}} \to \mathbb{T} \times \Omega, \qquad (t, \tilde{x}) \mapsto \left(t, x', \frac{x_3 + \eta(t, x')}{1 + \eta(t, x')} \right).$$

Moreover, utilizing ϕ_η we are finally able to describe the interaction of the fluid with the elastic plate at the boundary. On the upper part of the boundary we consider no-slip conditions, whereas on the lower part of the boundary we assume kinematic interface condition. That is, the velocity of the elastic structure must equal the fluid velocity at the fluid–solid interface. As the displacement from the elastic plate with respect to the reference configuration is described by η, the velocity of the displacement $\partial_t \eta$ only acts in normal direction of the reference configuration, that is, in x_3-direction. Mathematically speaking, the velocity of the plate and the fluid at the moving interface are described by $-(\partial_t \eta \circ \phi_\eta^{-1})e_3$ and $u_{|\Gamma_\eta}$, respectively, expressed in the current configuration. Hence,

$$u = -(\partial_t \eta \circ \phi_\eta^{-1})e_3 \qquad \text{on } \Gamma_\eta^{\mathbb{T}}.$$

This condition then reads

$$u \circ \phi_\eta = -\partial_t \eta e_3 \qquad \text{on } \mathbb{T} \times \omega \times \{0\}, \qquad (4.1.3)$$

on the reference configuration (where the coupled system is investigated).

The incompressibility of the fluid implies that

$$0 = \int\limits_{\Omega_\eta} \operatorname{div} u \, \mathrm{d}x = \int\limits_{\Omega} \frac{1}{1 + \eta} \big(\operatorname{div}(u \circ \phi_\eta) + \operatorname{div} \Psi \big) |\det \nabla \phi_\eta| \, \mathrm{d}x,$$

with Ψ given as

$$\Psi := \begin{pmatrix} \eta \tilde{u}' \\ (1 - x_3)\nabla'\eta \cdot \tilde{u}' \end{pmatrix}.$$

Observing that the determinant of the Jacobian matrix $\nabla\phi_\eta$ is given by

$$|\det \nabla\phi_\eta| = |1 + \eta| = 1 + \eta,$$

and utilizing the no-slip and kinematic boundary condition of $u \circ \phi_\eta$, we obtain

$$0 = \int_{\omega \times \{0\}} (u \circ \phi_\eta) \cdot (-e_3)\, \mathrm{d}x' = \int_{\omega \times \{0\}} \partial_t \eta\, \mathrm{d}x'.$$

It follows that

$$\int_\omega \eta(t, x')\, \mathrm{d}x' = C_{28}, \tag{4.1.4}$$

for some constant $C_{28} > 0$ independent of $t \in \mathbb{T}$. We add the condition (4.1.4) to our system of equations to solve. It will serve as a normalization condition on η. In principle, we could choose an arbitrary constant C_{28}. However, we choose the constant $C_{28} = 0$ in accordance with the choice of the reference configuration being the natural rest state.

The dynamics of the elastic plate are governed by the equation

$$\partial_t^2 \eta = \Phi_1 + \Phi_2$$

with Φ_1 and Φ_2 the intrinsic and external forces, respectively. More precisely, Φ_1 is the elastic stress which stems from the deflection of the elastic structure and is given by

$$\Phi_1 = -\Delta'^2 \eta + \mu_s \Delta' \partial_t \eta.$$

The external contribution Φ_2 is given by

$$\Phi_2 = f_\eta - \mathrm{T}_\eta.$$

Here f_η is a (given) external force, and T_η results from the fluid stress and is defined as

$$\mathrm{T}_\eta := e_3 \cdot \left((\mathrm{T}(u, p)\nu_t) \circ \phi_\eta\right)_{|x_3=0} = e_3 \cdot \left(\mathrm{T}(\tilde{u}, \tilde{p})_{|x_3=0} + \mathrm{S}_\eta\right)\tilde{\nu}_t \tag{4.1.5}$$

with $\tilde{u} := u \circ \phi_\eta$ and $\tilde{p} := p \circ \phi_\eta$. Recall that the fluid stress is given as in (2.1.2). The term S_η stems from the transformation ϕ_η, and as it consists of products of u and η, it does not occur in the study of the linearized system. Therefore, it will be determined later in Section 4.9. For more details on the derivation of the interaction of the fluid with the structure, we refer to [64, Section 1.2], [78, Chapter 4] and the introduction in [50].

Summarizing this information, we consider the free boundary problem

$$
\begin{cases}
\partial_t^2 \eta + \Delta'^2 \eta - \mu_s \Delta' \partial_t \eta = f_\eta - \mathrm{T}_\eta & \text{in } \mathbb{T} \times \omega, \\
\partial_t u - \mu_f \Delta u + (u \cdot \nabla)u + \nabla p = f_u & \text{in } \Omega_\eta^{\mathbb{T}}, \\
\operatorname{div} u = 0 & \text{in } \Omega_\eta^{\mathbb{T}}, \\
u(t, x', -\eta(t, x')) = -\partial_t \eta(t, x') e_3 & \text{on } \mathbb{T} \times \omega, \\
u(t, x', 1) = 0 & \text{on } \mathbb{T} \times \omega, \\
\displaystyle\int_\omega \eta(t, x')\, \mathrm{d}x' = 0.
\end{cases}
\tag{4.1.6}
$$

Due to the periodic boundary conditions in tangential direction we let $\omega := \mathbb{T}_0^2 := (\mathbb{R}/L\mathbb{Z})^2$ be the fluid-solid interface for fixed $L > 0$. Similarly to the Haar measure on $G = \mathbb{T} \times \mathbb{R}^n$, we normalize the Haar measure $\mathrm{d}y$ on $\mathbb{T}_0^2$ such that

$$
\int_{\mathbb{T}_0^2} v(y)\, \mathrm{d}y = \frac{1}{L^2} \int_{[0,L]^2} v(x')\, \mathrm{d}x'.
\tag{4.1.7}
$$

Since η has a vanishing mean value, we conclude from the first line in (4.1.6) by integration over $\mathbb{T}_0^2$ that any solution to the coupled system above has to obey the further condition

$$
\int_{\mathbb{T}_0^2} f_\eta\, \mathrm{d}x' = \int_{\mathbb{T}_0^2} \mathrm{T}_\eta\, \mathrm{d}x'.
\tag{4.1.8}
$$

4.2 Reformulation in a Reference Configuration

Our aim is to study (4.1.6) and show existence of a strong time-periodic solution satisfying (4.1.8) when excited periodically via the external forcing terms f_u and f_η. To achieve this goal, we consider the corresponding linearized system, and, based on the L^q estimates for this solution, utilize a fixed-point argument as in the case of nonlinear acoustics in Section 3.3 and Section 3.4. However, due to the free boundary, the domain depends on the unknown η, and therefore, some modifications are necessary before employing the contraction mapping principle. To be more precise, the transformation ϕ_η is employed to reduce the study of (4.1.6) to that of an equivalent system on the fixed domain Ω. Observe that by proceeding in the described way and employing ϕ_η, some further nonlinear terms appear in the formulation of (4.1.6) with respect to the reference configuration $\Omega = \mathbb{T}_0^2 \times (0,1)$. These terms can be examined in a similar manner as the nonlinear terms occurring in (4.1.6) by applying the embedding properties stated in Theorem 2.6.3 and the contraction mapping principle. Moreover, throughout this chapter the vanishing mean value condition of η will be included into the function space and $W_{(0)}^{s,q}(\mathbb{T}_0^2)$ denotes the set of all Sobolev functions η with vanishing mean value, *i.e.*,

$$W_{(0)}^{s,q}(\mathbb{T}_0^2) := \left\{ \eta \in W^{s,q}(\mathbb{T}_0^2) \; \Big| \; \int_{\mathbb{T}_0^2} \eta \, dx' = 0 \right\}, \qquad (4.2.1)$$

with $s \in \mathbb{N}_0$ and $q \in (1, \infty)$. Furthermore, for $m \in \mathbb{N}$ we utilize the notation

$$W_{(0)}^{m,2m,q}(\mathbb{T} \times \mathbb{T}_0^2) := W^{m,q}\big(\mathbb{T}; L^q(\mathbb{T}_0^2)\big) \cap L^q\big(\mathbb{T}; W_{(0)}^{2m,q}(\mathbb{T}_0^2)\big)$$

for the inhomogeneous Sobolev spaces. Hence, instead of investigating the free boundary problem (4.1.6) we study

$$\begin{cases} \partial_t^2 \eta + \Delta'^2 \eta - \Delta' \partial_t \eta = f_\eta - T_\eta & \text{in } \mathbb{T} \times \mathbb{T}_0^2, \\ \partial_t \tilde{u} - \Delta \tilde{u} + (\tilde{u} \cdot \nabla)\tilde{u} + \nabla \tilde{p} = \tilde{f}_u + R_f & \text{in } \mathbb{T} \times \Omega, \\ \operatorname{div} \tilde{u} = \tilde{R}_d & \text{in } \mathbb{T} \times \Omega, \qquad (4.2.2) \\ \tilde{u}(t, x', 0) = -\partial_t \eta(t, x') e_3 & \text{on } \mathbb{T} \times \mathbb{T}_0^2, \\ \tilde{u}(t, x', 1) = 0 & \text{on } \mathbb{T} \times \mathbb{T}_0^2, \end{cases}$$

with $\tilde{u} := u \circ \phi_\eta$, $\tilde{p} := p \circ \phi_\eta$ and where we have set without loss of generality $\mu_s = \mu_f = 1$. Note that a solution to (4.1.6) is given by $(u, p, \eta) = (\tilde{u} \circ \phi_\eta^{-1}, \tilde{p} \circ \phi_\eta^{-1}, \eta)$, if

$$(\tilde{u}, \tilde{p}, \eta) \in W^{1,2,q}(\mathbb{T} \times \Omega) \times L^q\big(\mathbb{T}; W^{1,q}(\Omega)\big) \times W_{(0)}^{2,4,q}(\mathbb{T} \times \mathbb{T}_0^2)$$

solves (4.2.2). Here, the further nonlinear terms R_f and $\tilde{R}_d$ on the right-hand side stem from the transformation, and do not appear in the investigation of the linearized system. Therefore, we will determine those terms later in Section 4.9. Further observe that by reformulating (4.1.6) on the reference configuration (*i.e.* by utilizing ϕ_η), the vanishing mean value condition on η and the compatibility conditions (4.1.8) do not change since these conditions are defined on the bottom part of the boundary.

As mentioned above the coupled system (4.2.2) can be treated by a fixed-point argument (see Chapter 3), which is strongly based on the L^q estimates established for the solution to the linearized system

$$\begin{cases} \partial_t^2 \eta + \Delta'^2 \eta - \Delta' \partial_t \eta = f_\eta - e_3 \cdot T_0(u,p) & \text{in } \mathbb{T} \times \mathbb{T}_0^2, \\ \partial_t u - \Delta u + \nabla p = f_u & \text{in } \mathbb{T} \times \Omega, \\ \operatorname{div} u = g & \text{in } \mathbb{T} \times \Omega, \qquad (4.2.3) \\ u(t,x',0) = -\partial_t \eta(t,x')e_3 & \text{on } \mathbb{T} \times \mathbb{T}_0^2, \\ u(t,x',1) = 0 & \text{on } \mathbb{T} \times \mathbb{T}_0^2, \end{cases}$$

with

$$T_0(u,p) := T(u,p)e_{3|x_3=0},$$

and T denoting the fluid stress tensor, see (2.1.2). We begin by proving uniqueness of a solution to (4.2.3) via a duality argument as for the damped wave equation in the proof of Lemma 3.2.6. In order to prove the existence of a time-periodic solution to the linearized problem, we first reduce the investigation of (4.2.3) to that of a coupled system with homogeneous divergence $g = 0$ by an application of the Bogovskiĭ operator. Hence, it suffices to consider the "reduced" system

$$\begin{cases} \partial_t^2 \eta + \Delta'^2 \eta - \Delta' \partial_t \eta = F - e_3 \cdot T_0(v,p) & \text{in } \mathbb{T} \times \mathbb{T}_0^2, \\ \partial_t v - \Delta v + \nabla p = f & \text{in } \mathbb{T} \times \Omega, \\ \operatorname{div} v = 0 & \text{in } \mathbb{T} \times \Omega, \qquad (4.2.4) \\ v(t,x',0) = -\partial_t \eta(t,x')e_3 & \text{on } \mathbb{T} \times \mathbb{T}_0^2, \\ v(t,x',1) = 0 & \text{on } \mathbb{T} \times \mathbb{T}_0^2, \end{cases}$$

A solution to this linear problem is given by $(\mathscr{F}_\mathbb{T}^{-1}[v_k], \mathscr{F}_\mathbb{T}^{-1}[p_k], \mathscr{F}_\mathbb{T}^{-1}[\eta_k])$ if (v_k, p_k, η_k) solves the resolvent problem

$$\begin{cases} -k^2\eta_k + \Delta'^2\eta_k - ik\Delta'\eta_k = F_k - e_3 \cdot T_0(v_k,p_k) & \text{in } \mathbb{T}_0^2, \\ ikv_k - \Delta v_k + \nabla p_k = f_k & \text{in } \Omega, \\ \operatorname{div} v_k = 0 & \text{in } \Omega, \\ v(x',0) = -ik\eta_k(x')e_3 & \text{on } \mathbb{T}_0^2, \\ v(x',1) = 0 & \text{on } \mathbb{T}_0^2, \end{cases}$$

for fixed $k \in \frac{2\pi}{\mathcal{T}}\mathbb{Z}$. Note that the term "resolvent problem" is not used in the classical way. However, throughout this thesis we will use this phrase for the above system, as it is used similarly for (classical) resolvent problems. The investigation hereof is subject of Subsection 4.5. There we prove the existence of a weak time-periodic solution which, by regularity theory, is as regular as the data allow. In Section 4.6 and Section 4.7 we determine the L^q estimates essential to carry out the fixed-point argument when investigating the nonlinear system. These L^q estimates will be established for the purely oscillatory and the stationary part separately. Section 4.6 is dedicated to the investigation of the purely oscillatory problem, whereas in Section 4.7 we study the stationary part. Furthermore, a solution to the corresponding steady-state part of (4.2.4) is established there. In the case of purely periodic forcing we will employ the results for solutions to the Stokes equations (4.0.3), deduced in Section 4.4. For stationary data, (4.2.4) can be solved by considering first the stationary Stokes equations and then the Bi-Laplacian one by one, as $\partial_t \eta_s = 0$. Finally, in Section 4.8 we combine all the results collected in Section 4.3 to Section 4.6 to deduce a solution to (4.2.3). Then we employ the contraction mapping principle and find a solution to (4.2.2), see Section 4.9.

Remark 4.2.1. Observe that any solenoidal solution u to the coupled fluid-structure interaction problem (4.2.4) on $\Omega = \mathbb{T}_0^2 \times \mathbb{R}_+$ or $\Omega = \mathbb{T}_0^2 \times (0,1)$ satisfies

$$e_3 \cdot (\mathrm{D}(u)_{|x_3=0}\, e_3) = \partial_{x_3} u_{3|x_3=0} = -(\partial_{x_1} u_{1|x_3=0} + \partial_{x_2} u_{2|x_3=0}) = 0,$$

due to the boundary condition $(4.2.4)_4$. Thus, the normal component of the fluid stress tensor at the moving boundary can be expressed in terms of the pressure field, that is

$$e_3 \cdot \mathrm{T}_0(u,p) = -p_{|x_3=0}.$$

4.3 Uniqueness

Subject of this section is the uniqueness of solutions to the coupled fluid-structure system (4.2.3), that is,

$$
\begin{cases}
\partial_t^2 \eta + \Delta'^2 \eta - \Delta' \partial_t \eta = f_\eta - e_3 \cdot \mathrm{T}_0(u,p) & \text{in } \mathbb{T} \times \mathbb{T}_0^2, \\
\partial_t u - \Delta u + \nabla p = f_u & \text{in } \mathbb{T} \times \Omega, \\
\operatorname{div} u = g & \text{in } \mathbb{T} \times \Omega, \qquad (4.3.1) \\
u_{|x_3=0} = -\partial_t \eta e_3 & \text{on } \mathbb{T} \times \mathbb{T}_0^2, \\
u_{|x_3=1} = 0 & \text{on } \mathbb{T} \times \mathbb{T}_0^2,
\end{cases}
$$

with fully inhomogeneous right-hand side

$$f_u \in \mathrm{L}^q(\mathbb{T} \times \Omega)^3,$$
$$g \in \mathrm{L}^q\big(\mathbb{T}; \mathrm{W}^{1,q}(\Omega)\big) \cap \mathrm{W}^{1,q}\big(\mathbb{T}; \dot{\mathrm{W}}^{-1,q}(\Omega)\big), \qquad (4.3.2)$$
$$f_\eta \in \mathrm{L}^q(\mathbb{T} \times \mathbb{T}_0^2).$$

Compared to the wave equations considered in the previous chapter, we will not decompose the problem into steady-state and purely periodic parts. Uniqueness will be proven by a duality argument as for the damped wave equation in the half space (see the proof of Lemma 3.2.6). Since this approach does not depend on the geometry of the underlying domain, we will consider arbitrary domains of the form

$$\Omega_\ell := \mathbb{T}_0^2 \times (0, \ell), \qquad \text{with } \ell \in (0, \infty]. \qquad (4.3.3)$$

We are going to show the following uniqueness assertion:

Lemma 4.3.1 (Uniqueness). *Let $q \in (1, \infty)$. Furthermore, let (f_u, g, f_η) be in the class (4.3.2) and*

$$(u, p, \eta) \in \mathrm{W}^{1,2,q}(\mathbb{T} \times \Omega_\ell)^3 \times \mathrm{L}^q\big(\mathbb{T}; \dot{\mathrm{W}}^{1,q}(\Omega_\ell)\big) \times \mathrm{W}^{2,4,q}_{(0)}(\mathbb{T} \times \mathbb{T}_0^2) \quad (4.3.4)$$

a solution to (4.3.1) in $\mathbb{T} \times \Omega_\ell$, with Ω_ℓ as in (4.3.3). Then (u, p, η) is unique in the class (4.3.4).

Proof. Let $(\tilde{u}, \tilde{p}, \tilde{\eta})$ be another solution to (4.3.1) with the same right-hand side (f_u, g, f_η). Then $(v, \mathfrak{p}, \zeta) := (u - \tilde{u}, p - \tilde{p}, \eta - \tilde{\eta})$ solves the coupled system with homogeneous data, that is,

$$\begin{cases} \partial_t^2 \zeta + \Delta'^2 \zeta - \Delta' \partial_t \zeta = -e_3 \cdot \mathrm{T}_0(v, \mathfrak{p}) & \text{in } \mathbb{T} \times \mathbb{T}_0^2, \\ \partial_t v - \Delta v + \nabla \mathfrak{p} = 0 & \text{in } \mathbb{T} \times \Omega_\ell, \\ \operatorname{div} v = 0 & \text{in } \mathbb{T} \times \Omega_\ell, \\ v_{|x_3=0} = -\partial_t \zeta e_3 & \text{on } \mathbb{T} \times \mathbb{T}_0^2, \\ v_{|x_3=1} = 0 & \text{on } \mathbb{T} \times \mathbb{T}_0^2. \end{cases} \qquad (4.3.5)$$

Furthermore, let $\psi \in \mathrm{L}^{q'}(\mathbb{T} \times \Omega_\ell)^3$ be arbitrary and consider

$$\begin{cases} \partial_t^2 \theta + \Delta'^2 \theta - \Delta' \partial_t \theta = -e_3 \cdot \mathrm{T}_0(\phi, \pi) & \text{in } \mathbb{T} \times \mathbb{T}_0^2, \\ \partial_t \phi - \Delta \phi + \nabla \pi = \psi & \text{in } \mathbb{T} \times \Omega_\ell, \\ \operatorname{div} \phi = 0 & \text{in } \mathbb{T} \times \Omega_\ell, \\ \phi_{|x_3=0} = -\partial_t \theta e_3 & \text{on } \mathbb{T} \times \mathbb{T}_0^2, \\ \phi_{|x_3=1} = 0 & \text{on } \mathbb{T} \times \mathbb{T}_0^2. \end{cases} \qquad (4.3.6)$$

A solution

$$(\phi, \pi, \theta) \in W^{1,2,q'}(\mathbb{T} \times \Omega_\ell)^3 \times L^{q'}\left(\mathbb{T}; \dot{W}^{1,q'}(\Omega_\ell)\right) \times W^{2,4,q'}_{(0)}(\mathbb{T} \times \mathbb{T}^2_0)$$

to (4.3.6) exists by Proposition 4.8.1 below. We define $\tilde{\phi}(t, x) := \phi(-t, x)$, $\tilde{\pi}(t, x) := \pi(-t, x)$ and $\tilde{\theta}(t, x') := \theta(-t, x')$, and observe that the identities

$$-\partial_t \tilde{\phi} - \Delta \tilde{\phi} + \nabla \tilde{\pi} = \tilde{\psi} \qquad \text{in } \mathbb{T} \times \Omega_\ell,$$

and

$$\partial_t^2 \tilde{\theta} + \Delta'^2 \tilde{\theta} + \Delta' \partial_t \tilde{\theta} = \tilde{\pi}_{|x_3=0} = -e_3 \cdot \mathrm{T}_0(\tilde{\phi}, \tilde{\pi}) \qquad \text{in } \mathbb{T} \times \mathbb{T}^2_0 \qquad (4.3.7)$$

hold, with $\tilde{\psi}(t, x) := \psi(-t, x)$. Observe further that the integrability properties of $(\tilde{\phi}, \tilde{\pi}, \tilde{\theta})$ are sufficient to carry out the following integration by parts:

$$\int\limits_{\mathbb{T} \times \Omega_\ell} v\, \tilde{\psi}\, \mathrm{d}(t, x) = \int\limits_{\mathbb{T} \times \Omega_\ell} v(-\partial_t \tilde{\phi} - \Delta \tilde{\phi} + \nabla \tilde{\pi})\, \mathrm{d}(t, x)$$

$$= \int\limits_{\mathbb{T} \times \Omega_\ell} (\partial_t v - \Delta v) \tilde{\phi}\, \mathrm{d}(t, x) + \int\limits_{\mathbb{T} \times \mathbb{T}^2_0} [v \cdot \nabla \tilde{\phi}\, e_3]_{|x_3=0}\, \mathrm{d}(t, x') \qquad (4.3.8)$$

$$- \int\limits_{\mathbb{T} \times \mathbb{T}^2_0} [\nabla v \tilde{\phi} \cdot e_3]_{|x_3=0}\, \mathrm{d}(t, x') - \int\limits_{\mathbb{T} \times \mathbb{T}^2_0} [v \tilde{\pi}]_{|x_3=0} \cdot e_3\, \mathrm{d}(t, x').$$

As $v_{|x_3=\ell}$ and $\tilde{\phi}_{|x_3=\ell}$ vanish on the upper boundary, we deduce that the boundary integrals on $\mathbb{T} \times \mathbb{T}^2_0 \times \{\ell\}$ vanish, and therefore do not occur in (4.3.8). Note further, since ϕ is solenoidal we also have that $\tilde{\phi}$ is divergence free. Therefore, we can exploit the vanishing divergence of $\tilde{\phi}$ combined with the boundary conditions (4.3.5)$_4$ - (4.3.5)$_5$ and (4.3.6)$_4$ - (4.3.6)$_5$ to find for the second and third boundary integral on the right-hand side

$$\int\limits_{\mathbb{T} \times \mathbb{T}^2_0} [v \cdot \nabla \tilde{\phi}\, e_3]_{|x_3=0}\, \mathrm{d}(t, x') = -\int\limits_{\mathbb{T} \times \mathbb{T}^2_0} (\partial_t \zeta) \partial_{x_3} \tilde{\phi}_{3\,|x_3=0}\, \mathrm{d}(t, x')$$

$$= \int\limits_{\mathbb{T} \times \mathbb{T}^2_0} (\partial_t \zeta)[\partial_{x_1} \tilde{\phi}_1 + \partial_{x_2} \tilde{\phi}_2]_{|x_3=0}\, \mathrm{d}(t, x') = 0,$$

and similarly

$$\int\limits_{\mathbb{T} \times \mathbb{T}^2_0} [\nabla v \tilde{\phi} \cdot e_3]_{|x_3=0}\, \mathrm{d}(t, x') = -\int\limits_{\mathbb{T} \times \mathbb{T}^2_0} [\partial_{x_3} v\, \partial_t \tilde{\theta}]_{|x_3=0}\, \mathrm{d}(t, x') = 0.$$

To study the last boundary integral in (4.3.8), we first note that the fluid stress tensor in (4.3.7) is given by $e_3 \cdot T_0(\tilde{\phi}, \tilde{\pi}) = -\tilde{\pi}_{|x_3=0}$, due to Remark 4.2.1. Therefore, we deduce from (4.3.7) that

$$- \int_{\mathbb{T} \times \mathbb{T}_0^2} [v\tilde{\pi}]_{|x_3=0} \cdot e_3 \, \mathrm{d}(t, x') = \int_{\mathbb{T} \times \mathbb{T}_0^2} \partial_t \zeta (\partial_t^2 \tilde{\theta} + \Delta'^2 \tilde{\theta} + \Delta' \partial_t \tilde{\theta}) \, \mathrm{d}(t, x').$$

Integration by parts (with respect to time and space) and the periodicity of ζ and θ yields

$$- \int_{\mathbb{T} \times \mathbb{T}_0^2} [v\tilde{\pi}]_{|x_3=0} \cdot e_3 \, \mathrm{d}(t, x') = \int_{\mathbb{T} \times \mathbb{T}_0^2} -(\partial_t^2 \zeta + \Delta'^2 \zeta - \Delta' \partial_t \zeta) \partial_t \tilde{\theta} \, \mathrm{d}(t, x')$$

$$= \int_{\mathbb{T} \times \mathbb{T}_0^2} e_3 \cdot T_0(v, \mathfrak{p}) \partial_t \tilde{\theta} \, \mathrm{d}(t, x') = - \int_{\mathbb{T} \times \mathbb{T}_0^2} (\mathfrak{p}\, \tilde{\phi} \cdot e_3)_{|x_3=0} \, \mathrm{d}(t, x')$$

$$= \int_{\mathbb{T} \times \Omega_\ell} \nabla \mathfrak{p} \cdot \tilde{\phi} \, \mathrm{d}(t, x) + \int_{\mathbb{T} \times \Omega_\ell} \mathfrak{p} \operatorname{div} \tilde{\phi} \, \mathrm{d}(t, x),$$

where the third equality follows from the identity

$$e_3 \cdot T_0(v, \mathfrak{p}) = -\mathfrak{p}_{|x_3=0}$$

and

$$[\tilde{\phi} \cdot e_3]_{|x_3=0} = \partial_t \tilde{\theta}.$$

Since $\tilde{\phi}$ is solenoidal, we finally obtain from (4.3.8)

$$\int_{\mathbb{T} \times \Omega_\ell} v \tilde{\psi} \, \mathrm{d}(t, x) = \int_{\mathbb{T} \times \Omega_\ell} (\partial_t v - \Delta v + \nabla \mathfrak{p}) \tilde{\phi} \, \mathrm{d}(t, x) = 0.$$

As this identity holds for all $\tilde{\psi}$, it follows that $v = 0$, and in turn we deduce from $(4.3.5)_2$ and $(4.3.5)_4$ that $\mathfrak{p}$ is a function that only depends on time and ζ only depends on $x' \in \mathbb{T}_0^2$. Hence, $(4.3.5)_1$ reduces to

$$\Delta'^2 \zeta(x') = \mathfrak{p}_{|x_3=0}(t) \tag{4.3.9}$$

which yields that $\mathfrak{p} \equiv C$. Utilizing (4.1.8) and the identity $e_3 \cdot T_0(v, \mathfrak{p}) = -\mathfrak{p}_{|x_3=0}$, we obtain

$$\int_{\mathbb{T}_0^2} \mathfrak{p} \, \mathrm{d}(t, x') = 0 \implies \mathfrak{p} = 0.$$

This further implies that the right-hand side in (4.3.9) vanishes, and by multiplying (4.3.9) by ζ and integrating by parts, we deduce that $\Delta'\zeta = 0$. Hence, ζ is unique up to a constant C which due to the vanishing mean value condition (4.1.4) is zero. $\qquad\square$

Remark 4.3.2. Note that the assertion of Lemma 4.3.1 still holds for $\ell = \infty$. In this case we further have to assume, that $(v, \mathfrak{p})$ obey a decay property. More precisely, we have to assume that

$$\lim_{x_3 \to \infty} v = 0 = \lim_{x_3 \to \infty} \mathfrak{p}.$$

4.4 The Time-Periodic Stokes Equations

As a linearization of the Navier-Stokes equations, the Stokes problem (4.0.3), that is

$$\begin{cases} \partial_t u - \Delta u + \nabla p = f & \text{in } \mathbb{T} \times \Omega, \\ \operatorname{div} u = g & \text{in } \mathbb{T} \times \Omega, \\ u_{|x_n=0} = h & \text{on } \mathbb{T} \times \partial\Omega, \end{cases} \qquad (4.4.1)$$

with $\Omega \subset \mathbb{R}^n$ and data $f \colon \mathbb{T} \times \Omega \to \mathbb{R}^n$, $g \colon \mathbb{T} \times \Omega \to \mathbb{R}$, and $h \colon \mathbb{T} \times \partial\Omega \to \mathbb{R}^n$, plays a fundamental role in the mathematical investigation of many problems in fluid dynamics. For example, problems in the field of time-periodic fluid-structure interaction are typically studied in an L^2 framework. A comparable L^q theory was (to our knowledge) only established recently by DENK and SAAL in [24], and, therefore, we believe that the L^q framework and the maximal regularity established in the following section will lead to further improvements in this field. Layer domains in particular often appear as flow domains in the study of fluid-structure interaction problems where a fluid flow interacts with a structure in a plate configuration. The nature of such free boundary problems typically leads to a system of fully inhomogeneous equations in a reference layer domain $\Omega = \mathbb{R}^{n-1} \times (0, 1)$ with the Stokes equations appearing in the linearization hereof; see for example Section 4.2 above or [20]. Understanding the dynamics of such fluid-structure systems under periodic forcing is important in the study of resonance; see for example [38]. To further develop such investigations, in the following we establish time-periodic maximal regularity in an L^q setting for the fully inhomogeneous Stokes equations (4.4.1). However, as the localization argument carried out to determine the L^q estimates for the fluid-structure interaction problem (4.2.4) in Section 4.6 only requires maximal L^q regularity in the half space framework, it suffices to study (4.4.1) in the half space, more precisely, in $\mathbb{T}_0^2 \times \mathbb{R}_+$ (see Subsection 4.4.2).

Although comprehensive L^q estimates of maximal regularity type are available in the whole space case [60], similar estimates in the more complicated half space case were only established recently by MAEKAWA and SAUER [69]. Yet the analysis in [69], does not include data $h \neq 0$. Extending these results to the case $h \neq 0$ is by no means trivial. Observe that the nature of the Stokes problem does not allow for the treatment of inhomogeneous boundary data by a simple lifting argument. Introduction of a vector field $H : \mathbb{T} \times \mathbb{R}_+^n \to \mathbb{R}^n$ that coincides with h on the boundary $\mathbb{T} \times \partial\mathbb{R}_+^n$, may reduce the problem to one of homogeneous boundary data by considering $\tilde{u} := u - H$ instead of u, but then it would be difficult to treat the additional term $\operatorname{div} H$ on the right-hand side of the equation $\operatorname{div} \tilde{u} = g - \operatorname{div}(H)$. Indeed, if a vector field H is chosen from the canonical trace space of the solution, the term $\operatorname{div} H$ is not admissible as right-hand side in [69].

In the following Subsection, we investigate the Stokes system with fully inhomogeneous data and boundary values in the half space. As our main result in this section we prove existence and *a priori* estimates of a solution to (4.4.1). In order to achieve this goal, a different approach than the reflection type argument used in [69] is employed. Instead, we use the Fourier transform $\mathscr{F}_{\mathbb{T} \times \mathbb{R}^{n-1}}$ to reduce the Stokes equations with fully inhomogeneous right-hand side to an ordinary differential equation in the variable x_n. The L^q estimates are then established for the stationary and purely oscillatory part separately with arguments based on Fourier multipliers and interpolation techniques. Without this decomposition, it seems impossible to establish optimal L^q estimates. Whereas the estimates for the resulting steady-state problem are well-known and can be found in contemporary literature (see for example [36] and the references herein) the estimates for the purely oscillatory part in the following are new and shall be established as described above.

Based on the results deduced in Subsection 4.4.1 for (4.4.1), the Stokes equations in the periodic half space $\mathbb{T}_0^2 \times \mathbb{R}_+$ shall be investigated in Subsection 4.4.2. As the Fourier transform $\mathscr{F}_{\mathbb{T}_0^2}$ is admissible in this framework, the existence of a solution as well as the L^q estimates follow by an application of the transference principle (Theorem 2.3.1).

Note that as we are utilizing the partial Fourier transform, multipliers depending on the Fourier variable $\xi' \in \mathbb{R}^{n-1}$ will occur in the representation of our solution. We will omit the superscript and write ξ instead of ξ' throughout this section.

The results presented in this Section, especially in Subsection 4.4.1, were already published in [15].

4.4.1 The Stokes System in the Half Space

This subsection is dedicated to the investigation of the time-periodic Stokes system

$$\begin{cases} \partial_t u - \Delta u + \nabla p = f & \text{in } \mathbb{T} \times \mathbb{R}^n_+, \\ \operatorname{div} u = g & \text{in } \mathbb{T} \times \mathbb{R}^n_+, \\ u = h & \text{on } \mathbb{T} \times \partial \mathbb{R}^n_+, \end{cases} \tag{4.4.2}$$

in the half space, with $n \geq 2$. Analogously to the damped wave equation in Section 3.2 we decompose (4.4.2) into a stationary and a purely oscillatory problem, which will be studied separately. Combining the results established for the separate problems we find the following theorem as the main result of this subsection.

Theorem 4.4.1. *Let $q \in (1, \infty)$ and $n \geq 2$. For all*

$$\begin{aligned} & f \in \mathrm{L}^q \big(\mathbb{T}; \mathrm{L}^q(\mathbb{R}^n_+) \big)^n, \\ & g \in \mathrm{L}^q \big(\mathbb{T}; \mathrm{W}^{1,q}(\mathbb{R}^n_+) \big) \cap \mathrm{W}^{1,q} \big(\mathbb{T}; \dot{\mathrm{W}}^{-1,q}(\mathbb{R}^n_+) \big), \\ & h \in \mathrm{W}^{1 - \frac{1}{2q}, 2 - \frac{1}{q}, q}(\mathbb{T} \times \mathbb{R}^{n-1})^n \end{aligned} \tag{4.4.3}$$

with

$$h_n \in \mathrm{W}^{1,q} \big(\mathbb{T}; \dot{\mathrm{W}}^{-\frac{1}{q}, q}(\mathbb{R}^{n-1}) \big) \tag{4.4.4}$$

there is a solution (u, p) to (4.4.2) with

$$\begin{aligned} & u_{\mathrm{s}} \in \dot{\mathrm{W}}^{2,q}(\mathbb{R}^n_+)^n, \\ & u_{\mathrm{tp}} \in \mathrm{W}^{1,2,q}_\perp(\mathbb{T} \times \mathbb{R}^n_+)^n, \\ & p \in \mathrm{L}^q \big(\mathbb{T}; \dot{\mathrm{W}}^{1,q}(\mathbb{R}^n_+) \big), \end{aligned} \tag{4.4.5}$$

which satisfies

$$\begin{aligned} \| \nabla^2 u_{\mathrm{s}} \|_{\mathrm{L}^q(\mathbb{R}^n_+)} & + \| \nabla p_{\mathrm{s}} \|_{\mathrm{L}^q(\mathbb{R}^n_+)} \\ & \leq C_{29} \big(\| f_{\mathrm{s}} \|_{\mathrm{L}^q(\mathbb{R}^n_+)} + \| g_{\mathrm{s}} \|_{\mathrm{W}^{1,q}(\mathbb{R}^n_+)} + \| h_{\mathrm{s}} \|_{\mathrm{W}^{2 - \frac{1}{q}, q}(\mathbb{R}^{n-1})} \big) \end{aligned} \tag{4.4.6}$$

and

$$\|u_{\mathrm{tp}}\|_{\mathrm{W}_\perp^{1,2,q}(\mathbb{T}\times\mathbb{R}_+^n)} + \|\nabla p_{\mathrm{tp}}\|_{\mathrm{L}_\perp^q(\mathbb{T};\mathrm{L}^q(\mathbb{R}_+^n))}$$

$$\leq C_{29}\left(\|f_{\mathrm{tp}}\|_{\mathrm{L}_\perp^q(\mathbb{T};\mathrm{L}^q(\mathbb{R}_+^n))} + \|g_{\mathrm{tp}}\|_{\mathrm{L}_\perp^q(\mathbb{T};\mathrm{W}^{1,q}(\mathbb{R}_+^n))\cap\mathrm{W}_\perp^{1,q}(\mathbb{T};\dot{\mathrm{W}}^{-1,q}(\mathbb{R}_+^n))}\right. \quad (4.4.7)$$

$$\left. + \|h_{\mathrm{tp}}\|_{\mathrm{W}_\perp^{1-\frac{1}{2q},2-\frac{1}{q},q}(\mathbb{T}\times\mathbb{R}^{n-1})} + \|h_{\mathrm{tp},n}\|_{\mathrm{W}_\perp^{1,q}(\mathbb{T};\dot{\mathrm{W}}^{-\frac{1}{q},q}(\mathbb{R}^{n-1}))}\right)$$

with $C_{29} = C_{29}(n, q, \mathcal{T}) > 0$. If $(\tilde{u}, \tilde{p})$ is another solution to (4.4.2) in the class (4.4.5), then $u_{\mathrm{tp}} = \tilde{u}_{\mathrm{tp}}$, $u_{\mathrm{s}} = \tilde{u}_{\mathrm{s}} + (a_1 x_n, \ldots, a_{n-1} x_n, 0)$ for some vector $a \in \mathbb{R}^{n-1}$, and $p = \tilde{p} + d(t)$ for some function d that depends on time only.

The two separated estimates (4.4.6) and (4.4.7) of different regularity type for the steady-state u_{s} and the purely oscillatory part u_{tp} of the solution, respectively, demonstrate the necessity of the decomposition. Observe that the purely oscillatory part u_{tp} of the solution is unique, whereas the steady-state part u_{s} is only unique up to some function ax_n, $a \in \mathbb{R}^{n-1}$. The first estimate (4.4.6) is well-known for the former problem, whence the main objective in the following will be establishing existence of a unique solution to the latter that satisfies (4.4.7).

Remark 4.4.2. For properties of solutions to the steady-state Stokes problem we refer the reader to [36]. The homogeneous Sobolev space $\dot{\mathrm{W}}^{2,q}(\mathbb{R}_+^n)$ used to characterize the solution u_{s} in (4.4.5), however, is defined slightly different in [36] compared to the definition (2.4.1) employed in this doctoral thesis. In [36] it is defined as the subspace $\{u \in \mathrm{L}_{loc}^1(\mathbb{R}_+^n) \mid \nabla^2 u \in \mathrm{L}^q(\mathbb{R}_+^n)\}$ of locally integrable functions, whereas we introduce it in (2.4.1) as a subspace of the distributions $Z'(\mathbb{R}^n)$. The assertions in Theorem 4.4.1, in particular estimate (4.4.6), remain valid in both cases. The latter definition has the advantage that the norm $\|\nabla^2 \cdot\|_{\mathrm{L}^q(\mathbb{R}_+^n)}$ appearing on the left-hand side of (4.4.6) turns $\dot{\mathrm{W}}^{2,q}(\mathbb{R}_+^n)$ into a Banach space, but to the disadvantage that elements of $\dot{\mathrm{W}}^{2,q}(\mathbb{R}_+^n)$ can only be interpreted as functions modulo polynomials.

Remark 4.4.3. Theorem 4.4.1 does not characterize the $\mathcal{T}$-time-periodic Stokes operator as a homeomorphism between the function spaces of the solutions (4.4.5) and the function spaces of the data (4.4.3)–(4.4.4). In particular, there are clearly functions (u, p) in the function spaces (4.4.5) for which the condition (4.4.4) we require of the purely oscillatory part of the inhomogeneous boundary data is not satisfied for $\mathrm{Tr}\,\mathcal{P}_\perp u_n$. The same phenomenon can be observed in the analysis of the initial-value Stokes problem in half spaces (see for example [77, Chapter 7.2]) and seems to manifest itself in a similar manner in the time-periodic case.

In order to prove Theorem 4.4.1 we employ $\mathcal{P}$ and $\mathcal{P}_\perp$ to decompose the solution to the Stokes problem into a steady-state and purely oscillatory part that solve different systems. More specifically, we observe that $(u, p) = (u_s + u_{tp}, p_s + p_{tp})$ is a solution to (4.4.2) if (u_s, p_s) is a solution to the steady-state problem

$$\begin{cases} -\Delta u_s + \nabla p_s = f_s & \text{in } \mathbb{R}_+^n, \\ \operatorname{div} u_s = g_s & \text{in } \mathbb{R}_+^n, \\ u_s = h_s & \text{on } \partial\mathbb{R}_+^n \end{cases} \tag{4.4.8}$$

and (u_{tp}, p_{tp}) is a solution to

$$\begin{cases} \partial_t u_{tp} - \Delta u_{tp} + \nabla p_{tp} = f_{tp} & \text{in } \mathbb{T} \times \mathbb{R}_+^n, \\ \operatorname{div} u_{tp} = g_{tp} & \text{in } \mathbb{T} \times \mathbb{R}_+^n, \\ u_{tp} = h_{tp} & \text{on } \mathbb{T} \times \partial\mathbb{R}_+^n. \end{cases} \tag{4.4.9}$$

The steady-state problem (4.4.8) is a classical Stokes problem, for which a comprehensive theory is available (see for example [36]). We therefore focus on the purely oscillatory problem (4.4.9), which only differs from (4.4.2) by having purely oscillatory data

$$f_{tp} \in \mathrm{L}_\perp^q(\mathbb{T} \times \mathbb{R}_+^n)^n,$$
$$g_{tp} \in \mathrm{L}_\perp^q\big(\mathbb{T}; \mathrm{W}^{1,q}(\mathbb{R}_+^n)\big) \cap \mathrm{W}_\perp^{1,q}\big(\mathbb{T}; \dot{\mathrm{W}}^{-1,q}(\mathbb{R}_+^n)\big),$$
$$h_{tp} \in \mathrm{W}_\perp^{1-\frac{1}{2q},2-\frac{1}{q},q}(\mathbb{T} \times \mathbb{R}^{n-1})^n.$$

Observe that (4.4.9) can be reduced to the Stokes system with homogeneous data and inhomogeneous boundary values (see for example the proof of Theorem 4.4.5). For this reason, we consider the case of non-homogeneous boundary values.

Proposition 4.4.4. *Let $q \in (1, \infty)$ and $n \geq 2$. For any vector field H with*

$$\begin{aligned} H &\in \mathrm{W}_\perp^{1-\frac{1}{2q},2-\frac{1}{q},q}(\mathbb{T} \times \mathbb{R}^{n-1})^n, \\ H_n &\in \mathrm{W}_\perp^{1,q}\big(\mathbb{T}; \dot{\mathrm{W}}^{-\frac{1}{q},q}(\mathbb{R}^{n-1})\big) \end{aligned} \tag{4.4.10}$$

there is a solution

$$\begin{aligned} u &\in \mathrm{W}_\perp^{1,2,q}(\mathbb{T} \times \mathbb{R}_+^n)^n, \\ p &\in \mathrm{L}_\perp^q\big(\mathbb{T}; \dot{\mathrm{W}}^{1,q}(\mathbb{R}_+^n)\big) \end{aligned} \tag{4.4.11}$$

to

$$\begin{cases} \partial_t u - \Delta u + \nabla p = 0 & in \ \mathbb{T} \times \mathbb{R}_+^n, \\ \operatorname{div} u = 0 & in \ \mathbb{T} \times \mathbb{R}_+^n, \\ u = H & on \ \mathbb{T} \times \partial\mathbb{R}_+^n \end{cases} \tag{4.4.12}$$

that satisfies

$$\|u\|_{W_\perp^{1,2,q}(\mathbb{T}\times\mathbb{R}_+^n)} + \|\nabla p\|_{L_\perp^q(\mathbb{T};L^q(\mathbb{R}_+^n))}$$
$$\leq C_{30}\left(\|H\|_{W_\perp^{1-\frac{1}{2q},2-\frac{1}{q},q}(\mathbb{T};L^q(\mathbb{R}^{n-1}))} + \|H_n\|_{W_\perp^{1,q}(\mathbb{T};\dot{W}^{-\frac{1}{q},q}(\mathbb{R}^{n-1}))}\right) \tag{4.4.13}$$

with $C_{30} = C_{30}(n, q, \mathcal{T}) > 0$.

Proof. We shall employ the partial Fourier transform $\mathscr{F}_{\mathbb{T}\times\mathbb{R}^{n-1}}$ to transform (4.4.12) into a system of ODEs. Letting $v := u' := (u_1, \ldots, u_{n-1})$, $w := u_n$, $\widehat{v} := \mathscr{F}_{\mathbb{T}\times\mathbb{R}^{n-1}}[v]$, and $\widehat{w} := \mathscr{F}_{\mathbb{T}\times\mathbb{R}^{n-1}}[w]$ in (4.4.12), we obtain an equivalent formulation of the system as a family of ODEs. More precisely, (4.4.12) is equivalent to the following ODE being satisfied for each (fixed) $(k, \xi) \in \frac{2\pi}{\mathcal{T}}\mathbb{Z} \times \mathbb{R}^{n-1}$:

$$\begin{cases} ik\widehat{v}(x_n) + |\xi|^2\widehat{v}(x_n) - \partial_{x_n}^2\widehat{v}(x_n) + i\xi\widehat{p}(x_n) = 0 & in \ \mathbb{R}_+, \\ ik\widehat{w}(x_n) + |\xi|^2\widehat{w}(x_n) - \partial_{x_n}^2\widehat{w}(x_n) + \partial_{x_n}\widehat{p}(x_n) = 0 & in \ \mathbb{R}_+, \\ i\xi \cdot \widehat{v}(x_n) + \partial_{x_n}\widehat{w}(x_n) = 0 & in \ \mathbb{R}_+, \\ (\widehat{v}(0), \widehat{w}(0)) = (\widehat{H'}, \widehat{H_n}). \end{cases} \tag{4.4.14}$$

To solve this ODE, we first consider the case $k \neq 0$. Taking divergence on both sides in $(4.4.12)_1$ and utilizing that $\operatorname{div} u = 0$, we find that $\Delta p = 0$ and thus $-|\xi|^2\widehat{p}(x_n) + \partial_{x_n}^2\widehat{p}(x_n) = 0$ in $\mathbb{R}_+$. Consequently,

$$\widehat{p}(x_n) = q_0(k, \xi)e^{-|\xi|x_n} \quad in \ \mathbb{R}_+ \tag{4.4.15}$$

for some function $q_0 : \frac{2\pi}{\mathcal{T}}\mathbb{Z} \times \mathbb{R}^{n-1} \to \mathbb{C}$. Observe that the representation formula (4.4.15) holds, as $e^{|\xi|x_n} \notin \mathscr{S}'(\mathbb{R}^{n-1})$ for any $x_n > 0$. Inserting (4.4.15) into (4.4.14), we find that

$$\partial_{x_n}^2\widehat{v} = (ik + |\xi|^2)\widehat{v} + i\xi q_0 e^{-|\xi|x_n},$$
$$\partial_{x_n}^2\widehat{w} = (ik + |\xi|^2)\widehat{w} - |\xi|q_0 e^{-|\xi|x_n}.$$

Since $k \neq 0$, the resolution hereof yields

$$\widehat{v}(x_n) = -\frac{\xi q_0(k, \xi)}{k} e^{-|\xi| x_n} + \alpha(k, \xi) e^{-\sqrt{|\xi|^2 + ik}\, x_n}, \qquad (4.4.16)$$

$$\widehat{w}(x_n) = \frac{|\xi| q_0(k, \xi)}{ik} e^{-|\xi| x_n} + \beta(k, \xi) e^{-\sqrt{|\xi|^2 + ik}\, x_n}, \qquad (4.4.17)$$

for some functions $\alpha : \frac{2\pi}{T}\mathbb{Z} \times \mathbb{R}^{n-1} \to \mathbb{C}^{n-1}$ and $\beta : \frac{2\pi}{T}\mathbb{Z} \times \mathbb{R}^{n-1} \to \mathbb{C}$. Note that these identities follow by exploiting that $e^{\sqrt{|\xi|^2 + ik}\, x_n} \notin \mathscr{S}'(\mathbb{R}^{n-1})$ for any $x_n > 0$. Utilizing $(4.4.14)_3$ and the boundary conditions $(4.4.14)_4$, we deduce

$$\alpha = \widehat{H'} + \frac{\xi q_0}{k}, \qquad \beta = \widehat{H_n} - \frac{|\xi| q_0}{ik}$$

and

$$q_0 = -i \left(|\xi| + \sqrt{|\xi|^2 + ik} \right) \frac{\xi}{|\xi|} \cdot \widehat{H'} + \left(\sqrt{|\xi|^2 + ik} + |\xi| \right) \widehat{H_n} + \frac{ik}{|\xi|} \widehat{H_n}. \qquad (4.4.18)$$

By $(4.4.15)$–$(4.4.18)$ a solution to $(4.4.14)$ is identified in the case $k \neq 0$. Since H is purely oscillatory, we have $\widehat{H'}(0, \xi) = \widehat{H_n}(0, \xi) = 0$, whence $(v, w, p) := (0, 0, 0)$ solves $(4.4.14)$ in the case $k = 0$. We thus obtain a formula for the solution to $(4.4.12)$:

$$v = \mathscr{F}_{\mathbb{T} \times \mathbb{R}^{n-1}}^{-1} \left[-\frac{\xi q_0}{k} e^{-|\xi| x_n} + \left(\widehat{H'} + \frac{\xi q_0}{k} \right) e^{-\sqrt{|\xi|^2 + ik}\, x_n} \right],$$

$$w = \mathscr{F}_{\mathbb{T} \times \mathbb{R}^{n-1}}^{-1} \left[\frac{|\xi| q_0}{ik} e^{-|\xi| x_n} + \left(\widehat{H_n} - \frac{|\xi| q_0}{ik} \right) e^{-\sqrt{|\xi|^2 + ik}\, x_n} \right], \qquad (4.4.19)$$

$$p = \mathscr{F}_{\mathbb{T} \times \mathbb{R}^{n-1}}^{-1} \left[q_0 e^{-|\xi| x_n} \right].$$

Formally at least, (v, w, p) as defined above is a solution to $(4.4.12)$. It remains to show that this solution is well-defined in the class $(4.4.11)$ for data in the class $(4.4.10)$. We start by considering data $H \in \mathcal{P}_\perp Z(\mathbb{T} \times \mathbb{R}^{n-1})^n$. The space $Z(\mathbb{T} \times \mathbb{R}^{n-1})$ is dense in

$$W^{1 - \frac{1}{2q}, q}\big(\mathbb{T}; L^q(\mathbb{R}^{n-1}) \big) \cap L^q\big(\mathbb{T}; W^{2 - \frac{1}{q}, q}(\mathbb{R}^{n-1}) \big) \cap W^{1, q}\big(\mathbb{T}; \dot{W}^{-\frac{1}{q}, q}(\mathbb{R}^{n-1}) \big),$$

which is not a trivial assertion since it entails the construction of an approximating sequence that converges simultaneously in Sobolev spaces of positive order and in homogeneous Sobolev spaces of negative order.

Nevertheless, it can be shown by a standard "cut-off" and mollifier technique; see [88, proof of Theorem 2.3.3 and Theorem 5.1.5]. Consequently, $\mathcal{P}_\perp Z(\mathbb{T} \times \mathbb{R}^{n-1})^n$ is dense in the class (4.4.10). Clearly, for purely oscillatory data H the solution given by (4.4.19) is also purely oscillatory. If we can therefore show (4.4.13) for arbitrary $H \in \mathcal{P}_\perp Z(\mathbb{T} \times \mathbb{R}^{n-1})^n$, the claim of the proposition will follow by a density argument.

We first examine the pressure term p (more specifically ∇p). The terms in (4.4.18) have different order of regularity, so we decompose $q_0 = q_1 + q_2$ by

$$q_1(k, \xi) := -i\left(|\xi| + \sqrt{|\xi|^2 + ik}\right) \frac{\xi}{|\xi|} \cdot \widehat{H'} + \left(\sqrt{|\xi|^2 + ik} + |\xi|\right) \widehat{H_n},$$

$$q_2(k, \xi) := \frac{ik}{|\xi|} \widehat{H_n},$$

$$(4.4.20)$$

and introduce the operators

$$\mathcal{G} : Z(\mathbb{T} \times \mathbb{R}^{n-1})^n \to \mathscr{S}(\mathbb{T} \times \mathbb{R}^n_+)^n,$$

$$\mathcal{G}(H) := \mathscr{F}^{-1}_{\mathbb{T} \times \mathbb{R}^{n-1}}\left[\xi q_1(k, \xi) e^{-|\xi|x_n}\right]$$

$$(4.4.21)$$

and

$$\mathcal{B} : Z(\mathbb{T} \times \mathbb{R}^{n-1}) \to \mathscr{S}(\mathbb{T} \times \mathbb{R}^n_+)^n,$$

$$\mathcal{B}(H_n) := \mathscr{F}^{-1}_{\mathbb{T} \times \mathbb{R}^{n-1}}\left[\xi q_2(k, \xi) e^{-|\xi|x_n}\right].$$

$$(4.4.22)$$

For $m \in \mathbb{N}_0$, observe that for any $x_n > 0$ the symbol $\xi \to (|\xi|x_n)^m\, e^{-|\xi|x_n}$ is an $\mathrm{L}^q(\mathbb{R}^{n-1})$-multiplier. Specifically, one verifies that

$$\sup_{x_n > 0} \sup_{\varepsilon \in \{0,1\}^{n-1}} \sup_{\xi \in \mathbb{R}^{n-1}} \left| \xi_1^{\varepsilon_1} \cdots \xi_{n-1}^{\varepsilon_{n-1}} \partial_{\xi_1}^{\varepsilon_1} \cdots \partial_{\xi_{n-1}}^{\varepsilon_{n-1}}\left[(|\xi|x_n)^m e^{-|\xi|x_n}\right]\right| < \infty,$$

whence it follows from the Multiplier Theorem of Marcinkiewicz (see Theorem 2.3.2) that the Fourier multiplier operator with symbol $\xi \to (|\xi|x_n)^m\, e^{-|\xi|x_n}$ is a bounded operator on $\mathrm{L}^q(\mathbb{R}^{n-1})$ with operator norm independent on x_n, that is,

$$\sup_{x_n > 0} \left\| \phi \mapsto \mathscr{F}^{-1}_{\mathbb{R}^{n-1}}\left[(|\xi|x_n)^m\, e^{-|\xi|x_n}\, \mathscr{F}_{\mathbb{R}^{n-1}}[\phi]\right] \right\|_{\mathscr{L}(\mathrm{L}^q(\mathbb{R}^{n-1}), \mathrm{L}^q(\mathbb{R}^{n-1}))} < \infty.$$

$$(4.4.23)$$

We can thus estimate

$$\|\mathscr{G}(H)\|_{L^\infty(\mathbb{R}_+;L^q_\perp(\mathbb{T};L^q(\mathbb{R}^{n-1})))} \le c_0 \left\| \mathscr{F}_{\mathbb{T}\times\mathbb{R}^{n-1}}^{-1}\left[\xi q_1(k,\xi)\right] \right\|_{L^q_\perp(\mathbb{T};L^q(\mathbb{R}^{n-1}))}$$

$$\le c_1 \left(\|H\|_{L^q_\perp(\mathbb{T};H^{2,q}(\mathbb{R}^{n-1}))} \right.$$

$$+ \left\| \mathscr{F}_{\mathbb{T}\times\mathbb{R}^{n-1}}^{-1}\left[M(k,\xi)\left(|\xi|^2+ik\right)\frac{\xi}{|\xi|}\widehat{H_n} \right] \right\|_{L^q_\perp(\mathbb{T};L^q(\mathbb{R}^{n-1}))}$$

$$\left. + \left\| \mathscr{F}_{\mathbb{T}\times\mathbb{R}^{n-1}}^{-1}\left[M(k,\xi)\left(|\xi|^2+ik\right)\frac{\xi\otimes\xi}{|\xi|^2}\widehat{H'} \right] \right\|_{L^q_\perp(\mathbb{T};L^q(\mathbb{R}^{n-1}))} \right),$$

where

$$M : \mathbb{R}\times\mathbb{R}^{n-1} \to \mathbb{C}, \quad M(\eta,\xi) := \frac{|\xi|}{\sqrt{|\xi|^2+i\eta}}.$$

Employing again the Marcinkiewicz Multiplier Theorem, we find that the symbol M is an $L^q(\mathbb{R};L^q(\mathbb{R}^{n-1}))$-multiplier. An application of the Transference Principle (Theorem 2.3.1) therefore implies that the restriction $M_{|\frac{2\pi}{\mathcal{T}}\mathbb{Z}\times\mathbb{R}^{n-1}}$ is an $L^q(\mathbb{T};L^q(\mathbb{R}^{n-1}))$-multiplier. We thus conclude

$$\|\mathscr{G}(H)\|_{L^\infty(\mathbb{R}_+;L^q_\perp(\mathbb{T};L^q(\mathbb{R}^{n-1})))} \le c_2 \|H\|_{L^q_\perp(\mathbb{T};H^{2,q}(\mathbb{R}^{n-1}))\cap H^{1,q}_\perp(\mathbb{T};L^q(\mathbb{R}^{n-1}))}.$$

$$(4.4.24)$$

This estimate shall serve as an interpolation endpoint. To obtain the opposite endpoint, we again employ (4.4.23) to estimate

$$\sup_{x_n>0} \|x_n \mathscr{G}(H)\|_{L^q_\perp(\mathbb{T};L^q(\mathbb{R}^{n-1}))} \le c_3 \|q_1\|_{L^q_\perp(\mathbb{T};L^q(\mathbb{R}^{n-1}))},$$

which implies

$$\|\mathscr{G}(H)\|_{L^{1,\infty}(\mathbb{R}_+;L^q_\perp(\mathbb{T};L^q(\mathbb{R}^{n-1})))} = \left\| \frac{1}{x_n}\|x_n\mathscr{G}(H)\|_{L^q_\perp(\mathbb{T};L^q(\mathbb{R}^{n-1}))} \right\|_{L^{1,\infty}(\mathbb{R}_+)}$$

$$\le c_4 \|q_1\|_{L^q_\perp(\mathbb{T};L^q(\mathbb{R}^{n-1}))}.$$

Recalling (4.4.20), we estimate

$$\|q_1\|_{L^q_\perp(\mathbb{T};L^q(\mathbb{R}^{n-1}))} \le c_5 \left(\|H\|_{L^q_\perp(\mathbb{T};H^{1,q}(\mathbb{R}^{n-1}))} \right.$$

$$+ \left\| \mathscr{F}_{\mathbb{T}\times\mathbb{R}^{n-1}}^{-1}\left[M_1(k,\xi)\cdot\left(|\xi|+|k|^{\frac{1}{2}}\right)\widehat{H'} \right] \right\|_{L^q_\perp(\mathbb{T};L^q(\mathbb{R}^{n-1}))}$$

$$\left. + \left\| \mathscr{F}_{\mathbb{T}\times\mathbb{R}^{n-1}}^{-1}\left[M_2(k,\xi)\left(|\xi|+|k|^{\frac{1}{2}}\right)\widehat{H_n} \right] \right\|_{L^q_\perp(\mathbb{T};L^q(\mathbb{R}^{n-1}))} \right)$$

with M_1 and M_2 defined as

$$M_1 : \mathbb{R} \times \mathbb{R}^{n-1} \to \mathbb{C}^{n-1}, \quad M_1(\eta, \xi) := \frac{\sqrt{|\xi|^2 + i\eta}}{|\xi| + |\eta|^{\frac{1}{2}}} \frac{\xi}{|\xi|},$$

and

$$M_2 : \mathbb{R} \times \mathbb{R}^{n-1} \to \mathbb{C}, \quad M_2(\eta, \xi) := \frac{\sqrt{|\xi|^2 + i\eta}}{|\xi| + |\eta|^{\frac{1}{2}}}.$$

Again, one can utilize the Marcinkiewicz Multiplier Theorem to show that both M_1 and M_2 are $L^q(\mathbb{R}; L^q(\mathbb{R}^{n-1}))$-multipliers, and subsequently obtain from the Transference Principle that their restrictions to $\frac{2\pi}{\mathcal{T}}\mathbb{Z} \times \mathbb{R}^{n-1}$ are $L^q(\mathbb{T}; L^q(\mathbb{R}^{n-1}))$-multipliers. Consequently, we find that

$$\|\mathscr{G}(H)\|_{L^{1,\infty}(\mathbb{R}_+; L_\perp^q(\mathbb{T}; L^q(\mathbb{R}^{n-1})))} \le c_6 \|H\|_{L_\perp^q(\mathbb{T}; H^{1,q}(\mathbb{R}^{n-1})) \cap H_\perp^{\frac{1}{2},q}(\mathbb{T}; L^q(\mathbb{R}^{n-1}))}. \tag{4.4.25}$$

By (4.4.24) and (4.4.25), the operator $\mathscr{G}$ extends uniquely to a bounded operator

$$\mathscr{G} : L^q(\mathbb{T}; H^{2,q}(\mathbb{R}^{n-1}))^n \cap H^{1,q}(\mathbb{T}; L^q(\mathbb{R}^{n-1}))^n$$
$$\to L^\infty(\mathbb{R}_+; L^q(\mathbb{T}; L^q(\mathbb{R}^{n-1})))^n,$$
$$\mathscr{G} : L^q(\mathbb{T}; H^{1,q}(\mathbb{R}^{n-1}))^n \cap H^{\frac{1}{2},q}(\mathbb{T}; L^q(\mathbb{R}^{n-1}))^n$$
$$\to L^{1,\infty}(\mathbb{R}_+; L^q(\mathbb{T}; L^q(\mathbb{R}^{n-1})))^n. \tag{4.4.26}$$

These extensions rely on the fact that $Z(\mathbb{T} \times \mathbb{R}^{n-1})$ is dense in the function spaces on the left-hand side above. We once more refer to [88, proof of Theorem 2.3.3 and Theorem 5.1.5] for a verification of this fact. Using the projection $\mathcal{P}_\perp$ on the left-hand side in (4.4.26), we obtain scales of the anisotropic Bessel potential spaces introduced in (2.5.6). Consequently, $\mathscr{G}$ is a bounded operator:

$$\mathscr{G} : H^2_{q,\perp}(\mathbb{T} \times \mathbb{R}^{n-1})^n \to L^\infty(\mathbb{R}_+; L_\perp^q(\mathbb{T}; L^q(\mathbb{R}^{n-1})))^n,$$
$$\mathscr{G} : H^1_{q,\perp}(\mathbb{T} \times \mathbb{R}^{n-1})^n \to L^{1,\infty}(\mathbb{R}_+; L_\perp^q(\mathbb{T}; L^q(\mathbb{R}^{n-1})))^n. \tag{4.4.27}$$

Utilizing Lemma 2.5.3, we find that

$$\left(H^1_{q,\perp}(\mathbb{T}\times\mathbb{R}^{n-1}), H^2_{q,\perp}(\mathbb{T}\times\mathbb{R}^{n-1})\right)_{1-\frac{1}{q},q} = \mathrm{B}^{2-\frac{1}{q}}_{qq,\perp}(\mathbb{T}\times\mathbb{R}^{n-1})$$

$$= \left(\mathrm{L}^q_\perp(\mathbb{T};\mathrm{L}^q(\mathbb{R}^{n-1})), H^2_{q,\perp}(\mathbb{T}\times\mathbb{R}^{n-1})\right)_{1-\frac{1}{2q},q}$$

$$= \left(\mathrm{L}^q_\perp(\mathbb{T};\mathrm{L}^q(\mathbb{R}^{n-1})), \mathrm{H}^{1,q}_\perp(\mathbb{T};\mathrm{L}^q(\mathbb{R}^{n-1}))\cap\mathrm{L}^q_\perp(\mathbb{T};\mathrm{H}^{2,q}(\mathbb{R}^{n-1}))\right)_{1-\frac{1}{2q},q}.$$

Where the last identity follows in view of (2.5.6). Employing [87, Theorem 1.12.1], we obtain for the intersection

$$\left(H^1_{q,\perp}(\mathbb{T}\times\mathbb{R}^{n-1}), H^2_{q,\perp}(\mathbb{T}\times\mathbb{R}^{n-1})\right)_{1-\frac{1}{q},q}$$

$$= \mathcal{P}_\perp\left(\mathrm{L}^q(\mathbb{T};\mathrm{L}^q(\mathbb{R}^{n-1})), \mathrm{H}^{1,q}(\mathbb{T};\mathrm{L}^q(\mathbb{R}^{n-1}))\right)_{1-\frac{1}{2q},q}$$

$$\cap\,\mathcal{P}_\perp\left(\mathrm{L}^q(\mathbb{T};\mathrm{L}^q(\mathbb{R}^{n-1})), \mathrm{L}^q(\mathbb{T};\mathrm{H}^{2,q}(\mathbb{R}^{n-1}))\right)_{1-\frac{1}{2q},q}$$

$$= \mathcal{P}_\perp\mathrm{W}^{1-\frac{1}{2q},q}(\mathbb{T};\mathrm{L}^q(\mathbb{R}^{n-1}))\cap\mathcal{P}_\perp\mathrm{L}^q(\mathbb{T};\mathrm{W}^{2-\frac{1}{q},q}(\mathbb{R}^{n-1}))$$

$$= \mathrm{W}^{1-\frac{1}{2q},2-\frac{1}{q},q}_\perp(\mathbb{T}\times\mathbb{R}^{n-1}).$$

One can employ [87, Theorem 1.12.1] to verify the interpolation of the intersection space in the first equality above. Moreover, real interpolation yields

$$\left(\mathrm{L}^{1,\infty}(\mathbb{R}_+;\mathrm{L}^q_\perp(\mathbb{T};\mathrm{L}^q(\mathbb{R}^{n-1}))), \mathrm{L}^\infty(\mathbb{R}_+;\mathrm{L}^q_\perp(\mathbb{T};\mathrm{L}^q(\mathbb{R}^{n-1})))\right)_{1-\frac{1}{q},q}$$
$$= \mathrm{L}^q(\mathbb{R}_+;\mathrm{L}^q_\perp(\mathbb{T};\mathrm{L}^q(\mathbb{R}^{n-1}))).$$

Recalling (4.4.27), we conclude that $\mathscr{G}$ extends uniquely to a bounded operator

$$\mathscr{G}: \mathrm{W}^{1-\frac{1}{2q},2-\frac{1}{q},q}_\perp(\mathbb{T}\times\mathbb{R}^{n-1})^n \to \mathrm{L}^q_\perp(\mathbb{T};\mathrm{L}^q(\mathbb{R}^n_+))^n. \qquad (4.4.28)$$

We now recall (4.4.22) and examine the operator $\mathscr{B}$. Utilizing (4.4.23) with $m=0$, we obtain

$$\|\mathscr{B}(H_n)\|_{\mathrm{L}^\infty(\mathbb{R}_+;\mathrm{L}^q_\perp(\mathbb{T};\mathrm{L}^q(\mathbb{R}^{n-1})))} \le c_7\|\mathscr{F}^{-1}_{\mathbb{T}\times\mathbb{R}^{n-1}}[\xi q_2(k,\xi)]\|_{\mathrm{L}^q_\perp(\mathbb{T};\mathrm{L}^q(\mathbb{R}^{n-1}))}$$
$$\le c_8\|H_n\|_{\mathrm{H}^{1,q}_\perp(\mathbb{T};\mathrm{L}^q(\mathbb{R}^{n-1}))}.$$
$$(4.4.29)$$

We again employ (4.4.23) to estimate

$$\sup_{x_n>0} \|x_n \mathscr{B}(H_n)\|_{\mathrm{L}^q_\perp(\mathbb{T};\mathrm{L}^q(\mathbb{R}^{n-1}))} \le c_9 \|H_n\|_{\mathrm{H}^{1,q}_\perp(\mathbb{T};\dot{\mathrm{H}}^{-1,q}(\mathbb{R}^{n-1}))},$$

which implies

$$\|\mathscr{B}(H_n)\|_{\mathrm{L}^{1,\infty}(\mathbb{R}_+;\mathrm{L}^q_\perp(\mathbb{T};\mathrm{L}^q(\mathbb{R}^{n-1})))} = \left\| \frac{1}{x_n} \|x_n \mathscr{B}(H_n)\|_{\mathrm{L}^q_\perp(\mathbb{T};\mathrm{L}^q(\mathbb{R}^{n-1}))} \right\|_{\mathrm{L}^{1,\infty}(\mathbb{R}_+)}$$
$$\le c_{10} \|H_n\|_{\mathrm{H}^{1,q}_\perp(\mathbb{T};\dot{\mathrm{H}}^{-1,q}(\mathbb{R}^{n-1}))}.$$

It follows that $\mathscr{B}$ extends to a bounded operator

$$\mathscr{B} : \mathrm{H}^{1,q}_\perp\big(\mathbb{T};\mathrm{L}^q(\mathbb{R}^{n-1})\big) \to \mathrm{L}^\infty\big(\mathbb{R}_+;\mathrm{L}^q_\perp(\mathbb{T};\mathrm{L}^q(\mathbb{R}^{n-1}))\big)^n,$$
$$\mathscr{B} : \mathrm{H}^{1,q}_\perp\big(\mathbb{T};\dot{\mathrm{H}}^{-1,q}(\mathbb{R}^{n-1})\big) \to \mathrm{L}^{1,\infty}\big(\mathbb{R}_+;\mathrm{L}^q_\perp(\mathbb{T};\mathrm{L}^q(\mathbb{R}^{n-1}))\big).^n$$

Real interpolation thus implies that $\mathscr{B}$ extends to a bounded operator

$$\mathscr{B} : \mathrm{W}^{1,q}_\perp\big(\mathbb{T};\dot{\mathrm{W}}^{-\frac{1}{q},q}(\mathbb{R}^{n-1})\big) \to \mathrm{L}^q_\perp\big(\mathbb{T};\mathrm{L}^q(\mathbb{R}^n_+)\big)^n. \qquad (4.4.30)$$

We now return to the solution formulas (4.4.19) and consider data $H \in \mathcal{P}_\perp Z(\mathbb{T} \times \mathbb{R}^{n-1})^n$. In this case, an application of (4.4.23) ensures that p is well-defined as an element in the function space $\mathrm{L}^q_\perp\big(\mathbb{T};\dot{\mathrm{H}}^{1,q}(\mathbb{R}^n_+)\big)$. By (4.4.28) and (4.4.30), we obtain

$$\|\nabla p\|_{\mathrm{L}^q_\perp(\mathbb{T};\mathrm{L}^q(\mathbb{R}^n_+))} = \|\mathscr{G}(H) + \mathscr{B}(H_n)\|_{\mathrm{L}^q_\perp(\mathbb{T};\mathrm{L}^q(\mathbb{R}^n_+))}$$
$$\le c_{11}\big(\|H\|_{\mathrm{W}^{1-\frac{1}{2q},2-\frac{1}{q},q}_\perp(\mathbb{T}\times\mathbb{R}^{n-1})} + \|H_n\|_{\mathrm{W}^{1,q}_\perp(\mathbb{T};\dot{\mathrm{W}}^{-\frac{1}{q},q}(\mathbb{R}^{n-1}))}\big) \qquad (4.4.31)$$

In a similar manner, it can be shown that (v,w) is well-defined as an element in the space $\mathrm{L}^q\big(\mathbb{T};\mathrm{H}^{2,q}(\mathbb{R}^n_+)\big) \cap \mathrm{H}^{1,q}\big(\mathbb{T};\mathrm{L}^q(\mathbb{R}^n_+)\big)$. To this end, one may consider the symbol

$$m : \mathbb{R} \times \mathbb{R}^{n-1} \to \mathbb{C}, \quad m(\eta,\xi) := \big(\sqrt{|\xi|^2 + i\eta}\, x_n\big)^m \mathrm{e}^{-\sqrt{|\xi|^2 + i\eta}\, x_n}$$

and verify that

$$\sup_{x_n>0} \sup_{\varepsilon\in\{0,1\}^n} \sup_{(\eta,\xi)\in\mathbb{R}\times\mathbb{R}^{n-1}} \left| \eta^{\varepsilon_0} \xi_1^{\varepsilon_1} \cdots \xi_{n-1}^{\varepsilon_{n-1}} \partial_\eta^{\varepsilon_0} \partial_{\xi_1}^{\varepsilon_1} \cdots \partial_{\xi_{n-1}}^{\varepsilon_{n-1}} m(\eta,\xi) \right| < \infty.$$

It follows that the Fourier multiplier operator corresponding to the symbol m is a bounded operator on $\mathrm{L}^q(\mathbb{R};\mathrm{L}^q(\mathbb{R}^{n-1}))$ with operator norm independent on x_n. An application of the Transference Principle (Theorem

2.3.1) therefore implies that the operator corresponding to the symbol $m_{|\frac{2\pi}{\mathcal{T}}\mathbb{Z}\times\mathbb{R}^{n-1}}$ is a bounded operator on $\mathrm{L}^q(\mathbb{T};\mathrm{L}^q(\mathbb{R}^{n-1}))$ with operator norm independent on x_n, that is,

$$\sup_{x_n>0}\left\|\phi\mapsto\mathscr{F}_{\mathbb{T}\times\mathbb{R}^{n-1}}\big[m(k,\xi)\mathscr{F}_{\mathbb{T}\times\mathbb{R}^{n-1}}[\phi]\big]\right\|_{\mathscr{L}(\mathrm{L}^q(\mathbb{T}\times\mathbb{R}^{n-1}),\mathrm{L}^q(\mathbb{T}\times\mathbb{R}^{n-1}))}<\infty. \tag{4.4.32}$$

With both (4.4.23) and (4.4.32) at our disposal, it is now straightforward to verify that $u:=(v,w)$ is well-defined as element in the space $\mathrm{L}^q\big(\mathbb{T};\mathrm{H}^{2,q}(\mathbb{R}^n_+)\big)\cap\mathrm{H}^{1,q}\big(\mathbb{T};\mathrm{L}^q(\mathbb{R}^n_+)\big)$. By construction, this choice of (u,p) is a solution to (4.4.12). Moreover, since $\mathcal{P}_\perp H=H$, also $u_{\mathrm{tp}}=u$. This means that u is a purely oscillatory solution in the aforementioned function space to the time-periodic heat equation in the half-space

$$\begin{cases}\partial_t u-\Delta u=-\nabla p & \text{in } \mathbb{T}\times\mathbb{R}^n_+,\\ \qquad\qquad u=H & \text{on } \mathbb{T}\times\partial\mathbb{R}^n_+.\end{cases}$$

By [61, Theorem 2.1] (see also [62, Theorem 1.4]), it is known that this problem has a unique purely oscillatory solution in the function space $\mathrm{L}^q_\perp\big(\mathbb{T};\mathrm{H}^{2,q}(\mathbb{R}^n_+)\big)\cap\mathrm{H}^{1,q}_\perp\big(\mathbb{T};\mathrm{L}^q(\mathbb{R}^n_+)\big)$, which satisfies

$$\begin{aligned}&\|u\|_{\mathrm{H}^{1,q}_\perp(\mathbb{T};\mathrm{L}^q(\mathbb{R}^n_+))\cap\mathrm{L}^q_\perp(\mathbb{T};\mathrm{H}^{2,q}(\mathbb{R}^n_+))}\\ &\qquad\leq c_{12}\Big(\|\nabla p\|_{\mathrm{L}^q_\perp(\mathbb{T};\mathrm{L}^q(\mathbb{R}^n_+))}+\|H\|_{\mathrm{W}^{1-\frac{1}{2q},q}_\perp(\mathbb{T};\mathrm{L}^q(\mathbb{R}^{n-1}))\cap\mathrm{L}^q(\mathbb{T};\mathrm{W}^{2-\frac{1}{q},q}(\mathbb{R}^{n-1}))}\Big).\end{aligned}$$

Combining this estimate with (4.4.31), we conclude (4.4.13). $\qquad\square$

Next, we consider the resolution of the fully non-homogeneous system (4.4.9) and establish L^q estimates. This step concludes the main result for the Stokes system if the system is excited by purely oscillatory forces.

Theorem 4.4.5. *Let $q\in(1,\infty)$ and $n\geq 2$. For all*

$$\begin{aligned}&f\in\mathrm{L}^q_\perp\big(\mathbb{T};\mathrm{L}^q(\mathbb{R}^n_+)\big)^n,\\ &g\in\mathrm{L}^q_\perp\big(\mathbb{T};\mathrm{W}^{1,q}(\mathbb{R}^n_+)\big)\cap\mathrm{W}^{1,q}_\perp\big(\mathbb{T};\dot{\mathrm{W}}^{-1,q}(\mathbb{R}^n_+)\big),\\ &h\in\mathrm{W}^{1-\frac{1}{2q},2-\frac{1}{q},q}_\perp\big(\mathbb{T}\times\mathbb{R}^{n-1}\big)^n\end{aligned}$$

with

$$h_n\in\mathrm{W}^{1,q}_\perp\big(\mathbb{T};\dot{\mathrm{W}}^{-\frac{1}{q},q}(\mathbb{R}^{n-1})\big) \tag{4.4.33}$$

there is a solution (u, p) to (4.4.9) with

$$(u, p) \in W_{\perp}^{1,2,q}\big(\mathbb{T} \times \mathbb{R}_{+}^{n}\big)^{n} \times L_{\perp}^{q}\big(\mathbb{T}; \dot{W}^{1,q}(\mathbb{R}_{+}^{n})\big), \qquad (4.4.34)$$

which satisfies

$$\begin{aligned} &\|u\|_{W_{\perp}^{1,2,q}(\mathbb{T} \times \mathbb{R}_{+}^{n})} + \|\nabla p\|_{L_{\perp}^{q}(\mathbb{T}; L^{q}(\mathbb{R}_{+}^{n}))} \\ &\leq C_{31}\big(\|f\|_{L_{\perp}^{q}(\mathbb{T}; L^{q}(\mathbb{R}_{+}^{n}))} + \|g\|_{L_{\perp}^{q}(\mathbb{T}; W^{1,q}(\mathbb{R}_{+}^{n})) \cap W_{\perp}^{1,q}(\mathbb{T}; \dot{W}^{-1,q}(\mathbb{R}_{+}^{n}))} \quad (4.4.35) \\ &\quad + \|h\|_{W_{\perp}^{1-\frac{1}{2q}, 2-\frac{1}{q}, q}(\mathbb{T} \times \mathbb{R}^{n-1})} + \|h_{n}\|_{W_{\perp}^{1,q}(\mathbb{T}; \dot{W}^{-\frac{1}{q}, q}(\mathbb{R}^{n-1}))}\big) \end{aligned}$$

with $C_{31} = C_{31}(n, q, \mathcal{T}) > 0$. Moreover, if $(\tilde{u}, \tilde{p})$ is another solution to (4.4.9) in the class (4.4.34), then $u = \tilde{u}$ and $p = \tilde{p} + d(t)$ for some function d that depends only on time.

Proof. Let $v \in W_{\perp}^{1,q}\big(\mathbb{T}; L^{q}(\mathbb{R}_{+}^{n})\big)^{n} \cap L_{\perp}^{q}\big(\mathbb{T}; W^{2,q}(\mathbb{R}_{+}^{n})\big)^{n}$ be the solution to the purely oscillatory time-periodic n-dimensional heat equation

$$\begin{cases} \partial_{t} v - \Delta v = f & \text{in } \mathbb{T} \times \mathbb{R}_{+}^{n}, \\ \qquad\quad v = 0 & \text{on } \mathbb{T} \times \partial \mathbb{R}_{+}^{n}. \end{cases} \qquad (4.4.36)$$

Existence of such a solution v that satisfies

$$\|v\|_{W_{\perp}^{1,q}(\mathbb{T}; L^{q}(\mathbb{R}_{+}^{n})) \cap L_{\perp}^{q}(\mathbb{T}; W^{2,q}(\mathbb{R}_{+}^{n}))} \leq c_{0}\|f\|_{L_{\perp}^{q}(\mathbb{T}; L^{q}(\mathbb{R}_{+}^{n}))}$$

follows from [61, Theorem 2.1]. Denote by $\mathbf{G}$ the extension of $g - \operatorname{div} v$ to $\mathbb{R}^{n}$ by even reflection in the x_{n} variable. Then $\mathbf{G} \in L_{\perp}^{q}\big(\mathbb{T}; W^{1,q}(\mathbb{R}^{n})\big)$, and by identifying $\dot{W}^{-1,q}(\mathbb{R}^{n})$ as the dual of $\dot{W}^{1,q'}(\mathbb{R}^{n})$ and recalling that $g \in W_{\perp}^{1,q}\big(\mathbb{T}; \dot{W}^{-1,q}(\mathbb{R}_{+}^{n})\big)$, one directly verifies that $\mathbf{G} \in W_{\perp}^{1,q}\big(\mathbb{T}; \dot{W}^{-1,q}(\mathbb{R}^{n})\big)$ with

$$\begin{aligned} \|\mathbf{G}\|_{W_{\perp}^{1,q}(\mathbb{T}; \dot{W}^{-1,q}(\mathbb{R}^{n})) \cap L_{\perp}^{q}(\mathbb{T}; W^{1,q}(\mathbb{R}^{n}))} &\leq c_{1}\big(\|g\|_{W_{\perp}^{1,q}(\mathbb{T}; \dot{W}^{-1,q}(\mathbb{R}_{+}^{n})) \cap L_{\perp}^{q}(\mathbb{T}; W^{1,q}(\mathbb{R}_{+}^{n}))} \\ &\quad + \|v\|_{W_{\perp}^{1,q}(\mathbb{T}; L^{q}(\mathbb{R}_{+}^{n})) \cap L_{\perp}^{q}(\mathbb{T}; W^{2,q}(\mathbb{R}_{+}^{n}))}\big). \end{aligned}$$

A solution to the purely oscillatory Stokes system

$$\begin{cases} \partial_{t} w - \Delta w + \nabla \pi = 0 & \text{in } \mathbb{T} \times \mathbb{R}^{n}, \\ \qquad\qquad\quad \operatorname{div} w = \mathbf{G} & \text{in } \mathbb{T} \times \mathbb{R}^{n} \end{cases} \qquad (4.4.37)$$

is obtained via the solution formulas

$$w := \mathscr{F}_{\mathbb{T} \times \mathbb{R}^{n}}^{-1}\left[\frac{-i\xi}{|\xi|^{2}} \mathscr{F}_{\mathbb{T} \times \mathbb{R}^{n}}[\mathbf{G}]\right], \quad \pi := \mathscr{F}_{\mathbb{T} \times \mathbb{R}^{n}}^{-1}\left[\frac{ik + |\xi|^{2}}{|\xi|^{2}} \mathscr{F}_{\mathbb{T} \times \mathbb{R}^{n}}[\mathbf{G}]\right].$$

$$(4.4.38)$$

From these formulas, we immediately obtain the estimate

$$\|w\|_{\mathrm{W}^{1,q}_{\perp}(\mathbb{T};\mathrm{L}^q(\mathbb{R}^n))\cap\mathrm{L}^q_{\perp}(\mathbb{T};\mathrm{W}^{2,q}(\mathbb{R}^n))} + \|\nabla\pi\|_{\mathrm{L}^q_{\perp}(\mathbb{T};\mathrm{L}^q(\mathbb{R}^n))}$$
$$\leq c_2\|\mathbf{G}\|_{\mathrm{L}^q_{\perp}(\mathbb{T};\mathrm{W}^{1,q}(\mathbb{R}^n))\cap\mathrm{W}^{1,q}_{\perp}(\mathbb{T};\dot{\mathrm{W}}^{-1,q}(\mathbb{R}^n))}.$$

By the symmetry of $\mathbf{G}$, the vector field $\tilde{w}$ obtained by odd reflection with respect to x_n of the nth component of w, that is,

$$\tilde{w}(t,x',x_n) := \big(w_1(t,x',-x_n),\dots,w_{n-1}(t,x',-x_n),-w_n(t,x',-x_n)\big),$$

is also a solution to (4.4.37) corresponding to the same pressure term π. This means that w and $\tilde{w}$ both solve the same time-periodic heat equation in the whole space $\mathbb{R}^n$. By [61, Theorem 2.1], $w = \tilde{w}$. It follows that $\mathrm{Tr}_0[w_n] = 0$.

Consequently, $H := h - \mathrm{Tr}_0[w]$ belongs to the space (4.4.10) (see Section 2.6 for a rigorous definition of the trace operator in this setting). Let $(U,\mathfrak{P})$ be the corresponding solution from Proposition 4.4.4 with boundary value H. It follows that $(u,p) := (U+w+v,\mathfrak{P}+\pi)$ is a solution to (4.4.9) in the class (4.4.34) satisfying (4.4.35).

It remains to show uniqueness, which follows from a standard duality argument. To this end, assume that (u,p) is a solution in the class (4.4.34) to the homogeneous Stokes problem

$$\begin{cases} \partial_t u - \Delta u + \nabla p = 0 & \text{in } \mathbb{T}\times\mathbb{R}^n_+, \\ \operatorname{div} u = 0 & \text{in } \mathbb{T}\times\mathbb{R}^n_+, \\ u = 0 & \text{on } \mathbb{T}\times\partial\mathbb{R}^n_+. \end{cases}$$

Let $\phi \in C^\infty_0(\mathbb{T}\times\mathbb{R}^n_+)^n$. With exactly the same arguments as above, one can establish existence of a solution

$$\psi \in \mathrm{W}^{1,q'}_{\perp}\big(\mathbb{T};\mathrm{L}^{q'}(\mathbb{R}^n_+)\big)^n \cap \mathrm{L}^{q'}_{\perp}\big(\mathbb{T};\mathrm{W}^{2,q'}(\mathbb{R}^n_+)\big)^n,$$
$$\eta \in \mathrm{L}^{q'}_{\perp}\big(\mathbb{T};\dot{\mathrm{W}}^{1,q'}(\mathbb{R}^n_+)\big),$$

to the adjoint Stokes problem

$$\begin{cases} \partial_t\psi + \Delta\psi + \nabla\eta = \phi & \text{in } \mathbb{T}\times\mathbb{R}^n_+, \\ \operatorname{div}\psi = 0 & \text{in } \mathbb{T}\times\mathbb{R}^n_+, \\ \psi = 0 & \text{on } \mathbb{T}\times\partial\mathbb{R}^n_+, \end{cases}$$

where $q\prime$ denotes the Hölder conjugate of q. Integration by parts yields

$$\int_{\mathbb{T}}\int_{\mathbb{R}^n_+} u\cdot\phi\,\mathrm{d}x\,\mathrm{d}t = \int_{\mathbb{T}}\int_{\mathbb{R}^n_+} u\cdot\big(\partial_t\psi+\Delta\psi+\nabla\eta\big)\,\mathrm{d}x\,\mathrm{d}t = 0.$$

Since this identity holds for all $\phi \in C_0^\infty(\mathbb{T} \times \mathbb{R}_+^n)^n$, it follows that $u = 0$. In turn, we deduce $\nabla p = 0$, whence $p \in L_\perp^q(\mathbb{T})$, that is, p depends only on time. $\qquad\square$

Finally it remains to consider the Stokes system (4.4.2) and prove Theorem 4.4.1 by combining Theorem 4.4.4 with known results for the corresponding stationary problem.

Proof of Theorem 4.4.1. Let f, g, h be vector fields in the class (4.4.3), with h satisfying (4.4.4). By [36, Theorem IV.3.2], the steady-state Stokes problem

$$\begin{cases} -\Delta u_{\mathrm{s}} + \nabla p_{\mathrm{s}} = f_{\mathrm{s}} & \text{in } \mathbb{R}_+^n, \\ \operatorname{div} u_{\mathrm{s}} = g_{\mathrm{s}} & \text{in } \mathbb{R}_+^n, \\ u_{\mathrm{s}} = h_{\mathrm{s}} & \text{on } \partial\mathbb{R}_+^n \end{cases}$$

admits a solution $(u_{\mathrm{s}}, p_{\mathrm{s}}) \in \dot{\mathrm{W}}^{2,q}(\mathbb{R}_+^n)^n \times \dot{\mathrm{W}}^{1,q}(\mathbb{R}_+^n)$ that satisfies

$$\begin{aligned} \|\nabla^2 u_{\mathrm{s}}\|_{\mathrm{L}^q(\mathbb{R}_+^n)} + \|\nabla p_{\mathrm{s}}\|_{\mathrm{L}^q(\mathbb{R}_+^n)} \leq c_0\big(&\|f_{\mathrm{s}}\|_{\mathrm{L}^q(\mathbb{R}_+^n)} + \|g_{\mathrm{s}}\|_{\mathrm{W}^{1,q}(\mathbb{R}_+^n)} \\ &+ \|h_{\mathrm{s}}\|_{\mathrm{W}^{2-\frac{1}{q},q}(\mathbb{R}^{n-1})}\big). \end{aligned}$$

$$(4.4.39)$$

By Theorem 4.4.5, the purely oscillatory Stokes problem (4.4.9) admits a solution $(u_{\mathrm{tp}}, p_{\mathrm{tp}})$ in the class (4.4.34) satisfying (4.4.35). Putting $(u, p) := (u_{\mathrm{s}} + u_{\mathrm{tp}}, p_{\mathrm{s}} + p_{\mathrm{tp}})$, we obtain a solution to (4.4.2) that satisfies (4.4.6) and (4.4.7). Finally, if $(\tilde{u}, \tilde{p})$ is another solution to (4.4.2) in the class (4.4.5), then $u_{\mathrm{tp}} = \mathcal{P}_\perp \tilde{u}$ by Theorem 4.4.5, and $u_{\mathrm{s}} = \mathcal{P}\tilde{u} + (a_1 x_n, \dots, a_{n-1} x_n, 0)$ for some vector $a \in \mathbb{R}^{n-1}$ by [36, Theorem IV.3.2]. It follows that $\nabla p = \nabla\tilde{p}$, and thus $p = \tilde{p} + d(t)$ for some function d that depends only on time. $\quad\square$

4.4.2 The Stokes System in the Periodic Half Space

Subject of this subsection is to transfer the results from the previous subsection to the so-called *periodic half space*

$$\Omega_+ := \mathbb{T}_0^2 \times \mathbb{R}_+, \qquad (4.4.40)$$

where $\mathbb{T}_0^2$ is the 2-dimensional torus $(\mathbb{R}/L\mathbb{Z})^2$ introduced in Section 4.1 for fixed $L > 0$. Note that in this subsection only purely oscillatory data are considered, since these results are only used to establish the L^q estimates for the solution to the coupled fluid-structure system under purely periodic forcing in Section 4.6. As the Fourier transform $\mathscr{F}_{\mathbb{T}\times\mathbb{T}_0^2}$ is admissible in the framework of periodic boundary conditions (in tangential directions), we can mimic the steps from the previous section to deduce:

Proposition 4.4.6. *Let $q \in (1,\infty)$ and let $h \in \mathrm{W}_\perp^{1-\frac{1}{2q},2-\frac{1}{q},q}(\mathbb{T}\times\mathbb{T}_0^2)^3$ satisfy condition*

$$h_n \in \mathrm{W}_\perp^{1,q}\big(\mathbb{T}; \dot{\mathrm{W}}^{-\frac{1}{q},q}(\mathbb{T}_0^2)\big). \tag{4.4.41}$$

Then the Stokes system

$$\begin{cases} \partial_t u - \Delta u + \nabla p = 0 & in\ \mathbb{T}\times\Omega_+, \\ \operatorname{div} u = 0 & in\ \mathbb{T}\times\Omega_+, \\ u|_{x_3=0} = h & on\ \mathbb{T}\times\mathbb{T}_0^2, \end{cases}$$

admits a solution

$$(u,p) \in \mathrm{W}_\perp^{1,2,q}(\mathbb{T}\times\Omega_+)^3 \times \mathrm{L}_\perp^q\big(\mathbb{T}; \dot{\mathrm{W}}^{1,q}(\Omega_+)\big) \tag{4.4.42}$$

and a constant $C_{32} = C_{32}(q,\mathcal{T},L) > 0$ such that the L^q estimate

$$\begin{aligned} \|u\|_{\mathrm{W}_\perp^{1,2,q}(\mathbb{T}\times\Omega_+)} &+ \|\nabla p\|_{\mathrm{L}_\perp^q(\mathbb{T}\times\Omega_+)} \\ &\leq C_{32}\big(\|h\|_{\mathrm{W}_\perp^{1-\frac{1}{2q},2-\frac{1}{q},q}(\mathbb{T}\times\mathbb{T}_0^2)} + \|h_n\|_{\mathrm{W}_\perp^{1,q}(\mathbb{T};\dot{\mathrm{W}}^{-\frac{1}{q},q}(\mathbb{T}_0^2))}\big) \end{aligned} \tag{4.4.43}$$

holds.

Proof. As mentioned above, we will utilize the partial Fourier transform $\mathscr{F}_{\mathbb{T}\times\mathbb{T}_0^2}$ to the Stokes equations and deduce in the same way as in the proof of Proposition 4.4.4 that for any fixed $(k,\xi) \in \frac{2\pi}{\mathcal{T}}\mathbb{Z} \times \big(\frac{2\pi}{L}\mathbb{Z}\big)^2$ a solution is given by

$$u' = \mathscr{F}_{\mathbb{T}\times\mathbb{T}_0^2}^{-1}\bigg[-\frac{\xi q_0}{k} e^{-|\xi|x_3} + \Big(\widehat{H'} + \frac{\xi q_0}{k}\Big) e^{-\sqrt{|\xi|^2+ik}\,x_3} \bigg],$$

$$u_3 = \mathscr{F}_{\mathbb{T}\times\mathbb{T}_0^2}^{-1}\bigg[\frac{|\xi|q_0}{ik} e^{-|\xi|x_3} + \Big(\widehat{H_n} - \frac{|\xi|q_0}{ik}\Big) e^{-\sqrt{|\xi|^2+ik}\,x_3} \bigg],$$

$$p = \mathscr{F}_{\mathbb{T}\times\mathbb{T}_0^2}^{-1}\big[q_0 e^{-|\xi|x_3} \big],$$

where q_0 is defined as

$$q_0 = -i\Big(|\xi| + \sqrt{|\xi|^2+ik}\Big)\frac{\xi}{|\xi|}\cdot\widehat{H'} + \Big(\sqrt{|\xi|^2+ik} + |\xi|\Big)\widehat{H_n} + \frac{ik}{|\xi|}\widehat{H_n},$$

see (4.4.19) and (4.4.18). Hence, following exactly the same steps as in the proof of Proposition 4.4.4, we obtain by an application of the transference principle (Theorem 2.3.1) that all the occurring terms are defined by $L^q(\mathbb{T} \times \mathbb{T}_0^2)$-multiplier, and therefore, (u, p) obeys the estimate (4.4.43). $\quad\square$

Utilizing the same reduction process as in the proof of Theorem 4.4.5, we finally find for the fully inhomogeneous Stokes system:

Theorem 4.4.7. *Let $q \in (1, \infty)$. For all*

$$f \in L_{\perp}^q(\mathbb{T} \times \Omega_+)^3,$$
$$g \in L_{\perp}^q\big(\mathbb{T}; W^{1,q}(\Omega_+)\big) \cap W_{\perp}^{1,q}\big(\mathbb{T}; \dot{W}^{-1,q}(\Omega_+)\big), \tag{4.4.44}$$
$$h \in W_{\perp}^{1-\frac{1}{2q}, 2-\frac{1}{q}, q}(\mathbb{T} \times \mathbb{T}_0^2)^3$$

satisfying (4.4.41), there is a solution (u, p) in the class (4.4.42) to the Stokes system

$$\begin{cases} \partial_t u - \Delta u + \nabla p = f & in\ \mathbb{T} \times \Omega_+, \\ \operatorname{div} u = g & in\ \mathbb{T} \times \Omega_+, \\ u_{|x_3=0} = h & on\ \mathbb{T} \times \mathbb{T}_0^2. \end{cases} \tag{4.4.45}$$

which satisfies

$$\|u\|_{W_{\perp}^{1,2,q}(\mathbb{T} \times \Omega_+)} + \|\nabla p\|_{L_{\perp}^q(\mathbb{T} \times \Omega_+)}$$
$$\leq C_{33} \big(\|f\|_{L_{\perp}^q(\mathbb{T} \times \Omega_+)} + \|g\|_{L_{\perp}^q(\mathbb{T}; W^{1,q}(\Omega_+)) \cap W_{\perp}^{1,q}(\mathbb{T}; \dot{W}^{-1,q}(\Omega_+))} \tag{4.4.46}$$
$$+ \|h\|_{W_{\perp}^{1-\frac{1}{2q}, 2-\frac{1}{q}, q}(\mathbb{T} \times \mathbb{T}_0^2)} + \|h_n\|_{W_{\perp}^{1,q}(\mathbb{T}; \dot{W}^{-\frac{1}{q},q}(\mathbb{T}_0^2))} \big)$$

with $C_{33} = C_{33}(q, \mathcal{T}, L) > 0$. Moreover, if $(\tilde{u}, \tilde{p})$ is another solution to (4.4.45) in $W_{\perp}^{1,2,q}(\mathbb{T} \times \Omega_+)^3 \times L_{\perp}^q\big(\mathbb{T}; \dot{W}^{1,q}(\Omega_+)\big)$, then $u = \tilde{u}$ and $p = \tilde{p} + d(t)$ for some function d that depends only on time.

Proof. The assertions will be proven by mimicking the steps in the proof of Theorem 4.4.5. For this reason, let v be a solution to the 3-dimensional heat equation

$$\partial_t v - \Delta v = F \qquad in\ \mathbb{T} \times \mathbb{T}_0^2 \times \mathbb{R},$$

and (w, π) be a solution to

$$\begin{cases} \partial_t w - \Delta w + \nabla \pi = 0 & in\ \mathbb{T} \times \mathbb{T}_0^2 \times \mathbb{R}, \\ \operatorname{div} w = \mathbf{G} & in\ \mathbb{T} \times \mathbb{T}_0^2 \times \mathbb{R}. \end{cases}$$

Here F and $\mathbf{G}$ denote the extensions of f and $g-\operatorname{div}(v_{|\mathbb{T}\times\Omega_+})$ to $\mathbb{T}\times\mathbb{T}_0^2\times\mathbb{R}$ by even reflection as used for the Stokes system in the half space. A solution to this problem is given by

$$v = \mathscr{F}_{\mathbb{T}\times\mathbb{T}_0^2\times\mathbb{R}}^{-1}\left[\frac{1}{|\xi|^2+ik}\mathscr{F}_{\mathbb{T}\times\mathbb{T}_0^2\times\mathbb{R}}[F]\right]$$

and (w,π) given as in (4.4.38). As the occurring terms in the representation formulas of v, w and π are $L^q(\mathbb{T}\times\mathbb{R}^3)$-multipliers, see for example [61, Theorem 2.1] and the proof of Theorem 4.4.5, we obtain from the transference principle that these are also $L^q(\mathbb{T}\times\mathbb{T}_0^2\times\mathbb{R})$-multipliers and obey the estimates

$$\|v\|_{W_\perp^{1,2,q}(\mathbb{T}\times\mathbb{T}_0^2\times\mathbb{R})} \le c_0\|F\|_{L_\perp^q(\mathbb{T}\times\mathbb{T}_0^2\times\mathbb{R})}$$

and

$$\|w\|_{W_\perp^{1,2,q}(\mathbb{T}\times\mathbb{T}_0^2\times\mathbb{R})} + \|\nabla\pi\|_{L_\perp^q(\mathbb{T}\times\mathbb{T}_0^2\times\mathbb{R})}$$
$$\le c_1\|G\|_{L_\perp^q(\mathbb{T};W^{1,q}(\mathbb{T}_0^2\times\mathbb{R}))\cap W_\perp^{1,q}(\mathbb{T};\dot{W}^{-1,q}(\mathbb{T}_0^2\times\mathbb{R}))}.$$

By the symmetry of F and $\mathbf{G}$, we deduce that v and w have the same reflection properties. That is,

$$\operatorname{Tr}_0\left[v_{|\mathbb{T}\times\Omega_+}\right] = -\operatorname{Tr}_0\left[v_{|\mathbb{T}\times\Omega_+}\right],$$
$$\operatorname{Tr}_0\left[w_{|\mathbb{T}\times\Omega_+}\right] = -\operatorname{Tr}_0\left[w_{|\mathbb{T}\times\Omega_+}\right],$$

and consequently $\tilde{v} := v_{|\mathbb{T}\times\Omega_+}$ and $\tilde{w} := w_{|\mathbb{T}\times\Omega_+}$ are solutions to

$$\begin{cases}\partial_t\tilde{v} - \Delta\tilde{v} = f & \text{in } \mathbb{T}\times\Omega_+,\\ \tilde{v} = 0 & \text{in } \mathbb{T}\times\partial\Omega_+,\end{cases}$$

and

$$\begin{cases}\partial_t\tilde{w} - \Delta\tilde{w} + \nabla\pi = 0 & \text{in } \mathbb{T}\times\Omega_+,\\ \nabla\cdot\tilde{w} = g - \operatorname{div}\tilde{v} & \text{in } \mathbb{T}\times\Omega_+,\\ \tilde{w}_3 = 0 & \text{on } \mathbb{T}\times\partial\Omega_+,\end{cases}$$

respectively. Moreover, $\tilde{v}$ and $\tilde{w}$ satisfy the L^q estimate (4.4.46). Finally, letting $(U,\mathfrak{P})$ be the solution to the Stokes system (4.4.45) with $f=0=g$ and boundary condition $h-\tilde{w}_{|\mathbb{T}\times\partial\Omega_+}$, which by Proposition 4.4.6 exists, we have a solution to (4.4.45) with fully inhomogeneous right-hand side by setting $(u,p) = (U+\tilde{v}+\tilde{w},\mathfrak{P}+\pi)$.

To complete the proof, it remains to show the uniqueness assertion. This can be established by the same duality argument as in the proof of Theorem 4.4.5. □

4.5 The Coupled Resolvent Problem

Subject of this section is the investigation of the resolvent-type problem

$$
\begin{cases}
-k^2\eta_k + \Delta'^2\eta_k - ik\Delta'\eta_k = F_k - e_3 \cdot \mathrm{T}_0(v_k, \mathfrak{p}_k) & \text{in } \mathbb{T}_0^2, \\
ikv_k - \Delta v_k + \nabla\mathfrak{p}_k = f_k & \text{in } \Omega, \\
\operatorname{div} v_k = 0 & \text{in } \Omega, \\
v_{k|x_3=0} = -ik\eta_k e_3 & \text{on } \mathbb{T}_0^2, \\
v_{k|x_3=1} = 0 & \text{on } \mathbb{T}_0^2,
\end{cases}
\tag{4.5.1}
$$

in the periodic layer $\Omega = \mathbb{T}_0^2 \times (0,1)$, for fixed $k \in \frac{2\pi}{T}\mathbb{Z}\setminus\{0\}$. Note that this resolvent problem is obtained by an application of the Fourier transform $\mathscr{F}_{\mathbb{T}}$ to the coupled fluid-structure system (4.2.4).To simplify reading, we omit the subscript k in this section as long as no confusion may arise. We begin by introducing the concept of weak solutions, which shall be established in the complex function space

$$
\mathcal{V}_k := \{(v,\eta) \in \mathcal{W}_\sigma^{1,2}(\Omega) \times \mathrm{W}_{(0)}^{2,2}(\mathbb{T}_0^2) \mid v_{|x_3=1} = 0,\ v_{|x_3=0} = -ik\eta\, e_3\},
$$

where we included the boundary conditions in the definition of the function space. Recall that $\mathcal{W}_\sigma^{1,2}(\Omega)$ consists of all solenoidal vector fields such that $v \cdot \nu$ has vanishing mean value on $\partial\Omega$, see (2.4.3), and $\mathrm{W}_{(0)}^{2,2}(\mathbb{T}_0^2)$ is the set of all $\mathrm{W}^{2,2}(\mathbb{T}_0^2)$ functions which have a vanishing mean value, see (4.2.1). In the investigation of (4.5.1) we consider the complexification of these function spaces.

Definition 4.5.1 (Weak Solution). *For $(f,F) \in \mathrm{W}^{-1,2}(\Omega)^3 \times \mathrm{W}^{-1,2}(\mathbb{T}_0^2)$ a pair $(v,\eta) \in \mathcal{V}_k$ is called a weak solution to (4.5.1) if and only if*

$$
ik\int_\Omega v\cdot\phi\,\mathrm{d}x + \int_\Omega \nabla v:\nabla\phi\,\mathrm{d}x + \int_{\mathbb{T}_0^2} ik\nabla'\eta\cdot(ik\nabla'\zeta)\,\mathrm{d}x'
$$
$$
+ \int_{\mathbb{T}_0^2} \Delta'\eta(ik\Delta'\zeta)\,\mathrm{d}x' - \int_{\mathbb{T}_0^2} k^2\eta(ik\zeta)\,\mathrm{d}x'
\tag{4.5.2}
$$
$$
= \langle F, ik\zeta\rangle_{\mathbb{T}_0^2} + \langle f, \phi\rangle_\Omega
$$

is satisfied for all $(\phi,\zeta) \in \mathcal{V}_k$.

Remark 4.5.2. Note that the weak formulation (4.5.2) stems from multiplying (4.5.1) with a test function (ϕ,ζ) and integration by parts as in the proof of Lemma 4.3.1.

Next we prove the following lemma.

Lemma 4.5.3 (Galerkin Basis). *There exists a basis*

$$\{(\psi_j, 0)\}_{j \in \mathbb{N}} \cup \{(\Psi_l, \Phi_l)\}_{l \in \mathbb{N}}$$

of $\mathcal{V}_k$ consisting of smooth functions, such that $(\psi_j)_{j \in \mathbb{N}}$ and $(\Phi_l)_{l \in \mathbb{N}}$ are bases of $W^{1,2}_{0,\sigma}(\Omega)$ and $W^{2,2}_{(0)}(\mathbb{T}^2_0)$, respectively. Moreover, $\Psi_{l|x_3=0} = ik\Phi_l e_3$.

Proof. A basis $(\psi_j)_{j \in \mathbb{N}}$ of smooth functions $W^{1,2}_{0,\sigma}(\Omega)$ exists, as the smooth functions are dense in $W^{1,2}_{0,\sigma}(\Omega)$ by definition, see Section 2.4. By the same argument, there is a basis $(\Phi_l)_{l \in \mathbb{N}}$ of $W^{2,2}_{(0)}(\mathbb{T}^2_0)$ consisting of smooth functions. Consequently, $\tilde{\Psi}_l := (1 - x_3)ik\Phi_l e_3$ is smooth and satisfies

$$\tilde{\Psi}_l(x', 0) = ik\Phi_l e_3 \qquad \text{and} \qquad \tilde{\Psi}_l(x', 1) = 0. \qquad (4.5.3)$$

However, $\tilde{\Psi}_l$ is not divergence free. Let $\mathcal{B}$ denote the Bogovskiĭ operator, see Subsection 2.7.2. Then, $\Psi_l = \tilde{\Psi}_l - \mathcal{B}(\nabla \cdot \tilde{\Psi}_l)$ is smooth and $\Psi_l \in \mathcal{V}_k$, as it is solenoidal and obeys (4.5.3). By the construction of $(\psi_j, 0)$ and (Ψ_l, Φ_l) one may check that these terms are linearly independent. Moreover, utilizing the properties of the Bogovskiĭ operator stated in Theorem 2.7.4, we deduce that

$$\begin{aligned} \|\Psi_l\|_{W^{1,2}(\Omega)} &\leq c_0 \big(\|\tilde{\Psi}_l\|_{W^{1,2}(\Omega)} + \|\mathcal{B}(\nabla \cdot \tilde{\Psi}_l)\|_{W^{1,2}(\Omega)} \big) \\ &\leq c_1 \big(\|\tilde{\Psi}_l\|_{W^{1,2}(\Omega)} + \|\nabla \cdot \tilde{\Psi}_l\|_{L^2(\Omega)} \big) \\ &\leq c_2 \|\tilde{\Psi}_l\|_{W^{1,2}(\Omega)} \leq c_3 \|\Phi_l\|_{W^{2,2}(\mathbb{T}^2_0)}. \end{aligned} \qquad (4.5.4)$$

In order to see that $\{(\psi_j, 0)\}_{j \in \mathbb{N}} \cup \{(\Psi_l, \Phi_l)\}_{l \in \mathbb{N}}$ constitute a basis of $\mathcal{V}_k$, let us consider $(u, \eta) \in \mathcal{V}_k$. As $(\Phi_l)_{l \in \mathbb{N}}$ is a basis of $W^{2,2}_{(0)}(\mathbb{T}^2_0)$, we have that

$$-\eta = \sum_{l=1}^\infty \alpha_l \Phi_l \qquad \Longrightarrow \qquad u_{|x_3=0} = -ik\eta e_3 = \sum_{l=1}^\infty \alpha_l \Psi_{l|x_3=0}$$

for some coefficients $\alpha_l \in \mathbb{C}$. Let $u_0 \in W^{1,2}_{0,\sigma}(\Omega)$ be the function such that

$$u = u_0 + \sum_{l=1}^\infty \alpha_l \Psi_l \in W^{1,2}_\sigma(\Omega).$$

Since ψ_j is a basis of $W^{1,2}_{0,\sigma}(\Omega)$, there are $\beta_j \in \mathbb{C}$ such that

$$u_0 = \sum_{j=1}^\infty \beta_j \psi_j \in W^{1,2}_{0,\sigma}(\Omega),$$

and thus

$$u = \sum_{j=1}^{\infty} \beta_j \psi_j + \sum_{l=1}^{\infty} \alpha_l \Psi_l.$$

To prove the assertion, both series on the right-hand side have to converge. Due to (4.5.4), this is true, as ψ_j and Φ_l are bases of $W_{0,\sigma}^{1,2}(\Omega)$ and $W_{(0)}^{2,2}(\mathbb{T}_0^2)$, respectively. Hence, $(u, \eta) \in \mathcal{V}_k$ can be expressed as a linear combination of $\{(\psi_j, 0)\}_{j \in \mathbb{N}} \cup \{(\Psi_l, \Phi_l)\}_{l \in \mathbb{N}}$. □

Proposition 4.5.4 (Existence of a Weak Solution). *For any data $(f, F) \in$ $W^{-1,2}(\Omega)^3 \times W^{-1,2}(\mathbb{T}_0^2)$ there exists a weak solution $(v, \eta) \in \mathcal{V}_k$ to (4.5.1). Furthermore, we find a constant $C_{34} = C_{34}(\mathcal{T}, L) > 0$, independent of k, such that*

$$\|v\|_{W^{1,2}(\Omega)} + \|ik\eta\|_{W^{1,2}(\mathbb{T}_0^2)} + \|\eta\|_{W^{2,2}(\mathbb{T}_0^2)} \\ \leq C_{34} \big(\|f\|_{W^{-1,2}(\Omega)} + \|F\|_{W^{-1,2}(\mathbb{T}_0^2)} \big) \quad (4.5.5)$$

holds.

Proof. The statement will be proved by utilizing a Galerkin approximation. Let us consider $(f, F) \in C_0^{\infty}(\Omega)^3 \times C_0^{\infty}(\mathbb{T}_0^2)$. Let further $\{(\psi_j, 0)\}_{j \in \mathbb{N}} \cup$ $\{(\Psi_l, \Phi_l)\}_{l \in \mathbb{N}}$ be the basis of $\mathcal{V}_k$ constructed in Lemma 4.5.3. Observe that

$$(v, \eta) \times (\phi, \zeta) \mapsto \langle (v, \eta), (\phi, \zeta) \rangle_{\mathcal{V}_k} := \langle \nabla v, \nabla \overline{\phi} \rangle + k^2 \langle \nabla' \eta, \nabla' \overline{\zeta} \rangle$$

defines a scalar product on the function space $\mathcal{W}_{\sigma}^{1,2}(\Omega) \times W_{(0)}^{1,2}(\mathbb{T}_0^2)$. Hence, employing Gram-Schmidt procedure to $\{(\psi_j, 0)\}_{j \in \mathbb{N}} \cup \{(\Psi_l, \Phi_l)\}_{l \in \mathbb{N}}$, we find an orthonormal basis of $\mathcal{W}_{\sigma}^{1,2}(\Omega) \times W_{(0)}^{1,2}(\mathbb{T}_0^2)$, and therefore of the subspace $\mathcal{V}_k$, which will be denoted by $\{(\psi_j, 0)\}_{j \in \mathbb{N}} \cup \{(\Psi_l, \Phi_l)\}_{l \in \mathbb{N}}$ as well. We are now looking for an approximate solution of the form

$$v^{m,n} = \sum_{j=1}^{m} \alpha_j \psi_j + \sum_{l=1}^{n} \beta_l \Psi_l, \qquad \eta^n = \sum_{l=1}^{n} \beta_l \Phi_l$$

satisfying for all $j \in \{1, \dots, m\}$ the discrete problem

$$ik \int_{\Omega} v^{m,n} \cdot \overline{\psi}_j \, \mathrm{d}x + \int_{\Omega} \nabla v^{m,n} : \nabla \overline{\psi}_j \, \mathrm{d}x = \langle f, \overline{\psi}_j \rangle_{\Omega} \quad (4.5.6)$$

and for all $l \in \{1, \ldots, n\}$

$$
ik \int_{\Omega} v^{m,n} \cdot \overline{\Psi}_l \, dx + \int_{\Omega} \nabla v^{m,n} : \nabla \overline{\Psi}_l \, dx + \int_{\mathbb{T}_0^2} ik \nabla' \eta^n \cdot (ik \nabla' \overline{\Phi}_l) \, dx'
$$

$$
+ \int_{\mathbb{T}_0^2} \Delta' \eta^n (ik \Delta' \overline{\Phi}_l) \, dx' - \int_{\mathbb{T}_0^2} k^2 \eta^n (ik \overline{\Phi}_l) \, dx' \qquad (4.5.7)
$$

$$
= \langle F, ik\overline{\Phi}_l \rangle_{\mathbb{T}_0^2} + \langle f, \overline{\Psi}_l \rangle_{\Omega},
$$

where α_j and β_l are constants. Testing with finitely many $(\overline{\psi}_j, 0)$, we obtain for $1 \le j \le m$ that

$$
ik \sum_{s=1}^{m} \alpha_s \langle \psi_s, \overline{\psi}_j \rangle + \sum_{s=1}^{m} \alpha_s \langle \nabla \psi_s, \nabla \overline{\psi}_j \rangle = \langle f, \overline{\psi}_j \rangle_{\Omega}.
$$

In view of the definition of $\langle \cdot, \cdot \rangle_{\mathcal{V}_k}$ and the choice of ψ_j, we find that

$$
ik \sum_{s=1}^{m} \alpha_s \langle \psi_s, \overline{\psi}_j \rangle + \sum_{s=1}^{m} \alpha_s \delta_{sj} \langle (\psi_s, 0), (\psi_j, 0) \rangle_{\mathcal{V}_k} = \langle f, \overline{\psi}_j \rangle_{\Omega}, \qquad (4.5.8)
$$

for all $j \in \{1, \ldots, m\}$. Similarly, for (Ψ_l, Φ_l) and $l \in \{1, \ldots, n\}$ we deduce that

$$
ik \left(\sum_{s=1}^{n} \beta_s (\langle \Psi_s, \overline{\Psi}_l \rangle + k^2 \langle \Phi_s, \overline{\Phi}_l \rangle) - \sum_{s=1}^{n} \beta_s \langle \Delta' \Phi_s, \Delta' \overline{\Phi}_l \rangle \right)
$$

$$
+ \sum_{s=1}^{n} \beta_s \delta_{sl} \langle (\Psi_s, \Phi_s), (\Psi_l, \Phi_l) \rangle_{\mathcal{V}_k} = \langle f, \overline{\Psi}_l \rangle_{\Omega} - ik \langle F, \overline{\Phi}_l \rangle_{\mathbb{T}_0^2}.
$$

$$
(4.5.9)
$$

Observe that (4.5.8) and (4.5.9) define a system of linear equations

$$
Mx = b \in \mathbb{C}^{m+n}
$$

with $x = (\alpha_1, \ldots, \alpha_m, \beta_1, \ldots, \beta_n) \in \mathbb{C}^{m+n}$, associated matrix representation

$$
M = I_{m+n} + ik N := I_{m+n} + ik \begin{pmatrix} N_1 & 0 \\ 0 & N_2 \end{pmatrix},
$$

and

$$b_j := \begin{cases} \langle f, \overline{\psi}_j \rangle, & \text{if } 1 \le j \le m, \\ \langle f, \overline{\Psi}_{j-m} \rangle - ik\langle F, \overline{\Phi}_{j-m} \rangle, & \text{if } m+1 \le j \le m+n. \end{cases}$$

Here, I_{m+n} denotes the identity matrix in $\mathbb{C}^{(m+n)\times(m+n)}$, and $N_1 \in \mathbb{C}^{m\times m}$ and $N_2 \in \mathbb{C}^{n\times n}$ are given by

$$(N_1)_{sj} = \langle \psi_s, \overline{\psi}_j \rangle$$
$$(N_2)_{sl} = \langle \Psi_s, \overline{\Psi}_l \rangle + k^2 \langle \Phi_s, \overline{\Phi}_l \rangle - \langle \Delta'\Phi_s, \Delta'\overline{\Phi}_l \rangle.$$

Since N is self-adjoint, it only has real eigenvalues and due to the identity

$$\left(\frac{1}{ik}I_{n+m} + N\right)x = \frac{1}{ik}Mx = \frac{1}{ik}\left(\langle f, \overline{\Psi}_l \rangle - ik\langle F, \overline{\Phi}_l \rangle\right)$$

we deduce that M is invertible, and therefore (4.5.8)–(4.5.9) admits a solution $(\alpha_1, \ldots, \alpha_m, \beta_1, \ldots, \beta_n) \in \mathbb{C}^{m+n}$. Next we multiply (4.5.8) with $\overline{\alpha}_j$ and (4.5.9) with $\overline{\beta}_l$. Summation over $j = 1, \ldots, m$ and $l = 1, \ldots, n$ in these equations, respectively, and addition of those two contributions, we obtain

$$ik\left(\|v^{m,n}\|^2_{L^2(\Omega)} + \|ik\eta^n\|^2_{L^2(\mathbb{T}_0^2)} - \|\Delta'\eta^n\|^2_{L^2(\mathbb{T}_0^2)}\right)$$
$$+ \|\nabla v^{m,n}\|^2_{L^2(\Omega)} + \|ik\nabla'\eta^n\|^2_{L^2(\mathbb{T}_0^2)} = \langle F, \overline{ik\eta^n}\rangle_{\mathbb{T}_0^2} + \langle f, v^{m,n}\rangle_\Omega.$$

Investigating the real and imaginary parts of this equation separately, we deduce for the real part that

$$\|\nabla v^{m,n}\|^2_{L^2(\Omega)} + \|ik\nabla'\eta^n\|^2_{L^2(\mathbb{T}_0^2)} \le c_0\left(\|f\|^2_{W^{-1,2}(\Omega)} + \|F\|^2_{W^{-1,2}(\mathbb{T}_0^2)}\right).$$

From Poincaré's inequality we conclude

$$\|v^{m,n}\|^2_{W^{1,2}(\Omega)} + \|ik\eta^n\|^2_{W^{1,2}(\mathbb{T}_0^2)} \le c_1\left(\|f\|^2_{W^{-1,2}(\Omega)} + \|F\|^2_{W^{-1,2}(\mathbb{T}_0^2)}\right)$$

with $c_0 > 0$ and $c_1 > 0$ independent of k. Taking the imaginary part of the above equality and making use of this estimate, we find that

$$\|\Delta'\eta^n\|^2_{L^2(\mathbb{T}_0^2)}$$
$$\le c_2\left(\|f\|^2_{W^{-1,2}(\Omega)} + \|F\|^2_{W^{-1,2}(\mathbb{T}_0^2)} + \|v^{m,n}\|^2_{W^{1,2}(\Omega)} + \|ik\eta^n\|^2_{W^{1,2}(\mathbb{T}_0^2)}\right)$$
$$\le c_3\left(\|f\|^2_{W^{-1,2}(\Omega)} + \|F\|^2_{W^{-1,2}(\mathbb{T}_0^2)}\right),$$

holds. Observe, that $c_3 > 0$ only depends on the period length $\mathcal{T} > 0$ and not on k. Hence, we obtain that $(v^{m,n}, \eta^n)$ satisfies the energy estimate

$$\|v^{m,n}\|_{\mathrm{W}^{1,2}(\Omega)} + \|ik\eta^n\|_{\mathrm{W}^{1,2}(\mathbb{T}_0^2)} + \|\eta^n\|_{\mathrm{W}^{2,2}(\mathbb{T}_0^2)}$$
$$\leq c_4 \big(\|f\|_{\mathrm{W}^{-1,2}(\Omega)} + \|F\|_{\mathrm{W}^{-1,2}(\mathbb{T}_0^2)} \big).$$
$$(4.5.10)$$

Finally, note that c_4 is independent of m, n. Due to (4.5.10), we find a weakly convergent subsequence of $(v^{n,n}, \eta^n)$ with weak limit $(v, \eta) \in \mathcal{V}_k$. Hence letting $n \to \infty$ in (4.5.6) and (4.5.7), we conclude a weak solution (v, η) to (4.5.1) satisfying the energy estimate (4.5.5). Due to the density of $C_0^\infty(\Omega)$ and $C_0^\infty(\mathbb{T}_0^2)$ in $\mathrm{W}^{-1,2}(\Omega)$ and $\mathrm{W}^{-1,2}(\mathbb{T}_0^2)$, respectively, we deduce from (4.5.5) that the assertion follows for any $(f, F) \in \mathrm{W}^{-1,2}(\Omega)^3 \times \mathrm{W}^{-1,2}(\mathbb{T}_0^2)$. □

In the next step we will construct a pressure field corresponding to the weak solution $(v, \eta) \in \mathcal{V}_k$. To this end we will proceed as in [20]. We first consider the Stokes resolvent problem

$$\begin{cases} ikv - \Delta v + \nabla \mathfrak{p} = f & \text{in } \Omega, \\ \operatorname{div} v = 0 & \text{in } \Omega, \\ v_{|x_3=0} = -ik\eta e_3 & \text{on } \mathbb{T}_0^2, \\ v_{|x_3=1} = 0 & \text{on } \mathbb{T}_0^2, \end{cases} \qquad (4.5.11)$$

where (v, η) is the weak solution to (4.5.1) from Proposition 4.5.4, and construct a pressure field $\mathfrak{p}$ corresponding to this system. Since this pressure field is unique up to an additional constant, we can modify $\mathfrak{p}$ such that $(v, \mathfrak{p}, \eta)$ is a solution to (4.5.1). Moreover, we show that this solution is as regular as the data allow.

Lemma 4.5.5 (Existence of a Pressure Field and Higher Regularity). *Let* $k \in \frac{2\pi}{\mathcal{T}}\mathbb{Z} \setminus \{0\}$, $r \in \mathbb{N}_0$, $(f, F) \in \mathrm{W}^{r,2}(\Omega)^3 \times \mathrm{W}^{r+\frac{1}{2},2}(\mathbb{T}_0^2)$ *and* $(v, \eta) \in \mathcal{V}_k$ *be the corresponding weak solution to (4.5.1) constructed in Proposition 4.5.4. Then there exists a pressure field* $\mathfrak{p}$ *such that* $(v, \mathfrak{p}, \eta) \in \mathrm{W}^{r+2,2}(\Omega)^3 \times \mathrm{W}^{r+1,2}(\Omega) \times \mathrm{W}^{r+\frac{9}{2},2}(\mathbb{T}_0^2)$ *solves (4.5.1) and satisfies*

$$\|v\|_{\mathrm{W}^{r+2,2}(\Omega)} + \|\mathfrak{p}\|_{\mathrm{W}^{r+1,2}(\Omega)} + \|\eta\|_{\mathrm{W}^{r+\frac{9}{2},2}(\mathbb{T}_0^2)}$$
$$\leq C_{35} \big(\|f\|_{\mathrm{W}^{r,2}(\Omega)} + \|F\|_{\mathrm{W}^{r+\frac{1}{2},2}(\mathbb{T}_0^2)} \big),$$
$$(4.5.12)$$

with $C_{35} = C_{35}(\mathcal{T}, L) > 0$ *independent of* k.

Proof. Let $(f, F) \in C_0^\infty(\Omega)^3 \times C_0^\infty(\mathbb{T}_0^2)$. Our starting point is the weak solution $(v, \eta) \in \mathcal{V}_k$. Let us consider the Stokes resolvent problem (4.5.11). Existence of a pressure field $\tilde{\mathfrak{p}} \in W^{1,2}(\Omega)$ follows similarly as in the proof of Theorem 1.2. in [31]. Moreover, regularity theory for the Stokes system implies that this solution is as regular as the data allow. For our purpose it suffices at this point in the proof to deduce $(v, \tilde{\mathfrak{p}}) \in W^{2,2}(\Omega) \times W^{1,2}(\Omega)$. Note that $\tilde{\mathfrak{p}}$ is only a pressure field corresponding to the Stokes equations (4.5.11), and obeys the weak formulation

$$ik \int_\Omega v \cdot \phi \, \mathrm{d}x + \int_\Omega \nabla v : \nabla \phi \, \mathrm{d}x - \int_\Omega \tilde{\mathfrak{p}} \, \mathrm{div}\, \phi \, \mathrm{d}x = \int_\Omega f \cdot \phi \, \mathrm{d}x - \int_{\mathbb{T}_0^2} \tilde{\mathfrak{p}} ik\zeta \, \mathrm{d}x'$$

for any test function $(\phi, \zeta) \in W^{1,2}(\Omega)^3 \times W_{(0)}^{2,2}(\mathbb{T}_0^2)$ with $\phi_{|x_3=1} = 0$ and $\phi_{|x_3=0} = -ik\zeta\, e_3$. Furthermore, it is well known that $(v, \tilde{\mathfrak{p}})$ has higher regularity if the data f are in function space of higher order, and

$$\|v\|_{W^{r+2,2}(\Omega)} + \|\tilde{\mathfrak{p}}\|_{W^{r+1,2}(\Omega)} \le c_0(\|f\|_{W^{r,2}(\Omega)} + \|\eta\|_{W^{r+\frac{3}{2},2}(\mathbb{T}_0^2)}), \quad (4.5.13)$$

if $\eta \in W^{r+\frac{3}{2},2}(\mathbb{T}_0^2)$. This estimate can be deduced by regularity theory for the Stokes system, see for example [36, Chapter 4].

To find a solution to (4.5.1), we recall the weak formulation (4.5.2) and observe that the pressure field corresponding to (4.5.1) obeys for any vector field $(\phi, \zeta) \in W^{1,2}(\Omega)^3 \times W_{(0)}^{2,2}(\mathbb{T}_0^2)$ satisfying $\phi_{|x_3=1} = 0$ and $\phi_{|x_3=0} = -ik\zeta\, e_3$ the identity

$$\Lambda((\phi, \zeta)) = \langle \tilde{\mathfrak{p}}, \nabla \cdot \phi \rangle, \quad (4.5.14)$$

with Λ given by

$$\begin{aligned}
\Lambda((\phi, \zeta)) := &\int_{\mathbb{T}_0^2} \Delta'\eta (ik\Delta'\zeta) \, \mathrm{d}x' + \int_{\mathbb{T}_0^2} ik\nabla'\eta \cdot (ik\nabla'\zeta) \, \mathrm{d}x' \\
&+ \int_\Omega \nabla v : \nabla \phi \, \mathrm{d}x - \int_{\mathbb{T}_0^2} k^2\eta (ik\zeta) \, \mathrm{d}x' + ik \int_\Omega v \cdot \phi \, \mathrm{d}x \\
&- \langle F, ik\zeta \rangle_{\mathbb{T}_0^2} - \langle f, \phi \rangle_\Omega.
\end{aligned}$$

Note that $(v, \tilde{\mathfrak{p}})$ is sufficiently regular to carry out integration by parts to determine from (4.5.14) the identity

$$0 = \Lambda((\phi, \zeta)) - \langle \tilde{\mathfrak{p}}, \nabla \cdot \phi \rangle$$

$$= \int_\Omega (ikv) \cdot \phi \, dx - \int_\Omega \Delta v \cdot \phi \, dx + \int_\Omega \nabla \tilde{\mathfrak{p}} \cdot \phi \, dx - \int_{\mathbb{T}_0^2} (\tilde{\mathfrak{p}}\phi)_{|x_3=0}(-e_3) \, dx'$$

$$- \int_{\mathbb{T}_0^2} \Delta'\eta(-ik\Delta'\zeta) \, dx' - \int_{\mathbb{T}_0^2} ik\nabla'\eta \cdot (-ik\nabla'\zeta) \, dx' + \int_{\mathbb{T}_0^2} k^2\eta(-ik\zeta) \, dx'$$

$$+ \langle F, -ik\zeta \rangle_{\mathbb{T}_0^2} - \langle f, \phi \rangle_\Omega,$$

where we used that v is solenoidal, $\phi_{|x_3=1} = 0$ and $\phi_{|x_3=0} = -ik\zeta e_3$, and thus

$$\int_{\partial\Omega} (\phi \cdot \nabla v) \cdot \nu \, dx' = \int_{\mathbb{T}_0^2} \partial_{x_3} v_{3|x_3=0}(ik\zeta) \, dx'$$

$$= -\int_{\mathbb{T}_0^2} (\partial_{x_1} u_{1|x_3=0} + \partial_{x_2} u_{2|x_3=0})(ik\zeta) \, dx' = 0.$$

Due to $(4.5.11)_1$, we obtain from the above identity

$$0 = -\int_{\mathbb{T}_0^2} \tilde{\mathfrak{p}}_{|x_3=0}(ik\zeta) \, dx' + \int_{\mathbb{T}_0^2} \Delta'\eta(ik\Delta'\zeta) \, dx' + \int_{\mathbb{T}_0^2} ik\nabla'\eta \cdot (ik\nabla'\zeta) \, dx'$$

$$- \int_{\mathbb{T}_0^2} k^2\eta(ik\zeta) \, dx' - \langle F, (ik\zeta) \rangle_{\mathbb{T}_0^2}$$

holds for any $\zeta \in W_{(0)}^{2,2}(\mathbb{T}_0^2)$. Observe that this is the weak formulation corresponding to the plate equation $(4.5.1)_1$, which is satisfied for any test function with vanishing mean value. However, as this identity has to hold for any test function $\zeta \in W^{2,2}(\mathbb{T}_0^2)$, which is not necessarily mean value free, it remains to modify $\tilde{\mathfrak{p}}$ such that

$$\int_{\mathbb{T}_0^2} F \, dx' = \int_{\mathbb{T}_0^2} e_3 \cdot \mathrm{T}_0(\tilde{v}, \tilde{\mathfrak{p}}) \, dx'. \tag{4.5.15}$$

For reasons of simplicity we will omit the subscript $x_3 = 0$ in the following. We utilize the notation

$$g_{(0)} = g - \int_{\mathbb{T}_0^2} g \, dy' \tag{4.5.16}$$

for an arbitrary function g (as long as the integral on the right-hand side is well defined). Then for $\zeta \in W^{2,2}(\mathbb{T}_0^2)$, (*i.e.*, ζ has not necessarily a vanishing mean value,) we obtain

$$0 = -\int_{\mathbb{T}_0^2} \tilde{\mathfrak{p}}(ik\zeta_{(0)}) \, \mathrm{d}x' + \int_{\mathbb{T}_0^2} \Delta'\eta(ik\Delta'\zeta_{(0)}) \, \mathrm{d}x' + \int_{\mathbb{T}_0^2} ik\nabla'\eta \cdot (ik\nabla'\zeta_{(0)}) \, \mathrm{d}x'$$
$$- \int_{\mathbb{T}_0^2} k^2\eta(ik\zeta_{(0)}) \, \mathrm{d}x' - \langle F, (ik\zeta_{(0)}) \rangle_{\mathbb{T}_0^2}.$$

By adding the three terms

$$\int_{\mathbb{T}_0^2} \tilde{\mathfrak{p}} \, \mathrm{d}x' \int_{\mathbb{T}_0^2} ik\zeta_{(0)} \, \mathrm{d}x' = \int_{\mathbb{T}_0^2} k^2\eta \, \mathrm{d}x' \int_{\mathbb{T}_0^2} ik\zeta_{(0)} \, \mathrm{d}x' = \int_{\mathbb{T}_0^2} F \, \mathrm{d}x' \int_{\mathbb{T}_0^2} ik\zeta_{(0)} \, \mathrm{d}x' = 0$$

and utilizing the fact that

$$\int_{\mathbb{T}_0^2} \tilde{\mathfrak{p}}_{(0)} \, \mathrm{d}x' \int_{\mathbb{T}_0^2} ik\zeta \, \mathrm{d}x' = k^2 \int_{\mathbb{T}_0^2} \eta_{(0)} \, \mathrm{d}x' \int_{\mathbb{T}_0^2} ik\zeta \, \mathrm{d}x' = \int_{\mathbb{T}_0^2} F_{(0)} \, \mathrm{d}x' \int_{\mathbb{T}_0^2} ik\zeta \, \mathrm{d}x' = 0,$$

from (4.1.4) we deduce the identity

$$0 = \int_{\mathbb{T}_0^2} \tilde{\mathfrak{p}}_{(0)}(-ik\zeta) \, \mathrm{d}x' - \int_{\mathbb{T}_0^2} \Delta'\eta(-ik\Delta'\zeta) \, \mathrm{d}x' - \int_{\mathbb{T}_0^2} ik\nabla'\eta \cdot (-ik\nabla'\zeta) \, \mathrm{d}x'$$
$$+ \int_{\mathbb{T}_0^2} k^2\eta(-ik\zeta) \, \mathrm{d}x' + \langle F_{(0)}, (-ik\zeta) \rangle_{\mathbb{T}_0^2}.$$

For this reason, we observe that $(\eta, \mathfrak{p})$ with

$$\mathfrak{p} := \tilde{\mathfrak{p}} - \int_{\mathbb{T}_0^2} \tilde{\mathfrak{p}} \, \mathrm{d}x' - \int_{\mathbb{T}_0^2} F \, \mathrm{d}x' \qquad (4.5.17)$$

is a weak solution to the plate equation, and a solution to (4.5.1) is given by $(v, \mathfrak{p}, \eta)$. A straightforward calculation shows that the compatibility condition (4.5.15) is obeyed. Moreover, we have $v \in W^{2,2}(\Omega)$, and in virtue of (4.5.17) it follows that $\mathfrak{p}$ is as regular as $\tilde{\mathfrak{p}}$, *i.e.*, $\mathfrak{p} \in W^{1,2}(\Omega)$. Observe that

$$m(\xi') := \frac{1}{|\xi'|^4 - k^2 + ik|\xi'|^2}$$

is bounded for any $\xi \in \mathbb{T}_0^2$ and fixed $k \in \frac{2\pi}{T}\mathbb{Z} \setminus \{0\}$, see for example (A.2.12), and a solution to $(4.5.1)_1$ is given by

$$\eta = \mathscr{F}_{\mathbb{T}_0^2}^{-1}\big[m\mathscr{F}_{\mathbb{T}_0^2}[\mathfrak{p}_{|x_3=0} + F]\big].$$

Therefore, we obtain from Plancherel's theorem and the transference principle (Theorem 2.3.1) that for any $r \in \mathbb{N}$ we have $\eta \in W_{(0)}^{r+\frac{9}{2},2}(\mathbb{T}_0^2)$ and

$$\|\eta\|_{W^{r+\frac{9}{2},2}(\mathbb{T}_0^2)} \leq c_1\big(\|\mathfrak{p}_{|x_3=0}\|_{W^{r+\frac{1}{2},2}(\mathbb{T}_0^2)} + \|F\|_{W^{r+\frac{1}{2},2}(\mathbb{T}_0^2)}\big), \qquad (4.5.18)$$

provided that $\mathfrak{p} \in W^{r+1,2}(\Omega)$. Observe that c_1 does not depend on k. Our starting point is the solution $(v, \mathfrak{p}, \eta) \in W^{2,2}(\Omega)^3 \times W^{1,2}(\Omega) \times W^{\frac{3}{2},2}(\mathbb{T}_0^2)$. For $r = 0$ we deduce in view of (4.5.5) and (4.5.13) that

$$\|\mathfrak{p}\|_{W^{1,2}(\Omega)} \leq c_2(\|f\|_{L^2(\Omega)} + \|\eta\|_{W^{\frac{3}{2},2}(\mathbb{T}_0^2)}) \leq c_3(\|f\|_{L^2(\Omega)} + \|F\|_{W^{-1,2}(\mathbb{T}_0^2)}),$$

and thus (4.5.18) yields

$$\begin{aligned}\|\eta\|_{W^{\frac{9}{2},2}(\mathbb{T}_0^2)} &\leq c_4(\|\mathfrak{p}_{|x_3=0}\|_{W^{\frac{1}{2},2}(\mathbb{T}_0^2)} + \|F\|_{W^{\frac{1}{2},2}(\mathbb{T}_0^2)}) \\ &\leq c_5(\|f\|_{L^2(\Omega)} + \|F\|_{W^{\frac{1}{2},2}(\mathbb{T}_0^2)})\end{aligned} \qquad (4.5.19)$$

Furthermore, this implies that $(v, \mathfrak{p})$ satisfies for $r = 3$

$$\|v\|_{W^{5,2}(\Omega)} + \|\mathfrak{p}\|_{W^{4,2}(\Omega)} \leq c_6(\|f\|_{W^{3,2}(\Omega)} + \|F\|_{W^{\frac{1}{2},2}(\mathbb{T}_0^2)}). \qquad (4.5.20)$$

Repeating the step (4.5.19), we deduce from (4.5.20) that $\eta \in W_{(0)}^{\frac{15}{2},2}(\mathbb{T}_0^2)$ and obeys

$$\|\eta\|_{W^{\frac{15}{2},2}(\mathbb{T}_0^2)} \leq c_7(\|f\|_{W^{3,2}(\Omega)} + \|F\|_{W^{\frac{7}{2},2}(\mathbb{T}_0^2)}).$$

In view of (4.5.13) this further implies that $(v, \mathfrak{p})$ has better regularity properties and satisfies

$$\|v\|_{W^{8,2}(\Omega)} + \|\mathfrak{p}\|_{W^{7,2}(\Omega)} \leq c_8(\|f\|_{W^{6,2}(\Omega)} + \|F\|_{W^{\frac{7}{2},2}(\mathbb{T}_0^2)}),$$

which again yields

$$\|\eta\|_{W^{\frac{21}{2},2}(\mathbb{T}_0^2)} \leq c_9(\|f\|_{W^{6,2}(\Omega)} + \|F\|_{W^{\frac{13}{2},2}(\mathbb{T}_0^2)}).$$

We can now boot-strap these steps to obtain (4.5.12). Since $C_0^\infty(\Omega)^3$ and $C_0^\infty(\mathbb{T}_0^2)$ are dense in $W^{r,2}(\Omega)^3$ and $W^{r+\frac{1}{2},2}(\mathbb{T}_0^2)$, respectively, we conclude from the energy estimate (4.5.12) that the assertion holds for any $(f, F) \in W^{r,2}(\Omega)^3 \times W^{r+\frac{1}{2},2}(\mathbb{T}_0^2)$. $\qquad\square$

Remark 4.5.6. Due to the representation formula for η one may check by mimicking the above steps that the estimate

$$\|v\|_{W^{r+2,2}(\Omega)} + \|\mathfrak{p}\|_{W^{r+1,2}(\Omega)} + \|\eta\|_{W^{r+4,2}(\mathbb{T}_0^2)}$$
$$\leq C_{36}\big(\|f\|_{W^{r,2}(\Omega)} + \|F\|_{W^{r,2}(\mathbb{T}_0^2)}\big),$$

holds for any $(f, F) \in W^{r,2}(\Omega) \times W^{r,2}(\mathbb{T}_0^2)$, which suffices for our purpose.

4.6 A priori Estimates

Subject of this section is the derivation of the L^q estimates satisfied by the solution to the linearized fluid-structure system

$$\begin{cases} \partial_t^2 \eta_{\mathrm{tp}} + \Delta'^2 \eta_{\mathrm{tp}} - \Delta' \partial_t \eta_{\mathrm{tp}} = F_{\mathrm{tp}} - e_3 \cdot T_0(u_{\mathrm{tp}}, p_{\mathrm{tp}}) & \text{in } \mathbb{T} \times \mathbb{T}_0^2, \\ \partial_t u_{\mathrm{tp}} - \Delta u_{\mathrm{tp}} + \nabla p_{\mathrm{tp}} = f_{\mathrm{tp}} & \text{in } \mathbb{T} \times \Omega, \\ \operatorname{div} u_{\mathrm{tp}} = 0 & \text{in } \mathbb{T} \times \Omega, \quad (4.6.1) \\ u_{\mathrm{tp}|x_3=0} = -\partial_t \eta_{\mathrm{tp}} e_3 & \text{on } \mathbb{T} \times \mathbb{T}_0^2, \\ u_{\mathrm{tp}|x_3=1} = 0 & \text{on } \mathbb{T} \times \mathbb{T}_0^2 \end{cases}$$

on $\Omega = \mathbb{T}_0^2 \times (0,1)$ under purely periodic forcing. We first consider (4.6.1) in the periodic half space $\mathbb{T}_0^2 \times (0,\infty)$ (Subsection 4.6.1) and introduce some L^q estimates obeyed by the pressure field (Subsection 4.6.2). Based on these estimates, we finally deduce in Subsection 4.6.3 the desired L^q estimates for the solution $(u_{\mathrm{tp}}, p_{\mathrm{tp}}, \eta_{\mathrm{tp}})$. For simplicity we omit the subscript tp in this section.

4.6.1 The Periodic Half Space

This section is devoted to the investigation of the coupled system

$$\begin{cases} \partial_t^2 \eta + \Delta'^2 \eta - \Delta' \partial_t \eta = h - e_3 \cdot T_0(v, \mathfrak{p}) & \text{in } \mathbb{T} \times \mathbb{T}_0^2, \\ \partial_t v - \Delta v + \nabla \mathfrak{p} = f & \text{in } \mathbb{T} \times \Omega_+, \\ \operatorname{div} v = g & \text{in } \mathbb{T} \times \Omega_+, \quad (4.6.2) \\ v_{|x_3=0} = -\partial_t \eta e_3 & \text{on } \mathbb{T} \times \mathbb{T}_0^2, \end{cases}$$

on the periodic half space $\Omega_+ := \mathbb{T}_0^2 \times \mathbb{R}_+$ with fully inhomogeneous purely oscillatory right-hand side

$$\begin{aligned} & f \in L_\perp^q(\mathbb{T} \times \Omega_+)^3, \\ & g \in L_\perp^q\big(\mathbb{T}; W^{1,q}(\Omega_+)\big) \cap W_\perp^{1,q}\big(\mathbb{T}; \dot{W}^{-1,q}(\Omega_+)\big), \quad (4.6.3) \\ & h \in L_\perp^q(\mathbb{T} \times \mathbb{T}_0^2). \end{aligned}$$

More precisely, the goal is to show that any solution to this system obeys the *a priori* estimate

$$\|v\|_{W^{1,2,q}_\perp(\mathbb{T}\times\Omega_+)} + \|\nabla\mathfrak{p}\|_{L^q_\perp(\mathbb{T}\times\Omega_+)} + \|\eta\|_{W^{2,4,q}_\perp(\mathbb{T}\times\mathbb{T}^2_0)} \leq C_{37}\big(\|f\|_{L^q_\perp(\mathbb{T}\times\Omega_+)}$$
$$+ \|g\|_{L^q_\perp(\mathbb{T};W^{1,q}(\Omega_+))\cap W^{1,q}_\perp(\mathbb{T};\dot{W}^{-1,q}(\Omega_+))} + \|h\|_{L^q_\perp(\mathbb{T}\times\mathbb{T}^2_0)}\big),$$

$$(4.6.4)$$

with $C_{37} = C_{37}(q,\mathcal{T},L) > 0$. Similarly to the previous subsection, we will omit the subscript tp as we are only examining purely periodic data in this subsection. As the solution to (4.6.2) is unique, see Lemma 4.3.1, we write $(v,\mathfrak{p},\eta)$ as a sum $(w+u,\pi+p,\eta)$ with (w,π) a solution to the Stokes system (4.4.45) with homogeneous boundary data $h = 0$ and (u,p,η) a solution to (4.6.2) with $f = 0 = g$ and h sufficiently chosen. Proceeding as in the proof of Proposition 4.4.4 we find a representation formula for (u,p,η) from which we deduce the desired L^q estimate, whereas an application of Theorem 4.4.7 yields that (w,π) obeys the *a priori* estimate (4.6.4). We begin by considering (4.6.2) with $f = 0 = g$ and show that (u,p,η) fulfills the L^q estimate (4.6.4). To this end we introduce the function space

$$\mathcal{W}^{1,2,q}_\sigma(\mathbb{T}\times\Omega) := W^{1,q}\big(\mathbb{T};L^q(\Omega)\big) \cap L^q\big(\mathbb{T};\mathcal{W}^{1,q}_\sigma(\Omega)\big).$$

Lemma 4.6.1. *Let* $q \in (1,\infty)$, $F \in L^q_\perp(\mathbb{T}\times\mathbb{T}^2_0)$ *and*

$$(u,p,\eta) \in \mathcal{W}^{1,2,q}_{\sigma,\perp}(\mathbb{T}\times\Omega_+)^3 \times L^q_\perp\big(\mathbb{T};\dot{W}^{1,q}(\Omega_+)\big) \times W^{2,4,q}_{(0),\perp}(\mathbb{T}\times\mathbb{T}^2_0)$$

be a corresponding solution to (4.6.2) *with* $f = 0$, $g = 0$ *and* $h = F$. *Then* (u,p,η) *satisfies* (4.6.4). *Moreover, there is a constant* $C_{38} = C_{38}(q,\mathcal{T},L) > 0$ *such that*

$$\|p\|_{L^q_\perp(\mathbb{T}\times\Omega_+)} \leq C_{38}\|F\|_{L^q_\perp(\mathbb{T}\times\mathbb{T}^2_0)} \qquad (4.6.5)$$

holds.

Proof. In order to prove the assertion we will mimic the arguments in the proof of Proposition 4.4.4 and utilize the Fourier transform $\mathscr{F}_{\mathbb{T}\times\mathbb{T}^2_0}$ to

$$\begin{cases} \partial_t^2\eta + \Delta'^2\eta - \Delta'\partial_t\eta = F - e_3\cdot\mathrm{T}_0(u,p) & \text{in } \mathbb{T}\times\mathbb{T}^2_0, \\ \partial_t u - \Delta u + \nabla p = 0 & \text{in } \mathbb{T}\times\Omega_+, \\ \mathrm{div}\, u = 0 & \text{in } \mathbb{T}\times\Omega_+, \\ u_{|x_3=0} = -\partial_t\eta e_3 & \text{on } \mathbb{T}\times\mathbb{T}^2_0. \end{cases} \qquad (4.6.6)$$

Letting

$$U := u', \quad \widehat{U} := \mathscr{F}_{\mathbb{T} \times \mathbb{T}_0^2}[U], \quad \widehat{\eta} := \mathscr{F}_{\mathbb{T} \times \mathbb{T}_0^2}[\eta],$$
$$V := u_3, \quad \widehat{V} := \mathscr{F}_{\mathbb{T} \times \mathbb{T}_0^2}[V], \quad \widehat{p} := \mathscr{F}_{\mathbb{T} \times \mathbb{T}_0^2}[p],$$

as well as

$$\mathfrak{m}_{\mathbb{T}} : \left(\frac{2\pi}{\mathcal{T}}\mathbb{Z}\right) \times \left(\frac{2\pi}{L}\mathbb{Z}\right)^2 \to \mathbb{C}, \quad \mathfrak{m}_{\mathbb{T}}(k, \xi') := |\xi'|^4 - k^2 + ik|\xi'|^2, \quad (4.6.7)$$

and utilizing that u being solenoidal implies

$$e_3 \cdot \mathrm{T}_0(u, p) = -p_{|x_3=0},$$

we deduce for each (fixed) $(k, \xi') \in \frac{2\pi}{\mathcal{T}}\mathbb{Z} \times \left(\frac{2\pi}{L}\mathbb{Z}\right)^2$

$$\begin{cases} \mathfrak{m}_{\mathbb{T}}\widehat{\eta} = \widehat{F} + \widehat{p}_{|x_3=0}, & \\ (|\xi'|^2 + ik)\widehat{U} - \partial_{x_3}^2\widehat{U} + i\xi'\,\widehat{p} = 0 & \text{in } \mathbb{R}_+, \\ (|\xi'|^2 + ik)\widehat{V} - \partial_{x_3}^2\widehat{V} + \partial_{x_3}\widehat{p} = 0 & \text{in } \mathbb{R}_+, \\ i\xi' \cdot \widehat{U} + \partial_{x_3}\widehat{V} = 0 & \text{in } \mathbb{R}_+, \\ \widehat{U}_{|x_3=0} = 0, & \\ \widehat{V}_{|x_3=0} = -ik\widehat{\eta}. & \end{cases} \quad (4.6.8)$$

In order to solve this system of ODEs, we first consider the system $(4.6.8)_2$ - $(4.6.8)_6$ and identify it as the system $(4.4.14)$ which was investigated in Section 4.4. For this system we have already seen that a solution is given by

$$U = \mathscr{F}_{\mathbb{T} \times \mathbb{T}_0^2}^{-1}\left[-\frac{\xi' q_0}{k}e^{-|\xi'|x_3} + \frac{\xi' q_0}{k}e^{-\sqrt{|\xi'|^2 + ik}\,x_3}\right],$$
$$V = \mathscr{F}_{\mathbb{T} \times \mathbb{T}_0^2}^{-1}\left[\frac{|\xi'| q_0}{ik}e^{-|\xi'|x_3} - \left(ik\widehat{\eta} + \frac{|\xi'| q_0}{ik}\right)e^{-\sqrt{|\xi'|^2 + ik}\,x_3}\right], \quad (4.6.9)$$
$$p = \mathscr{F}_{\mathbb{T} \times \mathbb{T}_0^2}^{-1}\left[q_0 e^{-|\xi'|x_3}\right],$$

see $(4.4.19)$, with q_0 as in $(4.4.18)$, *i.e.*,

$$q_0 = q_1\widehat{\eta} := \left[-ik\left(|\xi'| + \sqrt{|\xi'|^2 + ik}\right) + \frac{k^2}{|\xi'|}\right]\widehat{\eta} \quad (4.6.10)$$

In virtue of $(4.6.8)_1$ and $(4.6.9)$ we deduce

$$\mathfrak{m}_{\mathbb{T}}\widehat{\eta} - \widehat{p}_{|x_3=0} = \widehat{F} \Leftrightarrow \mathfrak{m}_{\mathbb{T}}\widehat{\eta} - q_0 = \widehat{F}$$

$$\Leftrightarrow \widehat{\eta} = \frac{|\xi'|}{|\xi'|\mathfrak{m}_{\mathbb{T}} - k^2 + ik|\xi'|\left(|\xi'| + \sqrt{|\xi'|^2 + ik}\right)}\widehat{F}$$

and therefore

$$\widehat{\eta} = \frac{\rho_{\mathbb{Z}}}{\mathfrak{m}_{\mathbb{T}}}\frac{\rho_{\mathbb{Z}}|\xi'|\mathfrak{m}_{\mathbb{T}}}{\mathcal{N}}\widehat{F} = \mathfrak{m}_{\mathbb{T}}M_{\mathbb{T}}\widehat{F} \qquad (4.6.11)$$

with $\mathfrak{m}_{\mathbb{T}}$ given as in $(4.6.7)$ and

$$\rho_{\mathbb{Z}}(k) := 1 - \delta_{\frac{2\pi}{\mathcal{T}}\mathbb{Z}}(k),$$

$$\mathcal{N}(k,\xi') := |\xi'|\mathfrak{m}_{\mathbb{T}}(k,\xi') - k^2 + ik|\xi'|\left(|\xi'| + \sqrt{|\xi'|^2 + ik}\right),$$

$$m_{\mathbb{T}}(k,\xi') := \rho_{\mathbb{Z}}(k)\mathfrak{m}_{\mathbb{T}}^{-1}(k,\xi'),$$

$$M_{\mathbb{T}}(k,\xi') := \frac{\rho_{\mathbb{Z}}(k)\,|\xi'|\mathfrak{m}_{\mathbb{T}}(k,\xi')}{\mathcal{N}(k,\xi')}.$$

Recall that by $\delta_{\frac{2\pi}{\mathcal{T}}\mathbb{Z}}$ we denote the delta distribution defined in $(2.3.1)$. Employing Proposition 4.4.6, we obtain that $(u,p) = (U,V,p)$ given as in $(4.6.9)$ obeys the L^q estimate $(4.4.43)$ with $h = -\partial_t\eta e_3$, that is

$$\|u\|_{\mathrm{W}_{\perp}^{1,2,q}(\mathbb{T}\times\Omega_+)} + \|\nabla p\|_{\mathrm{L}_{\perp}^q(\mathbb{T}\times\Omega_+)}$$
$$\leq c_0\left(\|\partial_t\eta\|_{\mathrm{W}_{\perp}^{1-\frac{1}{2q},2-\frac{1}{q},q}(\mathbb{T}\times\mathbb{T}_0^2)} + \|\partial_t\eta\|_{\mathrm{W}_{\perp}^{1,q}(\mathbb{T};\dot{\mathrm{W}}^{-\frac{1}{q},q}(\mathbb{T}_0^2))}\right)$$
$$\leq c_1\left(\|\eta\|_{\mathrm{W}_{\perp}^{2-\frac{1}{2q},4-\frac{1}{q},q}(\mathbb{T}\times\mathbb{T}_0^2)} + \|\partial_t\eta\|_{\mathrm{W}_{\perp}^{1,q}(\mathbb{T};\mathrm{L}^q(\mathbb{T}_0^2))}\right) \leq c_2\|\eta\|_{\mathrm{W}_{\perp}^{2,4,q}(\mathbb{T}\times\mathbb{T}_0^2)}$$

$$(4.6.12)$$

holds. Now let $(k,\xi') \in \frac{2\pi}{\mathcal{T}}\mathbb{Z} \times \left(\frac{2\pi}{L}\mathbb{Z}\right)^2$ and $(\alpha,\beta) \in \mathbb{N} \times \mathbb{N}^2$ with $|k| > \frac{\pi}{\mathcal{T}}$, $\alpha \leq 2$ and $|\beta| \leq 4$. Since $M_{\mathbb{T}}$, $m_{\mathbb{T}}$, $(ik)^\alpha m_{\mathbb{T}}$ and $(i\xi')^\beta m_{\mathbb{T}}$ are $\mathrm{L}^q(\mathbb{T}\times\mathbb{T}_0^2)$-multipliers, see Lemma A.2.3 and Lemma A.2.4, we conclude that

$$\|\eta\|_{\mathrm{W}_{\perp}^{2,4,q}(\mathbb{T}\times\mathbb{T}_0^2)} = \|\mathscr{F}_{\mathbb{T}\times\mathbb{T}_0^2}^{-1}\left[(1 + |k|^2 + |\xi'|^4)\widehat{\eta}\right]\|_{\mathrm{L}_{\perp}^q(\mathbb{T}\times\mathbb{T}_0^2)}$$
$$= \|\mathscr{F}_{\mathbb{T}\times\mathbb{T}_0^2}^{-1}\left[(m_{\mathbb{T}} + (|k|^2 m_{\mathbb{T}}) + (|\xi'|^4 m_{\mathbb{T}}))M_{\mathbb{T}}\widehat{F}\right]\|_{\mathrm{L}_{\perp}^q(\mathbb{T}\times\mathbb{T}_0^2)}$$
$$\leq c_3\|F\|_{\mathrm{L}_{\perp}^q(\mathbb{T}\times\mathbb{T}_0^2)}.$$

In view of (4.6.12), it remains to show (4.6.5), that is

$$\|p\|_{\mathrm{L}^q_\perp(\mathbb{T}\times\Omega_+)} \le c_4 \|F\|_{\mathrm{L}^q_\perp(\mathbb{T}\times\mathbb{T}^2_0)},$$

to complete the proof. For this purpose, we employ (4.4.23), that is, the boundedness of

$$\phi \mapsto \mathscr{F}^{-1}_{\mathbb{T}^2_0}\big[(|\xi'|x_3)^m \, e^{-|\xi'|x_3}\, \mathscr{F}_{\mathbb{T}^2_0}[\phi]\big]$$

on $\mathrm{L}^q_\perp(\mathbb{T}^2_0)$ for $m \in \mathbb{N}_0$, to deduce from the representation formulas (4.6.9), (4.6.10) and (4.6.11) that

$$
\begin{aligned}
\|p\|_{\mathrm{L}^q_\perp(\mathbb{T}\times\Omega_+)} &= \big\|\mathscr{F}^{-1}_{\mathbb{T}\times\mathbb{T}^2_0}\big[q_0 e^{-|\xi'|x_3}\big]\big\|_{\mathrm{L}^q(\mathbb{R}_+;\mathrm{L}^q_\perp(\mathbb{T}\times\mathbb{T}^2_0))}\\
&\le c_5 \big\|\mathscr{F}^{-1}_{\mathbb{T}\times\mathbb{T}^2_0}[q_0]\big\|_{\mathrm{L}^q_\perp(\mathbb{T}\times\mathbb{T}^2_0)} = c_5 \big\|\mathscr{F}^{-1}_{\mathbb{T}\times\mathbb{T}^2_0}[q_1\widehat{\eta}]\big\|_{\mathrm{L}^q_\perp(\mathbb{T}\times\mathbb{T}^2_0)}\\
&\le c_5\Big(\big\|\mathscr{F}^{-1}_{\mathbb{T}\times\mathbb{T}^2_0}\big[-ik|\xi'|m_\mathbb{T} M_\mathbb{T} \widehat{F}\big]\big\|_{\mathrm{L}^q_\perp(\mathbb{T}\times\mathbb{T}^2_0)} + \Big\|\mathscr{F}^{-1}_{\mathbb{T}\times\mathbb{T}^2_0}\Big[\frac{k^2}{\mathcal{N}}\widehat{F}\Big]\Big\|_{\mathrm{L}^q_\perp(\mathbb{T}\times\mathbb{T}^2_0)}\\
&\quad + \Big\|\mathscr{F}^{-1}_{\mathbb{T}\times\mathbb{T}^2_0}\Big[\frac{k^2 m_\mathbb{T} - ik|\xi'|^2 m_\mathbb{T}}{\sqrt{|\xi'|^2+ik}} M_\mathbb{T} \widehat{F}\Big]\Big\|_{\mathrm{L}^q_\perp(\mathbb{T}\times\mathbb{T}^2_0)}\Big).
\end{aligned}
$$

Note that from the estimates deduced in (A.2.31) one may conclude in a similar way as in the proof of Lemma A.2.4 that $\frac{k^2}{\mathcal{N}}$ is an $\mathrm{L}^q(\mathbb{T}\times\mathbb{T}^2_0)$-multiplier. Moreover, applying Marcinkiewicz multiplier theorem (Theorem 2.3.2) we first verify $(|\xi'|^2+ik)^{-1/2}$ as an $\mathrm{L}^q(\mathbb{R}\times\mathbb{R}^2)$-multiplier, and therefore the transference principle (Theorem 2.3.1) yields that it is also an $\mathrm{L}^q(\mathbb{T}\times\mathbb{T}^2_0)$-multiplier. Utilizing Lemma A.2.3 and Lemma A.2.4 we conclude (4.6.5). □

For the solution to the fully inhomogeneous system (4.6.2) we deduce:

Lemma 4.6.2 (L^q Estimate). *Let $q \in (1,\infty)$, (f,g,h) in the class* (4.6.3) *and $(v,\mathfrak{p},\eta)$ a corresponding solution to* (4.6.2) *in the class*

$$(v,\mathfrak{p},\eta) \in \mathrm{W}^{1,2,q}_\perp(\mathbb{T}\times\Omega_+)^3 \times \mathrm{L}^q_\perp\big(\mathbb{T};\dot{\mathrm{W}}^{1,q}(\Omega_+)\big) \times \mathrm{W}^{2,4,q}_{(0),\perp}(\mathbb{T}\times\mathbb{T}^2_0).$$

Then, $(v,\mathfrak{p},\eta)$ is unique and obeys the L^q estimate (4.6.4), *that is*

$$
\begin{aligned}
\|v\|_{\mathrm{W}^{1,2,q}_\perp(\mathbb{T}\times\Omega_+)} + \|\nabla\mathfrak{p}\|_{\mathrm{L}^q_\perp(\mathbb{T}\times\Omega_+)} + \|\eta\|_{\mathrm{W}^{2,4,q}_\perp(\mathbb{T}\times\mathbb{T}^2_0)} &\le C_{37}\big(\|f\|_{\mathrm{L}^q_\perp(\mathbb{T}\times\Omega_+)}\\
+ \|g\|_{\mathrm{L}^q_\perp(\mathbb{T};\mathrm{W}^{1,q}(\Omega_+))\cap\mathrm{W}^{1,q}_\perp(\mathbb{T};\dot{\mathrm{W}}^{-1,q}(\Omega_+))} &+ \|h\|_{\mathrm{L}^q_\perp(\mathbb{T}\times\mathbb{T}^2_0)}\big).
\end{aligned}
$$

Proof. The uniqueness assertion follows directly from Lemma 4.3.1 (for $\ell = \infty$). Hence we conclude that every solution to (4.6.2) can be expressed as a sum $(v, \mathfrak{p}, \eta) = (w + u, \pi + p, \eta)$, where (w, π) is a solution to the Stokes system

$$\partial_t w - \Delta w + \nabla \pi = f, \quad \operatorname{div} w = g \quad \text{and} \quad w_{|\mathbb{T} \times \partial \Omega_+} = 0 \qquad (4.6.13)$$

and (u, p, η) solves

$$\begin{cases} \partial_t^2 \eta + \Delta'^2 \eta - \Delta' \partial_t \eta = F - e_3 \cdot \mathrm{T}_0(u, p) & \text{in } \mathbb{T} \times \mathbb{T}_0^2, \\ \partial_t u - \Delta u + \nabla p = 0 & \text{in } \mathbb{T} \times \Omega_+, \\ \operatorname{div} u = 0 & \text{in } \mathbb{T} \times \Omega_+, \\ u_{|x_3 = 0} = -\partial_t \eta e_3 & \text{on } \mathbb{T} \times \mathbb{T}_0^2, \end{cases} \qquad (4.6.14)$$

with $F = h - e_3 \cdot \mathrm{T}_0(w, \pi)$. Utilizing Theorem 4.4.7 we obtain a solution $(w, \pi) \in \mathrm{W}_\perp^{1,2,q}(\mathbb{T} \times \Omega_+)^3 \times \mathrm{L}_\perp^q\big(\mathbb{T}; \dot{\mathrm{W}}^{1,q}(\Omega_+)\big)$ satisfying the L^q estimate

$$\begin{aligned} \|w\|_{\mathrm{W}_\perp^{1,2,q}(\mathbb{T} \times \Omega_+)} &+ \|\nabla \pi\|_{\mathrm{L}_\perp^q(\mathbb{T} \times \Omega_+)} \\ &\leq c_0\big(\|f\|_{\mathrm{L}_\perp^q(\mathbb{T} \times \Omega_+)} + \|g\|_{\mathrm{L}_\perp^q(\mathbb{T}; \mathrm{W}^{1,q}(\Omega_+)) \cap \mathrm{W}_\perp^{1,q}(\mathbb{T}; \dot{\mathrm{W}}^{-1,q}(\Omega_+))}\big). \end{aligned} \qquad (4.6.15)$$

Observe, that the pressure field π corresponding to the above Stokes system (4.6.13) is only unique up to a function that depends only on time. Let $\tilde{\pi}$ be defined as

$$\tilde{\pi} := \pi - \int_\Omega \pi \, \mathrm{d}y$$

with $\Omega = \mathbb{T}_0^2 \times (0, 1)$. Then $(w, \tilde{\pi})$ is still a solution to (4.6.13) satisfying (4.6.15). Moreover, $\tilde{\pi}_{|\Omega}$ is mean value free, and therefore Poincaré's inequality yields

$$\|\tilde{\pi}\|_{\mathrm{L}_\perp^q(\mathbb{T}; \mathrm{W}^{1,q}(\Omega))} \leq c_1 \|\nabla \tilde{\pi}\|_{\mathrm{L}_\perp^q(\mathbb{T} \times \Omega)} \leq c_2 \|\nabla \pi\|_{\mathrm{L}_\perp^q(\mathbb{T} \times \Omega_+)}. \qquad (4.6.16)$$

By considering $\tilde{\pi}$ instead of π, the right-hand side in (4.6.14) changes to $F = h - e_3 \cdot \mathrm{T}_0(w, \tilde{\pi})$. To complete the proof it remains to show that (u, p, η) satisfies the *a priori* estimate (4.6.4). From Lemma 4.6.1 it directly follows that (u, p, η) obeys the L^q estimates

$$\|u\|_{\mathrm{W}_\perp^{1,2,q}(\mathbb{T} \times \Omega_+)} + \|\nabla p\|_{\mathrm{L}_\perp^q(\mathbb{T}; \mathrm{L}^q(\Omega_+))} + \|\eta\|_{\mathrm{W}_\perp^{2,4,q}(\mathbb{T} \times \mathbb{T}_0^2)} \leq c_3 \|F\|_{\mathrm{L}_\perp^q(\mathbb{T} \times \mathbb{T}_0^2)}.$$

Utilizing (4.6.15) and the properties of the trace operator

$$\mathrm{Tr}_0 \colon \mathrm{L}_\perp^q\big(\mathbb{T}; \mathrm{W}^{1,q}(\Omega)\big) \to \mathrm{L}_\perp^q\big(\mathbb{T}; \mathrm{W}^{1 - \frac{1}{q}, q}(\mathbb{T}_0^2)\big), \qquad \phi \mapsto \phi_{|x_3 = 0}, \qquad (4.6.17)$$

stated in [36, Theorem II.4.3], we further estimate the norm on the right-hand side above as follows

$$\|F\|_{L^q_\perp(\mathbb{T}\times\mathbb{T}^2_0)} \le c_4\big(\|h\|_{L^q_\perp(\mathbb{T}\times\mathbb{T}^2_0)} + \|\nabla w_{|x_3=0}\|_{L^q_\perp(\mathbb{T}\times\mathbb{T}^2_0)} + \|\tilde{\pi}_{|x_3=0}\|_{L^q_\perp(\mathbb{T}\times\mathbb{T}^2_0)}\big)$$

$$\le c_5\big(\|h\|_{L^q_\perp(\mathbb{T}\times\mathbb{T}^2_0)} + \|\nabla w_{|x_3=0}\|_{L^q_\perp(\mathbb{T};W^{1-\frac{1}{q},q}(\mathbb{T}^2_0))}$$

$$+ \|\tilde{\pi}_{|x_3=0}\|_{L^q_\perp(\mathbb{T};W^{1-\frac{1}{q},q}(\mathbb{T}^2_0))}\big)$$

$$\le c_6\big(\|h\|_{L^q_\perp(\mathbb{T}\times\mathbb{T}^2_0)} + \|\nabla w\|_{L^q_\perp(\mathbb{T};W^{1,q}(\Omega_+))} + \|\tilde{\pi}\|_{L^q_\perp(\mathbb{T};W^{1,q}(\Omega))}\big).$$

Finally (4.6.16) yields

$$\|F\|_{L^q_\perp(\mathbb{T}\times\mathbb{T}^2_0)} \le c_7\big(\|h\|_{L^q_\perp(\mathbb{T}\times\mathbb{T}^2_0)} + \|w\|_{W^{1,2,q}_\perp(\Omega_+)} + \|\nabla\pi\|_{L^q_\perp(\mathbb{T}\times\Omega_+)}\big)$$

$$\le c_8\big(\|f\|_{L^q_\perp(\mathbb{T}\times\Omega_+)} + \|g\|_{L^q_\perp(\mathbb{T};W^{1,q}(\Omega_+))\cap W^{1,q}_\perp(\mathbb{T};\dot{W}^{-1,q}(\Omega_+))}$$

$$+ \|h\|_{L^q_\perp(\mathbb{T}\times\mathbb{T}^2_0)}\big),$$

which together with (4.6.15) completes the proof. □

4.6.2 Estimates of the Pressure Field

Before we consider the coupled fluid-structure system and establish L^q estimates, one further result on L^q estimates for the pressure field is needed. This subsection is dedicated to prove a lemma similar to [37, Lemma 5.4.] for the pressure field corresponding to (4.6.1). We show the following.

Lemma 4.6.3 (Pressure Field Estimates)*. Let $s \in (1, \infty)$. Let further $(f, F) \in L^s_\perp(\mathbb{T}\times\Omega)^3 \times L^s_\perp(\mathbb{T}\times\mathbb{T}^2_0)$ and*

$$(u, p, \eta) \in \mathcal{W}^{1,2,s}_{\sigma,\perp}(\mathbb{T}\times\Omega) \times L^s_\perp\big(\mathbb{T}; \dot{W}^{1,s}(\Omega)\big) \times W^{2,4,s}_{(0),\perp}(\mathbb{T}\times\mathbb{T}^2_0)$$

be a solution to (4.6.1). Then there exists a constant $C_{39} = C_{39}(\Omega, s) > 0$ such that

$$\|p(t,\cdot)\|_{L^{\frac{3}{2}s}(\Omega)} \le C_{39}\big(\|f(t,\cdot)\|_{L^s(\Omega)} + \|F(t,\cdot)\|_{L^s(\mathbb{T}^2_0)} + \|\nabla u(t,\cdot)\|_{L^s(\Omega)}$$

$$+ \|\nabla'\Delta'\eta(t,\cdot)\|_{L^s(\mathbb{T}^2_0)} + \|\nabla'\partial_t\eta(t,\cdot)\|_{L^s(\mathbb{T}^2_0)}$$

$$+ \|\nabla u(t,\cdot)\|^{\frac{s-1}{s}}_{L^s(\Omega)}\|\nabla u(t,\cdot)\|^{\frac{1}{s}}_{W^{1,s}(\Omega)}\big)$$

$$\text{(4.6.18)}$$

for a.e. $t \in \mathbb{T}$. Moreover, there is a constant $C_{40} = C_{40}(\Omega, s) > 0$ such that for a.e. $t \in \mathbb{T}$ the additional estimate

$$\|\nabla p(t,\cdot)\|_{L^s(\mathbb{T}^2_0\times(\frac{1}{3},\frac{2}{3}))} \le C_{40}\big(\|f(t,\cdot)\|_{L^s(\Omega)} + \|p(t,\cdot)\|_{L^s(\Omega)}\big) \quad \text{(4.6.19)}$$

holds.

Proof. In order to prove Lemma 4.6.3, we mimic the approach from [37, Proof of Lemma 5.4]. For simplicity, integrals over $\mathbb{T}_0^2 \times \{0\}$ are written as integrals of $\mathbb{T}_0^2$ in the following. Furthermore, we neglect the t-dependency of functions.

First consider an arbitrary $\phi \in C_0^\infty(\overline{\Omega})$, and observe that due to the periodicity of u and ϕ, as well as the boundary condition $(4.6.1)_4$, we obtain

$$\int_0^\mathcal{T} \int_\Omega \partial_t u \cdot \nabla\phi \, dx \psi \, dt = \int_0^\mathcal{T} \int_\Omega \operatorname{div} u\phi\partial_t\psi \, dx \, dt - \int_0^\mathcal{T} \int_{\mathbb{T}_0^2} u\,\phi \cdot (-e_3)\partial_t\psi \, dx' \, dt$$

$$= -\int_0^\mathcal{T} \int_{\mathbb{T}_0^2} \partial_t\eta \, \phi\partial_t\psi \, dx' \, dt = \int_0^\mathcal{T} \int_{\mathbb{T}_0^2} \partial_t^2\eta \, \phi \, dx'\psi \, dt$$

for any $\psi \in C_0^\infty(\mathbb{T})$. Thus

$$\int_\Omega \partial_t u \cdot \nabla\phi \, dx = \int_{\mathbb{T}_0^2} \partial_t^2\eta \, \phi \, dx' \qquad (4.6.20)$$

holds for a.e. $t \in \mathbb{T}$. Hence, by multiplication of $(4.6.1)_2$ with $\nabla\phi$, it follows that the pressure field p is a solution to the weak Poisson problem with homogeneous Robin boundary conditions on the bottom and homogeneous Neumann conditions on the other part of the boundary, *i.e.*, for any $\phi \in C_0^\infty(\overline{\Omega})$ the identity

$$\int_\Omega \nabla p \cdot \nabla\phi \, dx + \int_{\mathbb{T}_0^2} p\phi \, dx' = \int_\Omega f \cdot \nabla\phi \, dx + \int_\Omega \Delta u \cdot \nabla\phi \, dx$$

$$- \int_{\mathbb{T}_0^2} F\phi \, dx' + \int_{\mathbb{T}_0^2} \Delta'^2\eta\phi \, dx' - \int_{\mathbb{T}_0^2} \Delta'\partial_t\eta\phi \, dx'$$

$$(4.6.21)$$

holds. Existence of a unique solution to the Poisson problem with Robin boundary conditions follows as in [83] for the corresponding Poisson problem with Neumann boundary conditions.

To derive (4.6.18), let Φ be a solution to

$$\begin{cases} -\Delta\Phi = g & \text{in } \Omega, \\ \partial_\nu\Phi_{|x_3=0} + \Phi_{|x_3=0} = 0 & \text{on } \mathbb{T}_0^2, \\ \partial_\nu\Phi_{|x_3=1} = 0 & \text{on } \mathbb{T}_0^2 \end{cases} \qquad (4.6.22)$$

with $g \in C_0^\infty(\Omega)$ arbitrary. Then Φ obeys the weak formulation

$$\int_\Omega \nabla\Phi \cdot \nabla\psi \,\mathrm{d}x + \int_{\mathbb{T}_0^2} \Phi\psi \,\mathrm{d}x' = \int_\Omega g\psi \,\mathrm{d}x$$

for any $\psi \in C_0^\infty(\overline{\Omega})$. Existence of a solution to this problem obeying for any $q \in (1, \infty)$ the L^q estimate

$$\|\Phi\|_{W^{2,q}(\Omega)} \le c_0 \|g\|_{L^q(\Omega)} \tag{4.6.23}$$

follows by classical methods, one may mimic the proof in [83]. We can thus compute

$$\int_\Omega pg \,\mathrm{d}x = -\int_\Omega p\Delta\Phi \,\mathrm{d}x = \int_\Omega \nabla p \cdot \nabla\Phi \,\mathrm{d}x + \int_{\mathbb{T}_0^2} p\,\Phi \,\mathrm{d}x',$$

and in view of (4.6.21), this implies

$$\int_\Omega pg \,\mathrm{d}x = I_1 + \ldots + I_5,$$

with

$$I_1 := \int_\Omega f \cdot \nabla\Phi \,\mathrm{d}x, \quad I_2 := \int_\Omega \Delta u \cdot \nabla\Phi \,\mathrm{d}x, \quad I_3 := -\int_{\mathbb{T}_0^2} F\Phi \,\mathrm{d}x'$$
$$I_4 := \int_{\mathbb{T}_0^2} \Delta'^2\eta\,\Phi \,\mathrm{d}x', \quad I_5 := -\int_{\mathbb{T}_0^2} \Delta'\partial_t\eta\,\Phi \,\mathrm{d}x'.$$

The five integrals I_1–I_5 shall be estimated separately. But first observe that by Sobolev embedding

$$\|\Phi\|_{W^{1,s'}(\Omega)} \le c_1 \|\Phi\|_{W^{2,\frac{3s}{3s-2}}} \le c_2 \|g\|_{L^{\frac{3s}{3s-2}}(\Omega)}$$

holds. For the first integral, by Hölder's inequality we deduce

$$|I_1| \le \|f\|_{L^s(\Omega)} \|\nabla\Phi\|_{L^{s'}(\Omega)} \le c_3 \|f\|_{L^s(\Omega)} \|g\|_{L^{\frac{3s}{3s-2}}(\Omega)}.$$

To find a similar estimate for I_2, we twice integrate by parts to deduce

$$I_2 = \int_\Omega \partial_{x_j}^2 u_i \partial_{x_i}\Phi \,\mathrm{d}x = \int_{\partial\Omega} \partial_{x_j} u_i \,\partial_{x_i}\Phi\,\nu_j \,\mathrm{d}x' - \int_\Omega \partial_{x_j} u_i \,\partial_{x_j}\partial_{x_i}\Phi \,\mathrm{d}x$$
$$= \int_{\partial\Omega} (\partial_{x_j} u_i \,\partial_{x_i}\Phi\,\nu_j - \partial_{x_j} u_i \,\partial_{x_j}\Phi\,\nu_i) \,\mathrm{d}x', \tag{4.6.24}$$

where we utilized the Einstein summation convention and $\operatorname{div} u = 0$. Hence, applying first Hölder's inequality, Sobolev embeddings and a trace inequality (see for example [36, Theorem II.4.1]), this identity yields

$$
\begin{aligned}
|I_2| &\leq \left| \int_{\partial\Omega} \partial_{x_j} u_i \, \partial_{x_i} \Phi \, \nu_j - \partial_{x_j} u_i \, \partial_{x_j} \Phi \, \nu_i \, \mathrm{d}x' \right| \\
&\leq c_4 \|\nabla u\|_{\mathrm{L}^s(\partial\Omega)} \|\nabla\Phi\|_{\mathrm{L}^{s'}(\partial\Omega)} \leq c_5 \|\nabla u\|_{\mathrm{L}^s(\partial\Omega)} \|\nabla\Phi\|_{\mathrm{W}^{1,\frac{3s}{3s-2}}(\Omega)} \\
&\leq c_6 \left(\|\nabla u\|_{\mathrm{L}^s(\Omega)} + \|\nabla u\|_{\mathrm{L}^s(\Omega)}^{\frac{s-1}{s}} \|\nabla u\|_{\mathrm{W}^{1,s}(\Omega)}^{\frac{1}{s}} \right) \|g\|_{\mathrm{L}^{\frac{3s}{3s-2}}(\Omega)}.
\end{aligned}
$$

The estimates for the final three integrals will be established similarly to the estimates of I_2 by an application of the same trace inequality as above. It follows that

$$
\begin{aligned}
|I_3| &\leq \|F\|_{\mathrm{L}^s(\mathbb{T}_0^2)} \|\Phi\|_{\mathrm{L}^{s'}(\mathbb{T}_0^2)} \\
&\leq c_7 \|F\|_{\mathrm{L}^s(\mathbb{T}_0^2)} \|\Phi\|_{\mathrm{W}^{1,s'}(\Omega)} \leq c_8 \|F\|_{\mathrm{L}^s(\mathbb{T}_0^2)} \|g\|_{\mathrm{L}^{\frac{3s}{3s-2}}(\Omega)}
\end{aligned}
$$

holds. In order to utilize the same arguments as for I_2, we integrate by parts in I_4 and I_5 to find

$$
\begin{aligned}
|I_4| &= \left| \int_{\mathbb{T}_0^2} \Delta'^2 \eta \Phi \, \mathrm{d}x' \right| = \left| \int_{\mathbb{T}_0^2} \nabla' \Delta' \eta \cdot \nabla' \Phi \, \mathrm{d}x' \right| \\
&\leq c_9 \|\nabla' \Delta' \eta\|_{\mathrm{L}^s(\mathbb{T}_0^2)} \|\nabla\Phi\|_{\mathrm{L}^{s'}(\partial\Omega)} \leq c_{10} \|\nabla' \Delta' \eta\|_{\mathrm{L}^s(\mathbb{T}_0^2)} \|g\|_{\mathrm{L}^{\frac{3s}{3s-2}}(\Omega)}
\end{aligned}
$$

and

$$
\begin{aligned}
|I_5| &= \left| \int_{\mathbb{T}_0^2} \Delta' \partial_t \eta \Phi \, \mathrm{d}x' \right| = \left| \int_{\mathbb{T}_0^2} \nabla' \partial_t \eta \cdot \nabla' \Phi \, \mathrm{d}x' \right| \\
&\leq c_{11} \|\nabla' \partial_t \eta\|_{\mathrm{L}^s(\mathbb{T}_0^2)} \|\nabla\Phi\|_{\mathrm{L}^{s'}(\partial\Omega)} \leq c_{12} \|\nabla' \partial_t \eta\|_{\mathrm{L}^s(\mathbb{T}_0^2)} \|g\|_{\mathrm{L}^{\frac{3s}{3s-2}}(\Omega)}.
\end{aligned}
$$

Finally collecting the estimates deduced for I_1–I_5, it follows that

$$
\begin{aligned}
\left| \int_\Omega pg \, \mathrm{d}x \right| \leq c_{13} \Big(&\|f\|_{\mathrm{L}^s(\Omega)} + \|F\|_{\mathrm{L}^s(\mathbb{T}_0^2)} + \|\nabla u\|_{\mathrm{L}^s(\Omega)}^{\frac{s-1}{s}} \|\nabla u\|_{\mathrm{W}^{1,s}(\Omega)}^{\frac{1}{s}} \\
&+ \|\nabla u\|_{\mathrm{L}^s(\Omega)} + \|\nabla' \Delta' \eta\|_{\mathrm{L}^s(\mathbb{T}_0^2)} + \|\nabla' \partial_t \eta\|_{\mathrm{L}^s(\mathbb{T}_0^2)} \Big) \|g\|_{\mathrm{L}^{\frac{3s}{3s-2}}(\Omega)},
\end{aligned}
$$

and by using the duality $\left(\mathrm{L}^{\frac{3}{2}s}(\Omega) \right)' = \mathrm{L}^{\frac{3s}{3s-2}}(\Omega)$, we obtain (4.6.18).

To complete the proof it remains to show (4.6.19). For this purpose, let $\chi \in C_0^\infty(\mathbb{R})$ be a cut-off function such that

$$\chi(x_3) = 1 \quad \text{for } x_3 \in \left(\frac{1}{3}, \frac{2}{3}\right), \quad \text{and} \quad \chi(x_3) = 0 \quad \text{for } x_3 \in \mathbb{R} \setminus \left[\frac{1}{6}, \frac{5}{6}\right].$$

Set $\pi(x) := \chi(x_3) p(x)$ and observe from (4.6.1) that π is a solution to the weak Neumann problem

$$\int_\Omega \nabla \pi \cdot \nabla \phi \, dx = \int_\Omega f_1 \cdot \nabla \phi \, dx + \int_\Omega f_2 \, \phi \, dx \tag{4.6.25}$$

for all $\phi \in C_0^\infty(\overline{\Omega})$, with

$$f_1 := 2p\nabla\chi + f\chi \quad \text{and} \quad f_2 := p\Delta\chi + f \cdot \nabla\chi.$$

Note that to find (4.6.25), we have utilized that χ and $\frac{\partial\chi}{\partial\nu}$ vanish on $\partial\Omega$. Since $\operatorname{supp} f_1, \operatorname{supp} f_2 \subset \mathbb{T}_0^2 \times \left(\frac{1}{6}, \frac{5}{6}\right)$, we clearly have

$$\sup_{\|\nabla\phi\|_{\mathrm{L}^{s'}(\Omega)}=1} \left| \int_\Omega f_1 \cdot \nabla\phi \, dx \right| \leq c_{14}\left(\|f\|_{\mathrm{L}^s(\Omega)} + \|p\|_{\mathrm{L}^s(\Omega)}\right),$$

and by Poincaré's inequality

$$\sup_{\|\nabla\phi\|_{\mathrm{L}^{s'}(\Omega)}=1} \left| \int_\Omega f_2 \cdot \phi \, dx \right| = \sup_{\|\nabla\phi\|_{\mathrm{L}^{s'}(\Omega)}=1,\, \operatorname{supp}\phi \subset \mathbb{T}_0^2 \times (\frac{1}{6},\frac{5}{6})} \|f_2\|_{\mathrm{L}^s(\Omega)} \|\phi\|_{\mathrm{L}^{s'}(\Omega)}$$

$$\leq c_{15}\left(\|f\|_{\mathrm{L}^s(\Omega)} + \|p\|_{\mathrm{L}^s(\Omega')}\right).$$

Finally, (4.6.19) follows via a standard *a priori* estimate for the weak Neumann problem (4.6.25) and $p_{|\mathbb{T}_0^2 \times (\frac{1}{3},\frac{2}{3})} = \pi_{|\mathbb{T}_0^2 \times (\frac{1}{3},\frac{2}{3})}$. □

4.6.3 The Periodic Layer

Finally, this subsection is dedicated to the investigation of

$$\begin{cases} \partial_t^2\eta + \Delta'^2\eta - \Delta'\partial_t\eta = F - e_3 \cdot \mathrm{T}_0(u,p) & \text{in } \mathbb{T} \times \mathbb{T}_0^2, \\ \partial_t u - \Delta u + \nabla p = f & \text{in } \mathbb{T} \times \Omega, \\ \operatorname{div} u = 0 & \text{in } \mathbb{T} \times \Omega, \\ u_{|x_3=0} = -\partial_t\eta e_3 & \text{on } \mathbb{T} \times \mathbb{T}_0^2, \\ u_{|x_3=1} = 0 & \text{on } \mathbb{T} \times \mathbb{T}_0^2, \end{cases} \tag{4.6.26}$$

under purely periodic forcing, which means, in the case $\mathcal{P}f = 0 = \mathcal{P}F$. More precisely, we are interested in the L^q estimates satisfied by the solution to this system. For this purpose, we will extend the solution to the periodic half space and employ the results deduced in the previous two subsection in order to find a suitable L^q estimate. This procedure yields additional terms in the L^q estimate, which can subsequently be neglected by the same contradiction argument as in the case of nonlinear acoustics, see the proof of Lemma 3.2.9.

Lemma 4.6.4. *Let* $q \in (1, \infty)$, $(f, F) \in L^q_\perp(\mathbb{T} \times \Omega)^3 \times L^q_\perp(\mathbb{T} \times \mathbb{T}_0^2)$ *and* $(u, p, \eta) \in \mathcal{W}^{1,2,q}_{\sigma,\perp}(\mathbb{T} \times \Omega) \times L^q_\perp(\mathbb{T}; \dot{W}^{1,q}(\Omega)) \times W^{2,4,q}_{(0),\perp}(\mathbb{T} \times \mathbb{T}_0^2)$ *be a solution to* (4.6.26). *Then, there exists a constant* $C_{41} = C_{41}(q, \mathcal{T}, L) > 0$ *such that*

$$\|u\|_{W^{1,2,q}_\perp(\mathbb{T} \times \Omega)} + \|\nabla p\|_{L^q_\perp(\mathbb{T} \times \Omega)} + \|\eta\|_{W^{2,4,q}_\perp(\mathbb{T} \times \mathbb{T}_0^2)}$$
$$\leq C_{41}\big(\|f\|_{L^q_\perp(\mathbb{T} \times \Omega)} + \|F\|_{L^q_\perp(\mathbb{T} \times \mathbb{T}_0^2)}\big). \tag{4.6.27}$$

Proof. The L^q estimate (4.6.27) will be established by mimicking the steps in [37, Proof of Theorem 5.1] and employing Lemma 4.6.2. Let $\chi \in C^\infty(\mathbb{R})$ be a cut-off function such that

$$\chi(x_3) = \begin{cases} 1 & \text{if } x_3 \leq \frac{1}{3}, \\ 0 & \text{if } x_3 \geq \frac{2}{3}. \end{cases}$$

To find an appropriate extension of u to the upper half space $\Omega_+ = \mathbb{T}_0^2 \times \mathbb{R}_+$, observe that the components of the velocity field have to be extended differently. For this reason, the even and odd extensions of a function ϕ are defined as

$$\phi^e(t, x) := \begin{cases} \phi(t, x', x_3) & 0 \leq x_3 \leq 1, \\ (1 - \chi(2 - x_3))\phi(t, x', 2 - x_3) & 1 < x_3, \end{cases}$$
$$\phi^o(t, x) := \begin{cases} \phi(t, x', x_3) & 0 \leq x_3 \leq 1, \\ -(1 - \chi(2 - x_3))\phi(t, x', 2 - x_3) & 1 < x_3. \end{cases} \tag{4.6.28}$$

Observe, since η only depends on x_1 and x_2, it suffices to extend u and p. Letting

$$U := (u_1^e, u_2^e, u_3^o) \qquad \text{and} \qquad \mathfrak{P} := p^e,$$

we have extended (u, p, η) to a solution $(U, \mathfrak{P}, \eta)$ to the system

$$
\begin{cases}
\partial_t^2 \eta + \Delta'^2 \eta - \Delta' \partial_t \eta = F - e_3 \cdot \mathrm{T}_0(u, p) & \text{in } \mathbb{T} \times \mathbb{T}_0^2, \\
\partial_t U - \Delta U + \nabla \mathfrak{P} = \tilde{f} & \text{in } \mathbb{T} \times \Omega_+, \\
\operatorname{div} U = \tilde{g} & \text{in } \mathbb{T} \times \Omega_+, \\
U_{|x_3=0} = -\partial_t \eta e_3 & \text{on } \mathbb{T} \times \mathbb{T}_0^2,
\end{cases} \tag{4.6.29}
$$

where

$$
\tilde{f}_j(t, x) = \begin{cases}
f_j(t, x) & \text{if } x_3 < 1, \\
\varepsilon_j \big((1 - \chi(y)) f_j(t, x', y) - h_j(t, x) \big) & \text{if } x_3 > 1,
\end{cases} \tag{4.6.30}
$$

and

$$
\tilde{g}(t, x) = \begin{cases}
0 & \text{if } x_3 < 1, \\
-\chi'(y) \cdot u_3(t, x', y) & \text{if } x_3 > 1.
\end{cases}
$$

Here, the notation $y = 2 - x_3$ was employed. The change of sign in (4.6.30) is indicated by ε_j and stems form the different reflections utilized to define U. That is, ε_j is given by

$$
\varepsilon_j := \begin{cases}
1 & \text{if } j = 1, 2, \\
-1 & \text{if } j = 3.
\end{cases}
$$

Moreover, one may check (by a straightforward calculation) that the further term h_j in (4.6.30) is the j-th component of the vector field

$$
h(t, x) = \chi''(y) u(t, x', y) + 2\partial_{x_3} u(t, x', y) \chi'(y) + \chi'(y) p(t, x', y) e_3.
$$

Note that as $U(t, x', 0) = u(t, x', 0)$ and $\mathfrak{P}(t, x', 0) = p(t, x', 0)$, the boundary conditions at $x_3 = 0$ did not change.

Before exploiting Lemma 4.6.2 to find the desired L^q estimate (4.6.27), we require a further estimate. Due to the identity

$$
\chi' \Delta u_3 = \operatorname{div}[\chi' \nabla u_3] - \partial_{x_3} u_3 \chi'',
$$

we deduce by [36, Theorem III.3.4] and (2.7.7) that

$$
\begin{aligned}
\|\mathcal{B}(\chi' \Delta u_3)\|_{\mathrm{L}_\perp^q(\mathbb{T} \times \Omega)} &\leq \|\mathcal{B}(\operatorname{div}[\chi' \nabla u_3])\|_{\mathrm{L}_\perp^q(\mathbb{T} \times \Omega)} + \|\mathcal{B}(\partial_{x_3} u_3 \chi'')\|_{\mathrm{L}_\perp^q(\mathbb{T} \times \Omega)} \\
&\leq c_0 \big(\|\chi' \nabla u_3\|_{\mathrm{L}_\perp^q(\mathbb{T} \times \Omega)} + \|\partial_{x_3} u_3 \chi''\|_{\mathrm{L}_\perp^q(\mathbb{T} \times \Omega)} \big) \\
&\leq c_1(\chi) \|u\|_{\mathrm{L}_\perp^q(\mathbb{T}; \mathrm{W}^{1,q}(\Omega))}.
\end{aligned}
$$

$$\tag{4.6.31}$$

Letting now $V := \mathcal{B}(\chi' u_3)$, it follows that $\operatorname{supp} V \subset \mathbb{T} \times \mathbb{T}_0^2 \times \left(\frac{1}{3}, \frac{2}{3}\right)$ and (in view of $(4.6.26)_2$) that

$$
\begin{aligned}
\|\partial_t V\|_{\mathrm{L}_\perp^q(\mathbb{T}\times\Omega)} &= c_2 \|\mathcal{B}(\chi'\partial_t u_3)\|_{\mathrm{L}_\perp^q(\mathbb{T}\times\mathbb{T}_0^2\times(\frac{1}{3},\frac{2}{3}))} \\
&\le c_3\big(\|\mathcal{B}(\chi' f_3)\|_{\mathrm{L}_\perp^q(\mathbb{T}\times\mathbb{T}_0^2\times(\frac{1}{3},\frac{2}{3}))} + \|\mathcal{B}(\chi'\Delta u_3)\|_{\mathrm{L}_\perp^q(\mathbb{T}\times\mathbb{T}_0^2\times(\frac{1}{3},\frac{2}{3}))} \\
&\quad + \|\mathcal{B}(\chi'\partial_{x_3} p)\|_{\mathrm{L}_\perp^q(\mathbb{T}\times\mathbb{T}_0^2\times(\frac{1}{3},\frac{2}{3}))}\big) \\
&\le c_4(\chi)\big(\|f\|_{\mathrm{L}_\perp^q(\mathbb{T}\times\Omega)} + \|u\|_{\mathrm{L}_\perp^q(\mathbb{T};\mathrm{W}^{1,q}(\Omega))} + \|\nabla p\|_{\mathrm{L}_\perp^q(\mathbb{T}\times\mathbb{T}_0^2\times(\frac{1}{3},\frac{2}{3}))}\big).
\end{aligned}
$$

Now utilizing the pressure estimate (4.6.19), we obtain that V obeys the L^q estimate

$$
\|V\|_{\mathrm{W}_\perp^{1,q}(\mathbb{T};\mathrm{L}^q(\Omega))} \le c_5\big(\|f\|_{\mathrm{L}_\perp^q(\mathbb{T}\times\Omega)} + \|p\|_{\mathrm{L}_\perp^q(\mathbb{T}\times\Omega)} + \|u\|_{\mathrm{L}_\perp^q(\mathbb{T};\mathrm{W}^{1,q}(\Omega))}\big).
$$

Moreover, a further application of (2.7.7) yields

$$
\|V\|_{\mathrm{L}_\perp^q(\mathbb{T};\mathrm{W}^{2,q}(\Omega))} \le c_6 \|u\|_{\mathrm{L}_\perp^q(\mathbb{T};\mathrm{W}^{1,q}(\Omega))}
$$

In total, we thus obtain

$$
\|V\|_{\mathrm{W}_\perp^{1,2,q}(\mathbb{T}\times\Omega_+)} \le c_7\big(\|f\|_{\mathrm{L}_\perp^q(\mathbb{T}\times\Omega)} + \|u\|_{\mathrm{L}_\perp^q(\mathbb{T};\mathrm{W}^{1,q}(\Omega))} + \|p\|_{\mathrm{L}_\perp^q(\mathbb{T}\times\Omega)}\big). \quad (4.6.32)
$$

By setting

$$
\begin{aligned}
w &: \mathbb{T}\times\Omega_+ \to \mathbb{R}^3, & w &= U - V, \\
\pi &: \mathbb{T}\times\Omega_+ \to \mathbb{R}, & \pi &= \mathfrak{P},
\end{aligned}
$$

we obtain a solution $(w, \pi, \eta) \in \mathcal{W}_{\sigma,\perp}^{1,2,q}(\mathbb{T}\times\Omega_+) \times \mathrm{L}_\perp^q\big(\mathbb{T}; \dot{\mathrm{W}}^{1,q}(\Omega_+)\big) \times \mathrm{W}_{(0),\perp}^{2,4,q}(\mathbb{T}\times\mathbb{T}_0^2)$ to

$$
\begin{cases}
\partial_t^2 \eta + \Delta'^2 \eta - \Delta'\partial_t \eta = F - e_3 \cdot \mathrm{T}_0(w, \pi) & \text{in } \mathbb{T}\times\mathbb{T}_0^2, \\
\partial_t w - \Delta w + \nabla\pi = \tilde{f} - [\partial_t - \Delta]V & \text{in } \mathbb{T}\times\Omega_+, \\
\operatorname{div} w = 0 & \text{in } \mathbb{T}\times\Omega_+, \\
w|_{x_3=0} = -\partial_t\eta e_3 & \text{on } \mathbb{T}\times\mathbb{T}_0^2.
\end{cases}
$$

Moreover, this solution obeys the L^q estimate

$$
\begin{aligned}
\|w\|_{\mathrm{W}_\perp^{1,2,q}(\mathbb{T}\times\Omega_+)} &+ \|\nabla\pi\|_{\mathrm{L}_\perp^q(\mathbb{T}\times\Omega_+)} + \|\eta\|_{\mathrm{W}_\perp^{2,4,q}(\mathbb{T}\times\mathbb{T}_0^2)} \\
&\le C_{42}\big(\|\tilde{f}\|_{\mathrm{L}_\perp^q(\mathbb{T}\times\Omega_+)} + \|V\|_{\mathrm{W}_\perp^{1,2,q}(\mathbb{T}\times\Omega_+)} + \|F\|_{\mathrm{L}_\perp^q(\mathbb{T}\times\mathbb{T}_0^2)}\big),
\end{aligned}
$$

see Lemma 4.6.2. By exploiting (4.6.32), this implies

$$
\begin{aligned}
\|w\|_{\mathrm{W}^{1,2,q}_\perp(\mathbb{T}\times\Omega_+)} &+ \|\nabla\pi\|_{\mathrm{L}^q_\perp(\mathbb{T}\times\Omega_+)} + \|\eta\|_{\mathrm{W}^{2,4,q}_\perp(\mathbb{T}\times\mathbb{T}^2_0)} \\
&\le C_{43}\big(\|\tilde{f}\|_{\mathrm{L}^q_\perp(\mathbb{T}\times\Omega_+)} + \|f\|_{\mathrm{L}^q_\perp(\mathbb{T}\times\Omega)} + \|F\|_{\mathrm{L}^q_\perp(\mathbb{T}\times\mathbb{T}^2_0)} \qquad (4.6.33)\\
&\qquad + \|u\|_{\mathrm{L}^q_\perp(\mathbb{T};\mathrm{W}^{1,q}(\Omega))} + \|p\|_{\mathrm{L}^q_\perp(\mathbb{T}\times\Omega)}\big).
\end{aligned}
$$

In view of (4.6.30), we deduce

$$
\begin{aligned}
\|\tilde{f}\|_{\mathrm{L}^q_\perp(\mathbb{T}\times\Omega_+)} &\le c_8\big(\|f\|_{\mathrm{L}^q_\perp(\mathbb{T}\times\Omega)} + \|\tilde{f}\|_{\mathrm{L}^q_\perp(\mathbb{T}\times(1,\infty))}\big) \\
&\le \big(\|f\|_{\mathrm{L}^q_\perp(\mathbb{T}\times\Omega)} + \|\mathfrak{f}\|_{\mathrm{L}^q_\perp(\mathbb{T}\times\mathbb{T}^2_0\times(1,\infty))} + \|h\|_{\mathrm{L}^q_\perp(\mathbb{T}\times\mathbb{T}^2_0\times(1,\infty))}\big),
\end{aligned}
$$

with

$$
\mathfrak{f}(t,x) = (1 - \chi(2 - x_3))f(t,x',2-x_3)
$$

for all $(t,x) \in \mathbb{T}\times\mathbb{T}^2_0\times(1,\infty)$. Utilizing Hölder's inequality and shifting the coordinates in the second and third norm above, we obtain from the definition of h

$$
\|\tilde{f}\|_{\mathrm{L}^q_\perp(\mathbb{T}\times\Omega_+)} \le c_9(\chi)\big(\|f\|_{\mathrm{L}^q_\perp(\mathbb{T}\times\Omega)} + \|u\|_{\mathrm{L}^q_\perp(\mathbb{T};\mathrm{W}^{1,q}(\Omega))} + \|p\|_{\mathrm{L}^q_\perp(\mathbb{T}\times\Omega)}\big).
$$

Therefore, from $(u,p,\eta) = (w_{|\mathbb{T}\times\Omega}, \pi_{|\mathbb{T}\times\Omega}, \eta) + (V_{|\mathbb{T}\times\Omega},0,0)$, as well as the estimates (4.6.32) and (4.6.33) we conclude

$$
\begin{aligned}
\|u\|_{\mathrm{W}^{1,2,q}_\perp(\mathbb{T}\times\Omega)} &+ \|\nabla p\|_{\mathrm{L}^q_\perp(\mathbb{T}\times\Omega)} + \|\eta\|_{\mathrm{W}^{2,4,q}_\perp(\mathbb{T}\times\mathbb{T}^2_0)} \\
&\le c_{10}\big(\|f\|_{\mathrm{L}^q_\perp(\mathbb{T}\times\Omega)} + \|F\|_{\mathrm{L}^q_\perp(\mathbb{T}\times\mathbb{T}^2_0)} + \|u\|_{\mathrm{L}^q_\perp(\mathbb{T};\mathrm{W}^{1,q}(\Omega))} + \|p\|_{\mathrm{L}^q_\perp(\mathbb{T}\times\Omega)}\big).
\end{aligned}
$$

$$(4.6.34)$$

In order to complete the proof, it remains to omit the final two terms on the right-hand side. For this purpose, first observe that by Young's inequality we obtain

$$
\int_{\mathbb{T}} \big(\|\nabla u\|^{\frac{q-1}{q}}_{\mathrm{L}^q(\Omega)}\|\nabla u(t,\cdot)\|^{\frac{1}{q}}_{\mathrm{W}^{1,q}(\Omega)}\big)^q \, \mathrm{d}t
$$

$$
\le c_{11}(q)\varepsilon^{-\frac{1}{q-1}}\|\nabla u\|^q_{\mathrm{L}^q_\perp(\mathbb{T}\times\Omega)} + \varepsilon\|\nabla u\|^q_{\mathrm{L}^q_\perp(\mathbb{T};\mathrm{W}^{1,q}(\Omega))}
$$

for any $\varepsilon > 0$, and from (4.6.18) we conclude

$$
\begin{aligned}
\|p\|_{\mathrm{L}^q_\perp(\mathbb{T}\times\Omega)} &\le \|p\|_{\mathrm{L}^q_\perp(\mathbb{T};\mathrm{L}^{\frac{3}{2}q}(\Omega))} = \Big(\int_{\mathbb{T}}\|p(t,\cdot)\|^q_{\mathrm{L}^{\frac{3}{2}q}(\Omega)}\,\mathrm{d}t\Big)^{\frac{1}{q}} \\
&\le c_{12}(q,\varepsilon)\big(\|f\|_{\mathrm{L}^q_\perp(\mathbb{T}\times\Omega)} + \|F\|_{\mathrm{L}^q_\perp(\mathbb{T}\times\mathbb{T}^2_0)} + \|\nabla u\|_{\mathrm{L}^q_\perp(\mathbb{T}\times\Omega)} \\
&\qquad + \|\nabla'\Delta'\eta\|_{\mathrm{L}^q_\perp(\mathbb{T}\times\mathbb{T}^2_0)} + \|\nabla'\partial_t\eta\|_{\mathrm{L}^q_\perp(\mathbb{T}\times\mathbb{T}^2_0)}\big) + \varepsilon\|\nabla u\|_{\mathrm{L}^q_\perp(\mathbb{T};\mathrm{W}^{1,q}(\Omega))}.
\end{aligned}
$$

$$(4.6.35)$$

Note that c_{12} grows for $\varepsilon \to 0$. Hence, we deduce from (4.6.34) and (4.6.35) that

$$\|u\|_{W_\perp^{1,2,q}(\mathbb{T}\times\Omega)} + \|\nabla p\|_{L_\perp^q(\mathbb{T}\times\Omega)} + \|\eta\|_{W_\perp^{2,4,q}(\mathbb{T}\times\mathbb{T}_0^2)}$$
$$\leq c_{13}\big(\|f\|_{L_\perp^q(\mathbb{T}\times\Omega)} + \|F\|_{L_\perp^q(\mathbb{T}\times\mathbb{T}_0^2)} + \|u\|_{L_\perp^q(\mathbb{T};W^{1,q}(\Omega))}$$
$$+ \|\nabla'\Delta'\eta\|_{L_\perp^q(\mathbb{T}\times\mathbb{T}_0^2)} + \|\nabla'\partial_t\eta\|_{L_\perp^q(\mathbb{T}\times\mathbb{T}_0^2)}\big) + c_{14}\|\nabla u\|_{L_\perp^q(\mathbb{T};W^{1,q}(\Omega))},$$

where $c_{13} = c_{13}(q,\varepsilon) > 0$ grows as $\varepsilon \to 0$ and $c_{14} = c_{14}(q,\varepsilon) > 0$ tends to zero. Utilizing Ehrling's Lemma, we obtain

$$\|u\|_{L_\perp^q(\mathbb{T};W^{1,q}(\Omega))} \leq c_{15}(\delta)\|u\|_{L^q(\mathbb{T}\times\Omega)} + \delta\|u\|_{L^q(\mathbb{T};W^{2,q}(\Omega))},$$

which leads to

$$\|u\|_{W_\perp^{1,2,q}(\mathbb{T}\times\Omega)} + \|\nabla p\|_{L_\perp^q(\mathbb{T}\times\Omega)} + \|\eta\|_{W_\perp^{2,4,q}(\mathbb{T}\times\mathbb{T}_0^2)}$$
$$\leq c_{16}\big(\|f\|_{L_\perp^q(\mathbb{T}\times\Omega)} + \|F\|_{L_\perp^q(\mathbb{T}\times\mathbb{T}_0^2)} + \|u\|_{L_\perp^q(\mathbb{T}\times\Omega)}$$
$$+ \|\nabla'\Delta'\eta\|_{L_\perp^q(\mathbb{T}\times\mathbb{T}_0^2)} + \|\nabla'\partial_t\eta\|_{L_\perp^q(\mathbb{T}\times\mathbb{T}_0^2)}\big) + c_{17}\|u\|_{L_\perp^q(\mathbb{T};W^{2,q}(\Omega))},$$

with $\delta = \delta(\varepsilon) > 0$, $c_{16} = c_{16}(q,\delta,\varepsilon) > 0$ and $c_{17} = c_{17}(q,\delta,\varepsilon) > 0$. Choosing ε and δ sufficiently small, we finally deduce

$$\|u\|_{W_\perp^{1,2,q}(\mathbb{T}\times\Omega)} + \|\nabla p\|_{L_\perp^q(\mathbb{T}\times\Omega)} + \|\eta\|_{W_\perp^{2,4,q}(\mathbb{T}\times\mathbb{T}_0^2)}$$
$$\leq c_{18}\big(\|f\|_{L_\perp^q(\mathbb{T}\times\Omega)} + \|F\|_{L_\perp^q(\mathbb{T}\times\mathbb{T}_0^2)} + \|u\|_{L_\perp^q(\mathbb{T}\times\Omega)} \qquad (4.6.36)$$
$$+ \|\nabla'\Delta'\eta\|_{L_\perp^q(\mathbb{T}\times\mathbb{T}_0^2)} + \|\nabla'\partial_t\eta\|_{L_\perp^q(\mathbb{T}\times\mathbb{T}_0^2)}\big).$$

To complete the proof, we have to show that the lower order terms on the right-hand side in (4.6.36) can be dropped. For this purpose, observe that the embeddings

$$W_\perp^{1,2,q}(\mathbb{T}\times\Omega) \hookrightarrow L_\perp^q(\mathbb{T}\times\Omega),$$
$$W_\perp^{2,4,q}(\mathbb{T}\times\Omega) \hookrightarrow W_\perp^{1,q}\big(\mathbb{T};W^{1,q}(\Omega)\big),$$
$$W_\perp^{2,4,q}(\mathbb{T}\times\Omega) \hookrightarrow L_\perp^q\big(\mathbb{T};W^{3,q}(\Omega)\big),$$

are compact and that by Lemma 4.3.1 the solution to (4.6.26) with homogeneous right-hand side is necessarily zero. Hence, we can mimic the contradiction argument employed in the proof of Lemma 3.2.9 to conclude (4.6.27) $\qquad\qquad\qquad\square$

Alternatively, we could proceed analogously to the proof of Lemma 4.6.2 and reduce the investigation of (4.6.26) to that of the heat equation in the periodic layer (or the Stokes system) and the fluid-structure problem with homogeneous right-hand side in the Stokes equations. By mimicking the proof of Lemma 4.6.1 we find a representation formula for the solution, and Fourier multiplier theory and the transference principle lead to the desired L^q estimate.

4.7 The Stationary Linear System

In this section we investigate the steady-state part of the linearized fluid-structure system (4.2.4), that is,

$$
\begin{cases}
\Delta'^2 \eta_s = F_s - e_3 \cdot T_0(v_s, \mathfrak{p}_s) & \text{on } \mathbb{T}_0^2, \\
-\Delta v_s + \nabla \mathfrak{p}_s = f_s & \text{in } \Omega, \\
\operatorname{div} v_s = 0 & \text{in } \Omega, \\
v_s = 0 & \text{on } \partial\Omega.
\end{cases}
\tag{4.7.1}
$$

Here, we prove existence of a solution which satisfies some L^q estimates. For this purpose, observe that the steady-state Stokes equations and the Bi-Laplacian can be studied separately, as the coupling only includes $\mathfrak{p}_s$ and η_s. Subject of the following two subsections is the investigation of these two systems. Finally in Subsection 4.7.3 we combine these results to deduce existence of a solution to (4.7.1). The subscript s is omitted in this section for reasons of simplicity.

4.7.1 The Stationary Stokes System

This subsection is dedicated to the investigation of the stationary Stokes system

$$
\begin{cases}
-\Delta u + \nabla p = f & \text{in } \Omega, \\
\operatorname{div} u = 0 & \text{in } \Omega, \\
u = 0 & \text{on } \partial\Omega,
\end{cases}
\tag{4.7.2}
$$

in the periodic layer $\Omega = \mathbb{T}_0^2 \times (0,1)$. We show the following.

Lemma 4.7.1 (The Stationary Stokes Equations)**.** *Let $q \in (1, \infty)$ and $f \in L^q(\Omega)^3$. There exists a solution $(u, p) \in W^{2,q}_\sigma(\Omega) \times \dot{W}^{1,q}(\Omega)$ to (4.7.2). Furthermore, there is a constant $C_{44} = C_{44}(q, L) > 0$ such that*

$$
\|u\|_{W^{2,q}(\Omega)} + \|\nabla p\|_{L^q(\Omega)} \le C_{44} \|f\|_{L^q(\Omega)}.
\tag{4.7.3}
$$

Proof. In order to find a solution to (4.7.2) satisfying (4.7.3), we follow the approach in [1]. There, (4.7.2) was investigated in the infinite layer $\mathbb{R}^{n-1} \times (0,1)$ and the authors showed existence of a solution to the Stokes equations. More precisely, they investigated $u' = (u_1, u_2)$, u_3 and p separately, determined a representation formula for these terms and used Fourier multiplier theory to establish the L^q estimate (4.7.3). For example, u_3 is decomposed into the sum $u_3 = v + w$, with $\widehat{v} = \mathscr{F}_{\mathbb{R}^2}[v]$ a solution to the ordinary differential equation

$$(\partial_3^2 - |\xi'|^2)^2 \widehat{v} = |\xi'|^2 \widehat{f}_3 + i\xi' \cdot \partial_3 \widehat{f'} \qquad \text{in } (0,1),$$

where boundary conditions are neglected, and $\widehat{w} = \mathscr{F}_{\mathbb{R}^2}[w]$ solves

$$\begin{cases} (\partial_3^2 - |\xi'|^2)^2 \widehat{w} = 0 & \text{in } (0,1), \\ \widehat{w}_{|x_3=a} = -\widehat{v}_{|x_3=a}, \\ \partial_{x_3} \widehat{w}_{|x_3=a} = -\partial_{x_3} \widehat{v}_{|x_3=a}, \end{cases}$$

for $a = 0$ and 1. Here $\widehat{f}_3 = \mathscr{F}_{\mathbb{R}^2}[f_3]$ and $\widehat{f'} = \mathscr{F}_{\mathbb{R}^2}[f']$. From these systems a representation formula for v and w is established. As in [1] we are going to distinguish the cases $|\xi'| \leq 2$ and $|\xi'| > 1$. Therefore, let $\phi \in C_0^\infty(\mathbb{R}^2)$ be a cut-off function such that $\phi(\xi') = 1$ for $|\xi'| \leq 1$ and $\phi(\xi') = 0$ for $|\xi'| \geq 2$. By setting

$$v = v_1 + v_2 := (\phi\, v) + (1 - \phi)v$$

a representation formula for v_1 is given by [1, (2.8)]. Hence, combining [1, Lemma 2.3], Marcinkiewicz multiplier theorem (Theorem 2.3.2) and the transference principle (Theorem 2.3.1), we conclude analogously to [1] that

$$\|v_1\|_{2,q} \leq c_0 \|f\|_q \tag{4.7.4}$$

holds. In the case $|\xi'| > 1$, the velocity field v_2 can be written as

$$\widehat{v}_2 = \frac{|\xi'|^2}{|\xi'|^4} \widehat{f}_3^o - \sum_{k=1}^2 \frac{\xi'_k \xi'_3}{|\xi'|^4} \widehat{f}_k^e,$$

and we conclude similarly to the first case with [1, Lemma 2.4] that

$$\|v_2\|_{2,q} \leq c_1 \|f\|_q \tag{4.7.5}$$

holds. Therefore, v satisfies the L^q estimate (4.7.3). Note that by $\widehat{f_3^o}$ and $\widehat{f_k^e}$ we denote the odd and even extensions of $\widehat{f_3}$ and $\widehat{f_k}$ defined in (4.6.28), respectively. For w we proceed analogously and represent it as in [1, (2.15)]. Utilizing [1, Lemma 2.6] and [1, Lemma 2.7] in combination with Theorem 2.3.2 and Theorem 2.3.1, we obtain that w obeys the estimate

$$\|w\|_{2,q} \le c_2 \|f\|_q.$$

Hence, in view of (4.7.4) and (4.7.5) we deduce

$$\|u_3\|_{2,q} \le c_3 \|f\|_q. \tag{4.7.6}$$

To determine existence of u' and p satisfying (4.7.4), we can proceed similarly and obtain from the transference principle and the results from [1, Section 2.4] and [1, Section 2.3], respectively,

$$\|u'\|_{2,q} + \|\nabla p\|_q \le c_4 \|f\|_q.$$

In view of (4.7.6), this completes the proof. □

4.7.2 The Bi-Laplacian

In what follows, we are going to investigate

$$\Delta'^2 \eta = F \qquad \text{in } \mathbb{T}_0^2, \tag{4.7.7}$$

in the periodic whole space $\mathbb{T}_0^2 = (\mathbb{R}/L\mathbb{Z})^2$. Proceeding similarly as for the Stokes system, we show via Fourier multiplier theory existence of a solution to the Bi-Laplacian satisfies the L^q estimate

$$\|\eta\|_{\mathrm{W}^{4,q}(\mathbb{T}_0^2)} \le C_{45} \|F\|_{\mathrm{L}^q(\mathbb{T}_0^2)} \tag{4.7.8}$$

for $q \in (1, \infty)$. For this purpose recall that by $\mathrm{W}_{(0)}^{s,q}(\mathbb{T}_0^2)$ we denote the set

$$\mathrm{W}_{(0)}^{s,q}(\mathbb{T}_0^2) = \{u \in \mathrm{W}^{s,q}(\mathbb{T}_0^2) \mid \langle u, 1 \rangle = 0\},$$

for all $s \in \mathbb{R}$, with $\langle \cdot, \cdot \rangle$ being the L^2 scalar product on $\mathbb{T}_0^2$, see Section 2.1. Observe further that the condition $\langle u, 1 \rangle$ implies that the zeroth Fourier coefficient of u vanishes and vice versa.

Lemma 4.7.2 (The Bi-Laplacian). *Let $q \in (1, \infty)$. For any $F \in \mathrm{L}_{(0)}^q(\mathbb{T}_0^2)$ there exists a solution $\eta \in \mathrm{W}_{(0)}^{4,q}(\mathbb{T}_0^2)$ to (4.7.7) satisfying (4.7.8) with $C_{45} = C_{45}(q, L) > 0$.*

Proof. In what follows we will use $\widehat{\eta}$ and $\widehat{F}$ instead of $\mathscr{F}_{\mathbb{T}_0^2}[\eta]$ and $\mathscr{F}_{\mathbb{T}_0^2}[F]$, respectively. Utilizing the Fourier transform $\mathscr{F}_{\mathbb{T}_0^2}^{-1}$ to the Bi-Laplacian (4.7.7), we observe that

$$|\xi'|^4 \, \widehat{\eta}(\xi') = \widehat{F}(\xi')$$

and conclude that

$$\langle F, 1 \rangle = 0 \tag{4.7.9}$$

is necessary to obtain a solution. Hence, a solution to (4.7.7) is given by

$$\eta := \mathscr{F}_{\mathbb{T}_0^2}^{-1}\left[\frac{1 - \delta_{\frac{2\pi}{L}\mathbb{Z}}(\xi')}{|\xi'|^4} \widehat{F} \right],$$

with $\delta_{\frac{2\pi}{L}\mathbb{Z}}$ the Dirac measure on $(\frac{2\pi}{L}\mathbb{Z})^2$, see (2.3.1). Moreover, observe that the fourth order derivatives of η are given by

$$(\nabla')^4 \eta = \left(\mathscr{F}_{\mathbb{T}_0^2}^{-1}\left[(1 - \delta_{\frac{2\pi}{L}\mathbb{Z}}(\xi')) \frac{\xi_i'\xi_j'\xi_k'\xi_l'}{|\xi'|^4} \widehat{F} \right] \right)_{i,j,k,l=1,2}$$

where $i\frac{\xi_j'}{|\xi'|}$ denotes the symbol of the Riesz transform, see for example [44, Definition 4.1.13]. However, since the Riesz transform is bounded in $L^q(\mathbb{R}^n)$ for any $q \in (1,\infty)$ (see [44, Corollary 4.2.8]), we obtain from the transference principle (Theorem 2.3.1) that the Riesz transform is an $L^q(\mathbb{T}_0^2)$-multiplier. Hence the estimate

$$\|(\nabla')^4\eta\|_q \le c_0\|F\|_q$$

follows. Since η is mean value free, an application of Poincaré's inequality is admissible and yields (4.7.8). $\qquad\square$

4.7.3 The Stationary Linearized Fluid-Structure System

Subject of this final subsection is to combine the results collected in the previous subsections to show existence of a solution $(v, \mathfrak{p}, \eta)$ to

$$\begin{cases} \Delta'^2 \eta = F - e_3 \cdot \mathrm{T}_0(v,\mathfrak{p}) & \text{in } \mathbb{T}_0^2, \\ -\Delta v + \nabla \mathfrak{p} = f & \text{in } \Omega, \\ \operatorname{div} v = 0 & \text{in } \Omega, \\ v = 0 & \text{on } \partial\Omega, \end{cases} \tag{4.7.10}$$

obeying the L^q estimate

$$\|v\|_{W^{2,q}(\Omega)} + \|\nabla\mathfrak{p}\|_{L^q(\Omega)} + \|\eta\|_{W^{4,q}(\mathbb{T}_0^2)} \le C_{46}\big(\|f\|_{L^q(\Omega)} + \|F\|_{L^q(\mathbb{T}_0^2)}\big)$$
(4.7.11)

for all $q \in (1,\infty)$. We will show the following result.

Proposition 4.7.3. *Let $q \in (1,\infty)$ and $(f,F) \in L^q(\Omega)^3 \times L^q(\mathbb{T}_0^2)$. Then there is a solution $(v,\mathfrak{p},\eta) \in W_\sigma^{2,q}(\Omega) \times \dot{W}^{1,q}(\Omega) \times W_{(0)}^{4,q}(\mathbb{T}_0^2)$ to (4.7.10) satisfying (4.7.11) with $C_{46} = C_{46}(q) > 0$.*

Proof. As described at the beginning of this section, (4.7.10) can be studied by investigating the stationary Stokes system and the Bi-Laplacian successively. For this purpose, we employ Lemma 4.7.1 which yields existence of a solution $(v,\mathfrak{p}) \in W_\sigma^{2,q}(\Omega)^3 \times \dot{W}^{1,q}(\Omega)$ to the stationary Stokes equations obeying

$$\|v\|_{W^{2,q}(\Omega)} + \|\nabla\mathfrak{p}\|_{L^q(\Omega)} \le c_0\|f\|_{L^q(\Omega)}.$$
(4.7.12)

Furthermore, we know that $\mathfrak{p}$ is unique up to a constant. Recalling (4.1.7), *i.e.*,

$$\int_{\mathbb{T}_0^2} \phi(y)\,\mathrm{d}y = \frac{1}{L^2} \int_{[0,L]^2} \phi(x')\,\mathrm{d}x',$$

we deduce that $(v,\mathfrak{P})$ with

$$\mathfrak{P} := \mathfrak{p} - \int_{\mathbb{T}_0^2} F\,\mathrm{d}x' - \int_{\mathbb{T}_0^2} \mathfrak{p}_{|x_3=0}\,\mathrm{d}x,$$

is still a solution to the Stokes system

$$-\Delta v + \nabla\mathfrak{P} = f, \qquad \nabla \cdot v = 0 \quad \text{in } \Omega \qquad \text{and} \qquad v_{|\partial\Omega} = 0,$$

and satisfies (4.7.12). Observe further that by this choice of $\mathfrak{P}$ the right-hand side in (4.7.10)$_1$ has a vanishing mean on $\mathbb{T}_0^2$. Thus an application of Lemma 4.7.2 is admissible and yields the existence of a solution $\eta \in W_{(0)}^{4,q}(\mathbb{T}_0^2)$ satisfying

$$\|\eta\|_{W^{4,q}(\mathbb{T}_0^2)} \le \|F + \mathfrak{P}_{|x_3=0}\|_{L^q(\mathbb{T}_0^2)}$$
$$\le c_1\big(\|F\|_{L^q(\mathbb{T}_0^2)} + \|\mathfrak{p}_{(0)|x_3=0}\|_{W^{1-\frac{1}{q},q}(\mathbb{T}_0^2)}\big),$$

with $\mathfrak{p}_{(0)}$ as in (4.5.16). Note that this estimate follows, as $e_3 \cdot T_0(v, \mathfrak{P}) = -\mathfrak{P}_{|x_3=0}$. Utilizing the properties of the trace operator

$$\mathrm{Tr}_0 \colon \mathrm{L}^q\big(\mathbb{T}; \mathrm{W}^{1,q}(\Omega)\big) \to \mathrm{L}^q\big(\mathbb{T}; \mathrm{W}^{1-\frac{1}{q},q}(\mathbb{T}_0^2)\big), \qquad \phi \mapsto \phi_{|x_3=0},$$

stated in [36, Theorem II.4.3], we deduce via Poincaré's inequality that

$$\|\eta\|_{\mathrm{W}^{4,q}(\mathbb{T}_0^2)} \leq c_2\big(\|F\|_{\mathrm{L}^q(\mathbb{T}_0^2)} + \|\mathfrak{p}_{(0)}\|_{\mathrm{W}^{1,q}(\Omega)}\big) \leq c_3\big(\|F\|_{\mathrm{L}^q(\mathbb{T}_0^2)} + \|\nabla \mathfrak{p}\|_{\mathrm{L}^q(\Omega)}\big)$$

which in view of (4.7.12) yields (4.7.11) and completes the proof. $\qquad\square$

4.8 The Linear Fluid-Structure Problem

This section is devoted to the study of the linearized fluid-structure interaction problem (4.2.3) with fully inhomogeneous data, that is

$$\begin{cases} \partial_t^2 \eta + \Delta'^2 \eta - \Delta' \partial_t \eta = f_\eta - e_3 \cdot T_0(u, p) & \text{in } \mathbb{T} \times \mathbb{T}_0^2, \\ \partial_t u - \Delta u + \nabla p = f_u & \text{in } \mathbb{T} \times \Omega, \\ \operatorname{div} u = g & \text{in } \mathbb{T} \times \Omega, \\ u_{|x_3=0} = -\partial_t \eta e_3 & \text{on } \mathbb{T} \times \mathbb{T}_0^2, \\ u_{|x_3=1} = 0 & \text{on } \mathbb{T} \times \mathbb{T}_0^2. \end{cases} \qquad (4.8.1)$$

We are going to show existence of a unique time-periodic solution to this system. Moreover, L^q estimates will be established. As customary in the study of viscous flows, we will utilize the Bogovskiĭ operator $\mathcal{B}$ to eliminate the inhomogeneous divergence in (4.8.1). Consequently, we shall study

$$\begin{cases} \partial_t^2 \eta + \Delta'^2 \eta - \Delta' \partial_t \eta = F - e_3 \cdot T_0(v, \mathfrak{p}) & \text{in } \mathbb{T} \times \mathbb{T}_0^2, \\ \partial_t v - \Delta v + \nabla \mathfrak{p} = f & \text{in } \mathbb{T} \times \Omega, \\ \operatorname{div} v = 0 & \text{in } \mathbb{T} \times \Omega, \\ v_{|x_3=0} = -\partial_t \eta e_3 & \text{on } \mathbb{T} \times \mathbb{T}_0^2, \\ v_{|x_3=1} = 0 & \text{on } \mathbb{T} \times \mathbb{T}_0^2. \end{cases} \qquad (4.8.2)$$

Using the Bogovskiĭ operator to lift the right-hand side g in the divergence equation in (4.8.1), we obtain additional terms on the right-hand side in (4.8.2)$_1$ and (4.8.2)$_2$, which we include in the terms (f, F). The investigation of (4.8.2) is carried out in the following subsection.

4.8.1 The Linearized Fluid-Structure Problem with Homogeneous Divergence

In order to show existence of a time-periodic solution to (4.8.2) that obeys certain L^q estimates, we are going to combine the results from Section 4.5 to Section 4.7. More precisely, we will prove the following existence result:

Proposition 4.8.1. *Let $q \in (1, \infty)$. Then for any $f \in L^q(\mathbb{T} \times \Omega)^3$ and $F \in L^q(\mathbb{T} \times \mathbb{T}_0^2)$ the coupled system (4.8.2) admits a solution*

$$
\begin{aligned}
&v \in \mathcal{W}_\sigma^{1,2,q}(\mathbb{T} \times \Omega)^3, \\
&\mathfrak{p} \in L^q\big(\mathbb{T}; \dot{W}^{1,q}(\Omega)\big), \\
&\eta \in W_{(0)}^{2,4,q}(\mathbb{T} \times \mathbb{T}_0^2)
\end{aligned}
\tag{4.8.3}
$$

such that the estimate

$$
\begin{aligned}
\|v\|_{W^{1,2,q}(\mathbb{T} \times \Omega)} + \|\nabla \mathfrak{p}\|_{L^q(\mathbb{T} \times \Omega)} &+ \|\eta\|_{W^{2,4,q}(\mathbb{T} \times \mathbb{T}_0^2)} \\
&\leq C_{47}\big(\|f\|_{L^q(\mathbb{T} \times \Omega)} + \|F\|_{L^q(\mathbb{T} \times \mathbb{T}_0^2)}\big),
\end{aligned}
\tag{4.8.4}
$$

holds, with $C_{47} = C_{47}(q, \mathcal{T}) > 0$.

Proof. We start by decomposing (4.8.2) into a steady-state problem (4.7.1) and a purely oscillatory system (4.6.1) by setting $(v, \mathfrak{p}, \eta) = (v_s, \mathfrak{p}_s, \eta_s) + (v_{tp}, \mathfrak{p}_{tp}, \eta_{tp})$. Existence of a solution $(v_s, \mathfrak{p}_s, \eta_s)$ in the class (4.8.3) satisfying the L^q estimate (4.8.4) follows from Proposition 4.7.3. Hence, it remains to investigate (4.6.1), that is, (4.8.2) with purely periodic data. To show the existence of a solution hereto, we consider $(f_{tp}, F_{tp}) \in C_{0,\perp}^\infty(\mathbb{T} \times \Omega)^3 \times C_{0,\perp}^\infty(\mathbb{T} \times \mathbb{T}_0^2)$. We express the data (f_{tp}, F_{tp}) as a Fourier series

$$
f_{tp} = \sum_{k \in \frac{2\pi}{\mathcal{T}}\mathbb{Z}} f_k e^{ikt} \qquad \text{and} \qquad F_{tp} = \sum_{k \in \frac{2\pi}{\mathcal{T}}\mathbb{Z}} F_k e^{ikt},
$$

with Fourier coefficients $(f_k, F_k) \in C_0^\infty(\Omega)^3 \times C_0^\infty(\mathbb{T}_0^2)$. Since $\mathcal{P}f_{tp} = 0 = \mathcal{P}F_{tp}$, it follows that $(f_0, F_0) = (0, 0)$. For each $k \in \frac{2\pi}{\mathcal{T}}\mathbb{Z} \setminus \{0\}$ consider system (4.5.1), that is,

$$
\begin{cases}
-k^2 \eta_k + {\Delta'}^2 \eta_k - ik\Delta' \eta_k = F_k - e_3 \cdot \mathrm{T}_0(v_k, \mathfrak{p}_k) & \text{in } \mathbb{T}_0^2, \\
ikv_k - \Delta v_k + \nabla \mathfrak{p}_k = f_k & \text{in } \Omega, \\
\operatorname{div} v_k = 0 & \text{in } \Omega, \\
v_{k|x_3=0} = -ik\eta_k e_3 & \text{on } \mathbb{T}_0^2, \\
v_{k|x_3=1} = 0 & \text{on } \mathbb{T}_0^2.
\end{cases}
\tag{4.8.5}
$$

Existence of a solution $(v_k, \mathfrak{p}_k, \eta_k)$ to this resolvent type problem follows from Proposition 4.5.4 and Lemma 4.5.5 in combination with Remark 4.5.6. Furthermore, Lemma 4.5.5, or more precisely Remark 4.5.6, implies

$$\|v_k\|_{\mathrm{W}^{r+2,2}(\Omega)} + \|\mathfrak{p}_k\|_{\mathrm{W}^{r+1,2}(\Omega)} + \|\eta_k\|_{\mathrm{W}^{r+4,2}(\mathbb{T}_0^2)} \\ \leq c_0 \big(\|f_k\|_{\mathrm{W}^{r,2}(\Omega)} + \|F_k\|_{\mathrm{W}^{r,2}(\mathbb{T}_0^2)} \big) \tag{4.8.6}$$

with c_0 independent of k, and $r \in \mathbb{N}_0$ arbitrary. Observe that a solution to (4.6.1) is given by $(\mathscr{F}_{\mathbb{T}}^{-1}[\mathfrak{p}_k], \mathscr{F}_{\mathbb{T}}^{-1}[v_k], \mathscr{F}_{\mathbb{T}}^{-1}[\eta_k])$ where $(v_k, \mathfrak{p}_k, \eta_k)$ solves (4.8.5). Hence, choosing $r = 0$, we thus obtain by Parseval's theorem that the vector fields

$$v_{\mathrm{tp}} := \sum_{k \in \frac{2\pi}{T}\mathbb{Z}\backslash\{0\}} v_k e^{ikt}, \qquad \mathfrak{p}_{\mathrm{tp}} := \sum_{k \in \frac{2\pi}{T}\mathbb{Z}\backslash\{0\}} \mathfrak{p}_k e^{ikt},$$

$$\eta_{\mathrm{tp}} := \sum_{k \in \frac{2\pi}{T}\mathbb{Z}\backslash\{0\}} \eta_k e^{ikt},$$

are well defined as elements in the spaces $\mathrm{L}_\perp^2\big(\mathbb{T}; \mathcal{W}_\sigma^{2,2}(\Omega)\big)$, $\mathrm{L}_\perp^2\big(\mathbb{T}; \dot{\mathrm{W}}^{1,2}(\Omega)\big)$ and $\mathrm{L}_\perp^2\big(\mathbb{T}; \mathrm{W}_{(0)}^{4,2}(\mathbb{T}_0^2)\big)$, respectively. Since c_0 does not depend on k and the symbol ik coincides with $\mathscr{F}_{\mathbb{T}}[\partial_t]$, we conclude from Parseval's theorem by multiplying (4.8.6) with $(ik)^s e^{ikt}$, for some $s \in \mathbb{N}_0$, and summing up over $k \in \frac{2\pi}{T}\mathbb{Z} \backslash \{0\}$ that

$$\|v_{\mathrm{tp}}\|_{\mathrm{W}_\perp^{s,2}(\mathbb{T};\mathrm{W}^{r+2,2}(\Omega))} + \|\mathfrak{p}_{\mathrm{tp}}\|_{\mathrm{W}_\perp^{s,2}(\mathbb{T};\mathrm{W}^{r+1,2}(\Omega))} + \|\eta_{\mathrm{tp}}\|_{\mathrm{W}_\perp^{s,2}(\mathbb{T};\mathrm{W}^{r+4,2}(\mathbb{T}_0^2))} \\ \leq c_1 \big(\|f_{\mathrm{tp}}\|_{\mathrm{W}_\perp^{s,2}(\mathbb{T};\mathrm{W}^{r,2}(\Omega))} + \|F_{\mathrm{tp}}\|_{\mathrm{W}_\perp^{s,2}(\mathbb{T};\mathrm{W}^{r,2}(\Omega))} \big). \tag{4.8.7}$$

This implies $(v_{\mathrm{tp}}, \mathfrak{p}_{\mathrm{tp}}, \eta_{\mathrm{tp}}) \in \mathrm{W}_\perp^{s,2}\big(\mathbb{T}; \mathrm{W}^{r+2,2}(\Omega)\big) \times \mathrm{W}_\perp^{s,2}\big(\mathbb{T}; \mathrm{W}^{r+1,2}(\Omega)\big) \times \mathrm{W}_\perp^{s,2}\big(\mathbb{T}; \mathrm{W}^{r+4,2}(\Omega)\big)$. Hence, by choosing r and s sufficiently large and utilizing Sobolev's Embedding Theorem (see for example [2, 4.12 Theorem]) successively in time and space, we obtain that the solution $(v_{\mathrm{tp}}, \mathfrak{p}_{\mathrm{tp}}, \eta_{\mathrm{tp}})$ satisfies (4.8.3) for any $q \in (1, \infty)$. Due to (4.8.7), this holds also for $(f_{\mathrm{tp}}, F_{\mathrm{tp}}) \in \mathrm{L}_\perp^q(\mathbb{T} \times \Omega)^3 \times \mathrm{L}_\perp^q(\mathbb{T} \times \mathbb{T}_0^2)$ by a density argument. Moreover, Lemma 4.6.4 yields the validity of (4.8.4), and since $(v, \mathfrak{p}, \eta) = (v_\mathrm{s}, \mathfrak{p}_\mathrm{s}, \eta_\mathrm{s}) + (v_{\mathrm{tp}}, \mathfrak{p}_{\mathrm{tp}}, \eta_{\mathrm{tp}})$, this completes the proof. $\square$

4.8.2 The Fully Inhomogeneous Fluid-Structure System

At this point we have collected all information necessary to prove existence of a unique solution to the coupled fluid-structure system (4.8.1) with fully inhomogeneous right-hand side. This subsection is dedicated to the investigation of this problem in the periodic layer $\Omega = \mathbb{T}_0^2 \times (0,1)$. Moreover, L^q estimates shall be established. The main Theorem in the linear case then reads as follow:

Theorem 4.8.2. *Let $q \in (1, \infty)$. For all*

$$f_u \in \mathrm{L}^q(\mathbb{T} \times \Omega)^3,$$
$$g \in \mathrm{L}^q\big(\mathbb{T}; \mathrm{W}_{(0)}^{1,q}(\Omega)\big) \cap \mathrm{W}^{1,q}\big(\mathbb{T}; \dot{\mathrm{W}}^{-1,q}(\Omega)\big),$$
$$f_\eta \in \mathrm{L}^q(\mathbb{T} \times \mathbb{T}_0^2),$$

there is a unique solution (u, p, η) to (4.8.1) with

$$
\begin{aligned}
u &\in \mathrm{W}^{1,2,q}(\mathbb{T} \times \Omega)^3, \\
p &\in \mathrm{L}^q\big(\mathbb{T}; \dot{\mathrm{W}}^{1,q}(\Omega)\big), \\
\eta &\in \mathrm{W}_{(0)}^{2,4,q}(\mathbb{T} \times \mathbb{T}_0^2),
\end{aligned}
\tag{4.8.8}
$$

which satisfies the L^q estimate

$$
\begin{aligned}
&\|u\|_{\mathrm{W}^{1,2,q}(\mathbb{T} \times \Omega)} + \|\nabla p\|_{\mathrm{L}^q(\mathbb{T} \times \Omega)} + \|\eta\|_{\mathrm{W}^{2,4,q}(\mathbb{T} \times \mathbb{T}_0^2)} \\
&\quad \leq C_{48}\big(\|f_\eta\|_{\mathrm{L}^q(\mathbb{T} \times \mathbb{T}_0^2)} + \|f_u\|_{\mathrm{L}^q(\mathbb{T} \times \Omega)} + \|g\|_{\mathrm{L}^q(\mathbb{T}; \mathrm{W}^{1,q}(\Omega)) \cap \mathrm{W}^{1,q}(\mathbb{T}; \dot{\mathrm{W}}^{-1,q}(\Omega))}\big),
\end{aligned}
\tag{4.8.9}
$$

with $C_{48} = C_{48}(q, \mathcal{T}) > 0$.

Proof. As described at the beginning of this section, we first consider for every $t \in \mathbb{T}$ the divergence problem

$$
\begin{cases}
\operatorname{div} w = g & \text{in } \mathbb{T} \times \Omega, \\
\quad\;\; w = 0 & \text{on } \mathbb{T} \times \partial\Omega.
\end{cases}
\tag{4.8.10}
$$

By assumption

$$\int_\Omega g \, \mathrm{d}x - 0.$$

Hence, a solution to (4.8.10) is given by $w := \mathcal{B}(g)$ (see Theorem 2.7.4). Moreover, we deduce from Theorem 2.7.4 that $w(t, \cdot)$ obeys for a.e. $t \in \mathbb{T}$ the L^q estimate (2.7.7) with $l = 1$, that is,

$$\|w(t, \cdot)\|_{\mathrm{W}^{2,q}(\Omega)} \leq c_0 \|g(t, \cdot)\|_{\mathrm{W}^{1,q}(\Omega)},$$

and (2.7.5), *i.e.*,

$$\|w(t,\cdot)\|_{\mathrm{L}^q(\Omega)} \le c_1 |g(t,\cdot)|^*_{-1,q},$$

with $|\cdot|^*_{-1,q}$ defined as in (2.7.6). Introducing the vector-field $v := u - w$, the investigation of (4.8.1) reduces to that of the coupled system (4.8.2) with $f = f_u - (\partial_t - \Delta)w$ and $F = f_\eta - \partial_{x_3} w_3|_{x_3=0} = f_\eta - g_{|x_3=0}$ as $\partial_{x_1} w_1|_{x_3=0} = 0 = \partial_{x_2} w_2|_{x_3=0}$, and thus

$$\partial_{x_3} w_3|_{x_3=0} = \operatorname{div} w_{|x_3=0} = g_{|x_3=0}.$$

A solution hereto satisfying

$$\begin{aligned}\|v\|_{\mathrm{W}^{1,2,q}(\mathbb{T}\times\Omega)} + \|\nabla p\|_{\mathrm{L}^q(\mathbb{T}\times\Omega)} + \|\eta\|_{\mathrm{W}^{2,4,q}(\mathbb{T}\times\mathbb{T}_0^2)} \\ \le C_{49}\big(\|f\|_{\mathrm{L}^q(\mathbb{T}\times\Omega)} + \|F\|_{\mathrm{L}^q(\mathbb{T}\times\mathbb{T}_0^2)}\big),\end{aligned} \tag{4.8.11}$$

follows from Proposition 4.8.1. One may verify that the identity $\partial_t w = \mathcal{B}(\partial_t g)$ holds. Therefore, the right-hand side in (4.8.11) can be further estimated as follow:

$$\begin{aligned}\|f\|_{\mathrm{L}^q(\mathbb{T}\times\Omega)} &\le c_2\big(\|f_u\|_{\mathrm{L}^q(\mathbb{T}\times\Omega)} + \|\mathcal{B}(\partial_t g)\|_{\mathrm{L}^q(\mathbb{T}\times\Omega)} + \|w\|_{\mathrm{L}^q(\mathbb{T};\mathrm{W}^{2,q}(\Omega))}\big) \\ &\le c_3\bigg(\|f_u\|_{\mathrm{L}^q(\mathbb{T}\times\Omega)} + \bigg(\int_{\mathbb{T}} \big(|\partial_t g|^*_{-1,q}\big)^q \,\mathrm{d}t\bigg)^{\frac{1}{q}} + \bigg(\int_{\mathbb{T}} \|g\|^q_{\mathrm{W}^{1,q}(\Omega)} \,\mathrm{d}t\bigg)^{\frac{1}{q}}\bigg) \\ &= c_3\big(\|f_u\|_{\mathrm{L}^q(\mathbb{T}\times\Omega)} + \|g\|_{\mathrm{L}^q(\mathbb{T};\mathrm{W}^{1,q}(\Omega))\cap\mathrm{W}^{1,q}(\mathbb{T};\dot{\mathrm{W}}^{-1,q}(\Omega))}\big)\end{aligned}$$

and

$$\begin{aligned}\|F\|_{\mathrm{L}^q(\mathbb{T}\times\mathbb{T}_0^2)} &\le c_4\big(\|f_\eta\|_{\mathrm{L}^q(\mathbb{T}\times\mathbb{T}_0^2)} + \|g_{|x_3=0}\|_{\mathrm{L}^q(\mathbb{T}\times\mathbb{T}_0^2)}\big) \\ &\le c_5\big(\|f_\eta\|_{\mathrm{L}^q(\mathbb{T}\times\mathbb{T}_0^2)} + \|g_{|x_3=0}\|_{\mathrm{L}^q(\mathbb{T};\mathrm{W}^{1-\frac{1}{q},q}(\mathbb{T}_0^2))}\big) \\ &\le c_6\big(\|f_\eta\|_{\mathrm{L}^q(\mathbb{T}\times\mathbb{T}_0^2)} + \|g\|_{\mathrm{L}^q(\mathbb{T};\mathrm{W}^{1,q}(\Omega))}\big),\end{aligned}$$

where the last inequality follows from utilizing the trace operator

$$\mathrm{Tr}_0\colon \mathrm{L}^q\big(\mathbb{T}; \mathrm{W}^{1,q}(\Omega)\big) \to \mathrm{L}^q\big(\mathbb{T}; \mathrm{W}^{1-\frac{1}{q},q}(\mathbb{T}_0^2)\big), \qquad \phi \mapsto \phi_{|x_3=0},$$

stated in [36, Theorem II.4.3], as in the proof of Proposition 4.7.3. Hence, (v,p,η) obeys (4.8.9). Setting $u = v + w$, we obtain a solution to (4.8.2) satisfying (4.8.9). The uniqueness assertion follows from Lemma 4.3.1. $\qquad\square$

4.9 The Nonlinear Problem

Subject of the final section of this thesis is the investigation of the coupled fluid-structure system (4.1.6) on the moving domain Ω_η, that is,

$$
\begin{cases}
\partial_t^2 \eta + \Delta'^2 \eta - \Delta' \partial_t \eta = f_\eta - \mathrm{T}_\eta & \text{in } \mathbb{T} \times \mathbb{T}_0^2, \\
\partial_t u - \Delta u + (u \cdot \nabla)u + \nabla p = f_u & \text{in } \mathbb{T} \times \Omega_\eta^{\mathbb{T}}, \\
\operatorname{div} u = 0 & \text{in } \mathbb{T} \times \Omega_\eta^{\mathbb{T}}, \\
u(t, x', -\eta(t, x')) = -\partial_t \eta(t, x')e_3 & \text{on } \mathbb{T} \times \mathbb{T}_0^2, \\
u(t, x', 1) = 0 & \text{on } \mathbb{T} \times \mathbb{T}_0^2,
\end{cases}
\tag{4.9.1}
$$

where we have set without loss of generality $\mu_s = 1 = \mu_f$. Via the transformation ϕ_η defined in (4.1.2), *i.e.*,

$$
\phi_\eta \colon \mathbb{T} \times \Omega \to \mathbb{T} \times \Omega_\eta, \qquad (t, \tilde{x}) \mapsto (t, x', x_3 - (1 - x_3)\eta(t, x')),
$$

we have already seen that a solution to (4.9.1) is given by $u = \tilde{u} \circ \phi_\eta^{-1}$, $p = \tilde{p} \circ \phi_\eta^{-1}$ and η, where $(\tilde{u}, \tilde{p}, \eta)$ solves the nonlinear system

$$
\begin{cases}
\partial_t^2 \eta + \Delta'^2 \eta - \Delta' \partial_t \eta = f_\eta - \mathrm{T}_\eta & \text{in } \mathbb{T} \times \mathbb{T}_0^2, \\
\partial_t \tilde{u} - \Delta \tilde{u} + (\tilde{u} \cdot \nabla)\tilde{u} + \nabla \tilde{p} = \tilde{f}_u + \mathrm{R}_f & \text{in } \mathbb{T} \times \Omega, \\
\operatorname{div} \tilde{u} = \tilde{\mathrm{R}}_d & \text{in } \mathbb{T} \times \Omega, \\
\tilde{u}(t, x', 0) = -\partial_t \eta(t, x')e_3 & \text{on } \mathbb{T} \times \mathbb{T}_0^2, \\
\tilde{u}(t, x', 1) = 0 & \text{on } \mathbb{T} \times \mathbb{T}_0^2,
\end{cases}
\tag{4.9.2}
$$

see Section 4.2. Hence, it suffices to study (4.9.2) on the reference configuration $\Omega = \mathbb{T}_0^2 \times (0, 1)$ instead of the free boundary problem (4.9.1) on the moving domain $\Omega_\eta = \mathbb{T}_0^2 \times (-\eta, 1)$, where a fixed-point argument is not applicable, as the boundary depends on η. By employing ϕ_η we find an equivalent formulation on a reference configuration, but by proceeding in the described way we obtain the further nonlinear terms R_f and $\tilde{\mathrm{R}}_d$ on the right-hand side in (4.9.2), where derivatives of the inverse transformation

$$
\phi_\eta^{-1} \colon \mathbb{T} \times \Omega_\eta \to \mathbb{T} \times \Omega, \qquad (t, \tilde{x}) \mapsto \left(t, x', \frac{x_3 + \eta(t, x')}{1 + \eta(t, x')}\right)
$$

occur. Therefore, observe that the gradient of the inverse transformation is given by

$$
\nabla(\phi_\eta^{-1}) = \mathrm{I}_3 + \mathrm{E},
$$

with $I_3 \in \mathbb{R}^{3 \times 3}$ the unit matrix and E defined as

$$E := \begin{pmatrix} 0 & 0 & 0 \\ 0 & 0 & 0 \\ \frac{(1-x_3)\partial_{x_1}\eta}{(1+\eta)^2} & \frac{(1-x_3)\partial_{x_2}\eta}{(1+\eta)^2} & -\frac{\eta}{1+\eta} \end{pmatrix}.$$

Hence, we deduce

$$\nabla u = \nabla(\tilde{u} \circ \phi_\eta^{-1}) = \nabla \tilde{u} \circ \phi_\eta^{-1} + (\nabla \tilde{u} \circ \phi_\eta^{-1})E = (\nabla \tilde{u} \circ \phi_\eta^{-1})(I_3 + E),$$

and thus

$$\nabla u \circ \phi_\eta = \nabla \tilde{u}(I_3 + E \circ \phi_\eta).$$

Setting

$$\rho(x_3) := 1 - x_3 \qquad \text{and} \qquad \rho \circ \phi_\eta = (1+\eta)(1-x_3),$$

we obtain similarly

$$\partial_t u = \partial_t \tilde{u} \circ \phi_\eta^{-1} + (\partial_{x_3} \tilde{u} \circ \phi_\eta^{-1}) \frac{\rho \, \partial_t \eta}{(1+\eta)^2}$$

which further implies

$$\partial_t u \circ \phi_\eta = \partial_t \tilde{u} + \partial_{x_3} \tilde{u} \frac{\rho \, \partial_t \eta}{1+\eta}.$$

Recalling Section 4.1, the nonlinear term T_η is given by

$$T_\eta = e_3 \cdot \left(T(\tilde{u}, \tilde{p})_{|x_3=0} + S_\eta \right) \tilde{\nu}_t,$$

with normal vector defined as

$$\nu_t = \frac{1}{\sqrt{1 + |\nabla'\eta|^2}} \begin{pmatrix} \nabla'\eta \\ -1 \end{pmatrix} \qquad \text{and} \qquad \tilde{\nu}_t := \nu_t \circ \phi_\eta.$$

A solution to (4.9.2) shall be established by utilizing the contraction mapping principle and the L^q estimates deduced for the solution to the corresponding linearization in the previous section. For this reason, we write (4.9.2) as follows

$$\begin{cases} \partial_t^2 \eta + \Delta'^2 \eta + \Delta' \partial_t \eta = f_\eta - e_3 \cdot T_0(\tilde{u}, \tilde{p}) + R_\eta & \text{in } \mathbb{T} \times \mathbb{T}_0^2, \\ \partial_t \tilde{u} - \Delta \tilde{u} + \nabla \tilde{p} = \tilde{f}_u + \tilde{R}_f & \text{in } \mathbb{T} \times \Omega, \\ \text{div } \tilde{u} = \tilde{R}_d & \text{in } \mathbb{T} \times \Omega, \qquad (4.9.3) \\ \tilde{u}(t, x', 0) = -\partial_t \eta(t, x') e_3 & \text{on } \mathbb{T} \times \mathbb{T}_0^2, \\ \tilde{u}(t, x', 1) = 0 & \text{on } \mathbb{T} \times \mathbb{T}_0^2, \end{cases}$$

with right-hand side

$$\tilde{f}_u := f_u \circ \phi_\eta,$$
$$\tilde{R}_f(\tilde{u}, \tilde{p}, \eta) := R_f - (\tilde{u} \cdot \nabla)\tilde{u},$$
$$\tilde{R}_d(\tilde{u}, \eta) := \operatorname{div} R_d,$$
$$R_\eta(\tilde{u}, \tilde{p}, \eta) := -e_3 \cdot \left[(T(\tilde{u}, \tilde{p})_{|x_3=0}(\tilde{\nu}_t - e_3)) + S_\eta \tilde{\nu}_t \right],$$

and R_f, R_d and S_η given by

$$R_f(\tilde{u}, \tilde{p}, \eta) := \partial_{x_3}\tilde{u} \frac{\rho \partial_t \eta}{1 + \eta} + \sum_{j,k,l=1}^{3} \partial_{x_l} \partial_{x_k} \tilde{u}(E_{k,l} + E_{l,k} + E_{j,k}E_{j,l}) \circ \phi_\eta$$

$$+ \sum_{j,k=1}^{3} \partial_{x_k}\tilde{u}(\partial_{x_j}E_{j,k}) \circ \phi_\eta - \frac{\partial_{x_3}\tilde{p}}{1 + \eta} \begin{pmatrix} \rho\nabla'\eta \\ -\eta \end{pmatrix} + \nabla\tilde{u}(E \circ \phi_\eta)\tilde{u},$$

$$R_d(\tilde{u}) := - \begin{pmatrix} \eta\tilde{u}' \\ \rho\nabla'\eta \cdot \tilde{u}' \end{pmatrix},$$

$$S_\eta(\tilde{u}, \eta) := \left((\nabla\tilde{u}(E \circ \phi_\eta) + (\nabla\tilde{u}(E \circ \phi_\eta))^\top) \right).$$

Due to the boundary condition $(4.9.3)_4$, we conclude

$$\int_\Omega \tilde{R}_d \, \mathrm{d}x = \int_\Omega \operatorname{div} R_d \, \mathrm{d}x = \int_{\mathbb{T} \times \mathbb{T}_0^2} \nabla'\eta \cdot \tilde{u}'_{|x_3=0} \, \mathrm{d}x' = 0,$$

that is, $\tilde{R}_d$ has a vanishing mean on Ω.

Note that $E_{j,l}$ vanish if $j \neq 3$. Hence, the representation formula above can be simplified. Since the matrix E, more precise $E \circ \phi_\eta$, is investigated as a whole and not component-wise in what follows, we keep the additional zeros in the L^q norm. As long as no confusions may arise, we will omit the tilde and write u instead of $\tilde{u}$. In order to find appropriate L^q estimates for these nonlinear terms, we have to deduce estimate for the matrix E, especially for $E \circ \phi_\eta$. For this purpose, observe that the coefficients of $E \circ \phi_\eta$ are basically η or $\nabla'\eta$. Hence, we obtain the following result on boundedness of $E \circ \phi_\eta$:

Lemma 4.9.1. *Let $q \in (1, \infty)$. Then there exists an $\varepsilon_0 > 0$ and a constant $C_{50} > 0$ such that if η satisfies*

$$\|\eta\|_{W^{2,4,q}(\mathbb{T} \times \mathbb{T}_0^2)} \leq \varepsilon_0, \tag{4.9.4}$$

we get $E \circ \phi_\eta \in L^q(\mathbb{T} \times \Omega)$ and the estimate

$$\|E \circ \phi_\eta\|_{L^q(\mathbb{T} \times \Omega)} \leq C_{50} \|\eta\|_{L^q(\mathbb{T}; W^{1,q}(\mathbb{T}_0^2))}.$$

Proof. First of all note that by comparing E and ϕ_η, we obtain that

$$\mathrm{E}_{i,j} \circ \phi_\eta := \frac{1}{1+\eta} \begin{cases} 0 & \text{if } i \neq 3, \\ \rho \partial_{x_j} \eta & \text{if } i = 3 \text{ and } j \neq 3, \\ -\eta & \text{if } i = 3 = j. \end{cases}$$

Due to this observation we now have

$$\|\mathrm{E} \circ \phi_\eta\|_{\mathrm{L}^q(\mathbb{T}\times\Omega)} \leq \left\| \frac{\rho\nabla'\eta}{1+\eta} \right\|_{\mathrm{L}^q(\mathbb{T}\times\Omega)} + \left\| \frac{\eta}{1+\eta} \right\|_{\mathrm{L}^q(\mathbb{T}\times\Omega)}.$$

Utilizing the Soboloev embedding theorem (Theorem 2.6.3) from Section 2.6 with $m = 2$, $m_x = 0 = M_t$ and $\alpha = 2$, we find for any $q > 1$ that

$$\|\eta\|_{\mathrm{L}^\infty(\mathbb{T}\times\mathbb{T}_0^2)} \leq c_0 \|\eta\|_{\mathrm{W}^{2,4,q}(\mathbb{T}\times\mathbb{T}_0^2)},$$

and therefore

$$\left\| \frac{1}{1+\eta} \right\|_{\mathrm{L}^\infty(\mathbb{T}\times\mathbb{T}_0^2)} \leq \frac{1}{1 - \|\eta\|_{\mathrm{L}^\infty(\mathbb{T}\times\mathbb{T}_0^2)}} \leq \frac{1}{1 - c_0\|\eta\|_{\mathrm{W}^{2,4,q}(\mathbb{T}\times\mathbb{T}_0^2)}} \tag{4.9.5}$$
$$\leq \frac{1}{1 - c_0\varepsilon_0}.$$

Hence, choosing $\varepsilon_0 = \frac{1}{2c_0}$, we deduce

$$\left\| \frac{1}{1+\eta} \right\|_{\mathrm{L}^\infty(\mathbb{T}\times\mathbb{T}_0^2)} \leq 2 < \infty, \tag{4.9.6}$$

and as $x_3 \in (0,1)$, this implies that

$$\|\mathrm{E} \circ \phi_\eta\|_{\mathrm{L}^q(\mathbb{T}\times\Omega)} \leq c_1 \|\eta\|_{\mathrm{L}^q(\mathbb{T};\mathrm{W}^{1,q}(\mathbb{T}_0^2))},$$

and therefore the boundedness of $\mathrm{E} \circ \phi_\eta$. $\qquad\square$

The next step towards our goal is to find appropriate L^q estimates for the nonlinear terms R_η, $\tilde{\mathrm{R}}_f$ and $\tilde{\mathrm{R}}_d$ in the corresponding L^q norms. For this purpose, we will again employ the Sobolev embeddings stated in Theorem 2.6.3. In contrast to our approach in the case of nonlinear acoustics (see Section 3.3 and Section 3.4), we will not split the system, but examine the stationary and purely oscillatory problem in one step. For the nonlinear terms we find the following L^q estimates.

Lemma 4.9.2 (L^q Estimates). *Let $\varepsilon_0 > 0$ and $q \in (2, \infty)$. Then for any $u \in \mathrm{W}^{1,2,q}(\mathbb{T} \times \Omega)^3$, $p \in \mathrm{L}^q\big(\mathbb{T}; \dot{\mathrm{W}}^{1,q}(\Omega)\big)$ and $\eta \in \mathrm{W}^{2,4,q}(\mathbb{T} \times \mathbb{T}_0^2)$ satisfying (4.9.4), the nonlinear terms obey the L^q estimates*

$$\|\tilde{\mathrm{R}}_f\|_{\mathrm{L}^q(\mathbb{T}\times\Omega)} \leq C_{51}\big((1+\varepsilon_0)\|u\|_{\mathrm{W}^{1,2,q}(\mathbb{T}\times\Omega)} + \|\nabla p\|_{\mathrm{L}^q(\mathbb{T}\times\Omega)}$$
$$+ \|u\|_{\mathrm{W}^{1,2,q}(\mathbb{T}\times\Omega)}^2\big)\|\eta\|_{\mathrm{W}^{2,4,q}(\mathbb{T}\times\mathbb{T}_0^2)}, \tag{4.9.7}$$

$$\|\tilde{\mathrm{R}}_d\|_{\mathrm{L}^q(\mathbb{T};\mathrm{W}^{1,q}(\Omega))\cap\mathrm{W}^{1,q}(\mathbb{T};\dot{\mathrm{W}}^{-1,q}(\Omega))} \leq C_{52}\|u\|_{\mathrm{W}^{1,2,q}(\mathbb{T}\times\Omega)}\|\eta\|_{\mathrm{W}^{2,4,q}(\mathbb{T}\times\mathbb{T}_0^2)}, \tag{4.9.8}$$

$$\|\mathrm{R}_\eta\|_{\mathrm{L}^q(\mathbb{T}\times\mathbb{T}_0^2)} \leq C_{53}(1+\varepsilon_0)\big(\|\eta\|_{\mathrm{W}^{2,4,q}(\mathbb{T}\times\mathbb{T}_0^2)}\|u\|_{\mathrm{W}^{1,2,q}(\mathbb{T}\times\Omega)}$$
$$+ \|u\|_{\mathrm{W}^{1,2,q}(\mathbb{T}\times\Omega)} + \|p\|_{\mathrm{L}^q(\mathbb{T};\mathrm{W}^{1,q}(\Omega))}\big). \tag{4.9.9}$$

Proof. We begin by investigating $\tilde{\mathrm{R}}_f$ and observe that by Hölder's inequality and (4.9.6) we determine

$$\|\tilde{\mathrm{R}}_f\|_{\mathrm{L}^q(\mathbb{T}\times\Omega)} \leq c_0\Big(\|\nabla^2 u : \mathrm{E} \circ \phi_\eta\|_{\mathrm{L}^q(\mathbb{T}\times\Omega)} + \|\nabla^2 u : (\mathrm{E} \circ \phi_\eta)^2\|_{\mathrm{L}^q(\mathbb{T}\times\Omega)}$$
$$+ \|\nabla p \nabla\eta\|_{\mathrm{L}^q(\mathbb{T}\times\Omega)} + \|\nabla p\, \eta\|_{\mathrm{L}^q(\mathbb{T}\times\Omega)} + \Big\|\sum_{j,k=1}^3 \partial_{x_k} u(\partial_{x_j}\mathrm{E}_{j,k}) \circ \phi_\eta\Big\|_{\mathrm{L}^q(\mathbb{T}\times\Omega)}$$
$$+ \|\nabla u(\mathrm{E} \circ \phi_\eta)u\|_{\mathrm{L}^q(\mathbb{T}\times\Omega)} + \|(u \cdot \nabla)u\|_{\mathrm{L}^q(\mathbb{T}\times\Omega)} + \|\nabla u\, \partial_t\eta\|_{\mathrm{L}^q(\mathbb{T}\times\Omega)}\Big). \tag{4.9.10}$$

Recall that by $\nabla\eta$ we denote the vector $(\nabla'\eta, 0)$. Utilizing Lemma 4.9.1 and Theorem 2.6.3 with $m = 2$, $M_t = 0$, $M_x = 1$ and $\alpha = \frac{3}{2}$ ($n = 2$), we obtain for $q \in (2, \infty)$ the L^q estimate

$$\|\nabla^2 u : \mathrm{E} \circ \phi_\eta\|_{\mathrm{L}^q(\mathbb{T}\times\Omega)} + \|\nabla^2 u : (\mathrm{E} \circ \phi_\eta)^2\|_{\mathrm{L}^q(\mathbb{T}\times\Omega)}$$
$$\leq c_1(1 + \|\eta\|_{\mathrm{L}^\infty(\mathbb{T};\mathrm{W}^{1,\infty}(\mathbb{T}_0^2))})\|\nabla^2 u\|_{\mathrm{L}^q(\mathbb{T}\times\Omega)}\|\eta\|_{\mathrm{L}^\infty(\mathbb{T};\mathrm{W}^{1,\infty}(\mathbb{T}_0^2))} \tag{4.9.11}$$
$$\leq c_2(1 + \|\eta\|_{\mathrm{W}^{2,4,q}(\mathbb{T}\times\mathbb{T}_0^2)})\|u\|_{\mathrm{W}^{1,2,q}(\mathbb{T}\times\Omega)}\|\eta\|_{\mathrm{W}^{2,4,q}(\mathbb{T}\times\mathbb{T}_0^2)}$$

and

$$\|\nabla p\, \nabla\eta\|_{\mathrm{L}^q(\mathbb{T}\times\Omega)} + \|\nabla u(\mathrm{E} \circ \phi_\eta)u\|_{\mathrm{L}^q(\mathbb{T}\times\Omega)} + \|\nabla p\, \eta\|_{\mathrm{L}^q(\mathbb{T}\times\Omega)}$$
$$\leq c_3\big(\|\nabla p\|_{\mathrm{L}^q(\mathbb{T}\times\Omega)} + \|(u \cdot \nabla)u\|_{\mathrm{L}^q(\mathbb{T}\times\Omega)}\big)\|\nabla\eta\|_{\mathrm{L}^\infty(\mathbb{T};\mathrm{W}^{1,\infty}(\mathbb{T}_0^2))}$$
$$\leq c_4\big(\|p\|_{\mathrm{L}^q(\mathbb{T};\mathrm{W}^{1,q}(\Omega))} + \|\nabla u\|_{\mathrm{L}^\infty(\mathbb{T};\mathrm{L}^q(\Omega))}\|u\|_{\mathrm{L}^q(\mathbb{T};\mathrm{L}^\infty(\Omega))}\big)\|\eta\|_{\mathrm{W}^{2,4,q}(\mathbb{T}\times\mathbb{T}_0^2)}. \tag{4.9.12}$$

To see that the L^∞-norms occurring on the right-hand side above are finite, employ the Sobolev embedding theorem with $m = 1$, $M_t = 0$, $M_x = 1$ and $\alpha = 1$, as well as with $m = 1$, $M_t = 0 = M_x$ and $\alpha = 0$ to get

$$\|\nabla u\|_{L^\infty(\mathbb{T};L^q(\Omega))} + \|u\|_{L^q(\mathbb{T};L^\infty(\Omega))} \leq c_5 \|u\|_{W^{1,2,q}(\mathbb{T}\times\Omega)}. \tag{4.9.13}$$

Furthermore, for η and $\partial_t\eta$ we find for $q \in (2, \infty)$ the L^q estimates

$$\|\eta\|_{L^\infty(\mathbb{T}\times\mathbb{T}_0^2)} + \|\partial_t\eta\|_{L^\infty(\mathbb{T}\times\mathbb{T}_0^2)} \leq c_6 \|\eta\|_{W^{2,4,q}(\mathbb{T}\times\mathbb{T}_0^2)} \tag{4.9.14}$$

via an application of Theorem 2.6.3 with $m = 2$, $M_t = 0 = M_x$ and $\alpha = 2$, and $m = 2$, $M_t = 1$, $M_x = 0$ and $\alpha = 1$, respectively. This implies

$$\|\nabla u\, \partial_t\eta\|_{L^q(\mathbb{T}\times\Omega)} \leq c_7 \|u\|_{W^{1,2,q}(\mathbb{T}\times\Omega)} \|\eta\|_{W^{2,4,q}(\mathbb{T}\times\mathbb{T}_0^2)}. \tag{4.9.15}$$

Note that the remaining term in (4.9.10) can be estimated as follow:

$$\left\|\sum_{j,k=1}^{3} \partial_{x_k} u (\partial_{x_j} \mathrm{E}_{j,k}) \circ \phi_\eta\right\|_{L^q(\mathbb{T}\times\Omega)} \leq c_8 \|\nabla u\|_{L^q(\mathbb{T}\times\Omega)} \|\eta\|_{L^\infty(\mathbb{T};W^{2,\infty}(\mathbb{T}_0^2))}$$

$$\leq c_9 \|u\|_{W^{1,2,q}(\mathbb{T}\times\Omega)} \|\eta\|_{W^{2,4,q}(\mathbb{T}\times\mathbb{T}_0^2)}. \tag{4.9.16}$$

Collecting the estimates (4.9.11) – (4.9.16) we obtain with view to (4.9.10) that $\tilde{\mathrm{R}}_f$ obeys (4.9.7).

Observe that Sobolev's embedding theorem (Theorem 2.6.3) yields for $m = 2$, $M_t = 0$, $M_x = 2$ and $\alpha = 1$ that

$$\|\eta\|_{L^\infty(\mathbb{T};W^{2,\infty}(\mathbb{T}_0^2))} \leq c_{10} \|\eta\|_{W^{2,4,q}(\mathbb{T}\times\mathbb{T}_0^2)},$$

and for $m = 2$, $M_t = 1$, $M_x = 0$ and $\alpha = 1$ that

$$\|\eta\|_{W^{1,\infty}(\mathbb{T};L^\infty(\mathbb{T}_0^2))} \leq c_{11} \|\eta\|_{W^{2,4,q}(\mathbb{T}\times\mathbb{T}_0^2)}$$

for $q > 2$. Note further that for a.e. $t \in \mathbb{T}$

$$\|\operatorname{div} \mathrm{R}_d(t, \cdot)\|_{\dot{W}^{-1,q}(\Omega)} = \sup_{\psi \in \dot{W}^{1,q'}(\Omega),\, \|\nabla\psi\|_{L^{q'}(\Omega)}=1} |\langle \operatorname{div} \mathrm{R}_d(t, \cdot), \psi(t, \cdot)\rangle|$$

$$= \sup_{\psi \in \dot{W}^{1,q'}(\Omega),\, \|\nabla\psi\|_{L^{q'}(\Omega)}=1} |\langle \mathrm{R}_d(t, \cdot), \nabla\psi(t, \cdot)\rangle|$$

$$\leq \sup_{\psi \in \dot{W}^{1,q'}(\Omega),\, \|\nabla\psi\|_{L^{q'}(\Omega)}=1} \|\mathrm{R}_d\|_{L^q(\Omega)} \|\nabla\phi\|_{L^{q'}(\Omega)} = \|\mathrm{R}_d\|_{L^q(\Omega)}$$

holds. Hence the nonlinear term $\tilde{R}_d$ appearing in the divergence equations fulfills

$$
\begin{aligned}
\|\tilde{R}_d\|_{\mathrm{L}^q(\mathbb{T};\mathrm{W}^{1,q}(\Omega))\cap\mathrm{W}^{1,q}(\mathbb{T};\dot{\mathrm{W}}^{-1,q}(\Omega))} & \\
&= \|\operatorname{div} R_d\|_{\mathrm{L}^q(\mathbb{T};\mathrm{W}^{1,q}(\Omega))\cap\mathrm{W}^{1,q}(\mathbb{T};\dot{\mathrm{W}}^{-1,q}(\Omega))} \\
&\leq \|R_d\|_{\mathrm{W}^{1,2,q}(\mathbb{T}\times\Omega)} \leq c_{12}\|u\|_{\mathrm{W}^{1,2,q}(\mathbb{T}\times\Omega)}\|\eta\|_{\mathrm{W}^{1,2,\infty}(\mathbb{T}\times\mathbb{T}_0^2)} \\
&\leq c_{13}\|u\|_{\mathrm{W}^{1,2,q}(\mathbb{T}\times\Omega)}\|\eta\|_{\mathrm{W}^{2,4,q}(\mathbb{T}\times\mathbb{T}_0^2)},
\end{aligned}
\tag{4.9.17}
$$

and therefore (4.9.8).

Finally it remains to show (4.9.9) to complete the proof. This L^q estimate holds due to

$$
\begin{aligned}
|\nu_t - e_3| &\leq \left|\left(\begin{array}{c}\nabla'\eta \\ -1-\sqrt{1+|\nabla\eta|^2}\end{array}\right)\right| \leq \sqrt{|\nabla'\eta|^2 + \left|1+\sqrt{1+|\nabla\eta|^2}\right|^2} \\
&\leq 2(1+|\nabla\eta|) \leq 2(1+\varepsilon_0),
\end{aligned}
$$

as well as the estimates

$$
\begin{aligned}
\|S_\eta\tilde{\nu}_t\|_{\mathrm{L}^q(\mathbb{T}\times\mathbb{T}_0^2)} &\leq 2\|\nabla u(\mathrm{E}\circ\phi_\eta)\|_{\mathrm{L}^q(\mathbb{T}\times\mathbb{T}_0^2)}\|\tilde{\nu}_t\|_{\mathrm{L}^\infty(\mathbb{T}\times\mathbb{T}_0^2)} \\
&\leq c_{14}(1+\varepsilon_0)\|\nabla u\nabla\eta\|_{\mathrm{L}^q(\mathbb{T}\times\mathbb{T}_0^2)} \\
&\leq c_{15}(1+\varepsilon_0)\|\eta\|_{\mathrm{W}^{2,4,q}(\mathbb{T}\times\mathbb{T}_0^2)}\|u\|_{\mathrm{W}^{1,2,q}(\mathbb{T}\times\Omega)}
\end{aligned}
$$

and

$$
\begin{aligned}
\|T_0(u,p)\|_{\mathrm{L}^q(\mathbb{T}\times\mathbb{T}_0^2)} &\leq c_{16}\big(\|\nabla u\|_{\mathrm{L}^q(\mathbb{T}\times\mathbb{T}_0^2)} + \|p\|_{\mathrm{L}^q(\mathbb{T}\times\mathbb{T}_0^2)}\big) \\
&\leq c_{17}\big(\|u\|_{\mathrm{W}^{1,2,q}(\mathbb{T}\times\Omega)} + \|p\|_{\mathrm{L}^q(\mathbb{T};\mathrm{W}^{1,q}(\Omega))}\big)
\end{aligned}
$$

for S_η and the fluid stress tensor T. $\qquad\square$

Now we have collected all information to prove existence of a $\mathcal{T}$-time-periodic solution to (4.9.1). For this reason, we will utilize a fixed-point argument, namely the contraction mapping principle, which is highly based on the theory for the linearized equations deduced in the previous section. The main result of this chapter reads as follow.

Theorem 4.9.3. *Let $q \in (2,\infty)$. There is an $\varepsilon > 0$ such that for all*

$$
\begin{aligned}
f_u &\in \mathrm{L}^q(\mathbb{T}\times\Omega)^3, \\
f_\eta &\in \mathrm{L}^q(\mathbb{T}\times\mathbb{T}_0^2),
\end{aligned}
$$

satisfying the smallness assumption

$$\|f_u\|_{\mathrm{L}^q(\mathbb{T}\times\Omega)} + \|f_\eta\|_{\mathrm{L}^q(\mathbb{T}\times\mathbb{T}_0^2)} \le \varepsilon,$$

we find a solution (u,p,η) to (4.9.2) in the class

$$(u,p,\eta) \in \mathrm{W}^{1,2,q}(\mathbb{T}\times\Omega)^3 \times \mathrm{L}^q\big(\mathbb{T};\dot{\mathrm{W}}^{1,q}(\Omega)\big) \times \mathrm{W}^{2,4,q}_{(0)}(\mathbb{T}\times\mathbb{T}_0^2).$$

Proof. We will employ the contraction mapping principle which is based on the L^q estimates deduced for the linear system

$$\begin{cases} \partial_t^2\eta + \Delta'^2\eta + \Delta'\partial_t\eta + e_3\cdot\mathrm{T}_0(u,p) = f_\eta & \text{in } \mathbb{T}\times\mathbb{T}_0^2, \\ \partial_t u - \Delta u + \nabla p = f_u & \text{in } \mathbb{T}\times\Omega, \\ \operatorname{div} u = g & \text{in } \mathbb{T}\times\Omega, \\ u_{|x_3=0} + \partial_t\eta e_3 = 0 & \text{on } \mathbb{T}\times\mathbb{T}_0^2, \\ u_{|x_3=1} = 0 & \text{on } \mathbb{T}\times\mathbb{T}_0^2, \end{cases} \quad (4.9.18)$$

where g is a linearization of $\tilde{\mathrm{R}}_d$. For this reason, let

$$\mathcal{S}\colon \mathcal{Y}_q \to \mathcal{X}_q$$

be the solution operator corresponding to this linear system, with

$$\mathcal{X}_q := \mathrm{W}^{1,2,q}(\mathbb{T}\times\Omega)^3 \times \mathrm{L}^q\big(\mathbb{T};\dot{\mathrm{W}}^{1,q}(\Omega)\big) \times \mathrm{W}^{2,4,q}_{(0)}(\mathbb{T}\times\mathbb{T}_0^2),$$

and

$$\mathcal{Y}_q := \mathrm{L}^q(\mathbb{T}\times\Omega)^3 \times \mathrm{L}^q\big(\mathbb{T};\mathrm{W}^{1,q}_{(0)}(\Omega)\big) \cap \mathrm{W}^{1,q}\big(\mathbb{T};\dot{\mathrm{W}}^{-1,q}(\Omega)\big) \times \mathrm{L}^q(\mathbb{T}\times\mathbb{T}_0^2).$$

From Theorem 4.8.2 we deduce the existence of a solution, and consequently conclude that $\mathcal{S}$ is invertible and bounded. Hence, as a solution to the linear system (4.9.18) we further obtain from Theorem 4.8.2, or more precisely from (4.8.9), that (u,p,η) satisfies the L^q estimate

$$\|u\|_{\mathrm{W}^{1,2,q}(\mathbb{T}\times\Omega)} + \|\nabla p\|_{\mathrm{L}^q(\mathbb{T}\times\Omega)} + \|\eta\|_{\mathrm{W}^{2,4,q}(\mathbb{T}\times\mathbb{T}_0^2)} \le c_0\varepsilon,$$

with c_0 independent of ε, and therefore we assume that η obeys the smallness condition from Lemma 4.9.2 if ε is sufficiently small. The solution to (4.9.2) shall be obtained as a fixed point of the mapping

$$\mathcal{F}\colon \mathcal{X}_q \to \mathcal{X}_q,$$
$$\mathcal{F}(u,p,\eta) := \mathcal{S}\big(f_u + \tilde{\mathrm{R}}_f, \tilde{\mathrm{R}}_d, f_\eta + \mathrm{R}_\eta\big).$$

To show that $\mathcal{F}$ possesses a fixed point, we let $r > 0$ and consider some $(u, p, \eta) \in \mathcal{X}_q \cap B_r$, where B_r denotes the closed ball with radius r. Due to Lemma 4.9.2, we obtain

$$\|\tilde{R}_f\|_{L^q(\mathbb{T}\times\Omega)} \leq c_1\big((1+\varepsilon)r + r^2\big)r,$$

$$\|\tilde{R}_d\|_{L^q(\mathbb{T};W^{1,q}(\Omega))\cap W^{1,q}(\mathbb{T};\dot{W}^{-1,q}(\Omega))} \leq c_2 r^2,$$

$$\|R_\eta\|_{L^q(\mathbb{T}\times\mathbb{T}_0^2)} \leq c_3(1+\varepsilon)r^2,$$

and therefore

$$\|\mathcal{F}\|_{\mathcal{X}_q} \leq \|\mathcal{S}\|\Big(\|f_u\|_{L^q(\mathbb{T}\times\Omega)} + \|\tilde{R}_f\|_{L^q(\mathbb{T}\times\Omega)} + \|R_\eta\|_{L^q(\mathbb{T}\times\mathbb{T}_0^2)}$$

$$+ \|\tilde{R}_d\|_{L^q(\mathbb{T};W^{1,q}(\Omega))\cap W^{1,q}(\mathbb{T};\dot{W}^{-1,q}(\Omega))} + \|f_\eta\|_{L^q(\mathbb{T}\times\mathbb{T}_0^2)}\Big)$$

$$\leq c_4(\varepsilon + r^2 + r^3),$$

since $\mathcal{S}$ is a homeomorphism. Choosing $r = \sqrt{\varepsilon}$ and ε sufficiently small, we have $c_4(\varepsilon + r^2 + r^3) \leq r$. Hence, $\mathcal{F}$ becomes a self-mapping on B_r. To complete the proof, it remains to show that $\mathcal{F}$ is a contraction mapping. For this purpose observe first that

$$\left\|\nabla u \partial_t \eta \frac{1}{1+\eta} - \nabla v \partial_t \zeta \frac{1}{1+\zeta}\right\|_{L^q(\mathbb{T}\times\Omega)}$$

$$\leq \|\nabla u\|_{L^q(\mathbb{T}\times\Omega)}\|\partial_t\eta\|_{L^\infty(\mathbb{T}\times\mathbb{T}_0^2)}\left\|\frac{1}{1+\eta} - \frac{1}{1+\zeta}\right\|_{L^\infty(\mathbb{T}\times\mathbb{T}_0^2)}$$

$$+ \|\nabla u(\partial_t\eta - \partial_t\zeta) + (\nabla u - \nabla v)\partial_t\zeta\|_{L^q(\mathbb{T}\times\Omega)}\left\|\frac{1}{1+\zeta}\right\|_{L^\infty(\mathbb{T}\times\mathbb{T}_0^2)}$$

holds, and due to

$$\|\eta\|_{L^\infty(\mathbb{T}\times\mathbb{T}_0^2)} + \|\partial_t\eta\|_{L^\infty(\mathbb{T}\times\mathbb{T}_0^2)} \leq c_5\|\eta\|_{W^{2,4,q}(\mathbb{T}\times\mathbb{T}_0^2)}$$

(see Theorem 2.6.3 with $m = 2$, $M_t = 1$, $M_x = 0$, $\alpha = 1$, and $m = 2$, $M_t = 0$, $M_x = 0$, $\alpha = 2$, respectively,) we thus conclude from (4.9.4)

$$\left\|\nabla u \partial_t \eta \frac{1}{1+\eta} - \nabla v \partial_t \zeta \frac{1}{1+\zeta}\right\|_{L^q(\mathbb{T}\times\Omega)}$$

$$\leq c_6\varepsilon\big(\|\eta - \zeta\|_{W^{2,4,q}(\mathbb{T}\times\mathbb{T}_0^2)} + \|u - v\|_{W^{1,2,q}(\mathbb{T}\times\Omega)}\big).$$

Similarly it follows that

$$\left\|\frac{\partial_{x_3} p}{1+\eta}\begin{pmatrix}\rho\nabla'\eta \\ -\eta\end{pmatrix} - \frac{\partial_{x_3}\mathfrak{p}}{1+\zeta}\begin{pmatrix}\rho\nabla'\zeta \\ -\zeta\end{pmatrix}\right\|_{L^q(\mathbb{T}\times\Omega)}$$

$$\leq c_7\varepsilon\big(\|\eta - \zeta\|_{W^{2,4,q}(\mathbb{T}\times\mathbb{T}_0^2)} + \|p - \mathfrak{p}\|_{L^q(\mathbb{T};\dot{W}^{1,q}(\Omega))}\big).$$

The further terms occurring in the representation formula of $\tilde{\mathrm{R}}_f$ can be investigated analogously, such that

$$\|\tilde{\mathrm{R}}_f(u,p,\eta) - \tilde{\mathrm{R}}_f(v,\mathfrak{p},\zeta)\|_{\mathrm{L}^q(\mathbb{T}\times\Omega)} \leq c_8\varepsilon\|(u,p,\eta) - (v,\mathfrak{p},\zeta)\|$$

holds, with

$$\|(u,p,\eta) - (v,\mathfrak{p},\zeta)\|$$
$$= \|u - v\|_{\mathrm{W}^{1,2,q}(\mathbb{T}\times\Omega)} + \|p - \mathfrak{p}\|_{\mathrm{L}^q(\mathbb{T};\dot{\mathrm{W}}^{1,q}(\Omega))} + \|\eta - \zeta\|_{\mathrm{W}^{2,4,q}(\mathbb{T}\times\mathbb{T}_0^2)}.$$

For $\tilde{\mathrm{R}}_d$ a straightforward calculation yields

$$\|\tilde{\mathrm{R}}_d(u,p,\eta) - \tilde{\mathrm{R}}_d(v,\mathfrak{p},\zeta)\|_{\mathrm{L}^q(\mathbb{T};\mathrm{W}^{1,q}(\Omega))\cap\mathrm{W}^{1,q}(\mathbb{T};\dot{\mathrm{W}}^{-1,q}(\Omega))}$$
$$\leq c_9\varepsilon\|(u,p,\eta) - (v,\mathfrak{p},\zeta)\|.$$

In the case of R_η, we further have to employ the properties of the trace operator

$$\mathrm{Tr}_0\colon \mathrm{L}_\perp^q\big(\mathbb{T};\mathrm{W}^{1,q}(\Omega)\big) \to \mathrm{L}_\perp^q\big(\mathbb{T};\mathrm{W}^{1-\frac{1}{q},q}(\mathbb{T}_0^2)\big), \qquad \phi \mapsto \phi_{|x_3=0},$$

defined in (4.6.17), to deduce that

$$\|\mathrm{R}_\eta(u,p,\eta) - \mathrm{R}_\eta(v,\mathfrak{p},\zeta)\|_{\mathrm{L}^q(\mathbb{T}\times\mathbb{T}_0^2)} \leq \|\mathrm{S}_\eta(u,p,\eta) - \mathrm{S}_\eta(v,\mathfrak{p},\zeta)\|_{\mathrm{L}^q(\mathbb{T}\times\mathbb{T}_0^2)}$$
$$+ \|\mathrm{T}_0(u,p)\tilde{\nu}_t(\eta) - \mathrm{T}_0(v,\mathfrak{p})\tilde{\nu}_t(\zeta)\|_{\mathrm{L}^q(\mathbb{T}\times\mathbb{T}_0^2)}$$
$$\leq c_{10}\varepsilon\|(u,p,\eta) - (v,\mathfrak{p},\zeta)\|.$$

Hence, collecting the estimates deduced for R_f, $\tilde{\mathrm{R}}_d$ and R_η, we obtain that

$$\|\mathcal{F}(u,p,\eta) - \mathcal{F}(v,\mathfrak{p},\zeta)\|_{\mathcal{X}_q} \leq c_{11}\|\mathcal{S}\|\Big(\|\tilde{\mathrm{R}}_f(u,p,\eta) - \tilde{\mathrm{R}}_f(v,\mathfrak{p},\zeta)\|_{\mathrm{L}^q(\mathbb{T}\times\Omega)}$$
$$+ \|\mathrm{R}_\eta(u,p,\eta) - \mathrm{R}_\eta(v,\mathfrak{p},\zeta)\|_{\mathrm{L}^q(\mathbb{T}\times\mathbb{T}_0^2)}$$
$$+ \|\tilde{\mathrm{R}}_d(u,p,\eta) - \tilde{\mathrm{R}}_d(v,\mathfrak{p},\zeta)\|_{\mathrm{L}^q(\mathbb{T};\mathrm{W}^{1,q}(\Omega))\cap\mathrm{W}^{1,q}(\mathbb{T};\dot{\mathrm{W}}^{-1,q}(\Omega))}\Big)$$
$$\leq c_{12}\varepsilon\|(u,p,\eta) - (v,\mathfrak{p},\zeta)\|.$$

Therefore, choosing ε sufficiently small $\mathcal{F}$ becomes a contracting self-mapping. By the contraction mapping principle, existence of a fixed point for $\mathcal{F}$ follows and completes the proof. □

Summarizing the results deduced in this chapter, we have shown existence of a solution to the nonlinear coupled system (4.9.2). However, as the transformation ϕ_η maps a solution to this system on a solution to the fluid-structure problem (4.9.1), we have found a solution to the free boundary problem which is given by $(u,v,\eta) = (\tilde{u}\circ\phi_\eta^{-1}, \tilde{u}\circ\phi_\eta^{-1}, \eta)$. Note that η only exists on the fluid-solid interface and is already defined on the reference configuration, hence ϕ_η has no effect on η.

Appendix

A.1 Some Bounded Functions

In this section we consider some of the functions appearing in the investigation of the L^q estimates in this thesis, and show that these are bounded. Therefore, throughout this section we consider $\eta \in \mathbb{R}$ and assume it to satisfy $|\eta| > \frac{\pi}{\mathcal{T}}$.

A.1.1 Functions in the Context of Nonlinear Acoustics

This subsection is dedicated to prove that the terms $I_1 - I_4$ given as in (3.2.13), that is,

$$
\begin{aligned}
I_1 &:= \frac{1}{(|\xi|^2 - \eta^2)^2 + \eta^2 |\xi|^4} \\
I_2 &:= \frac{\eta^2}{(|\xi|^2 - \eta^2)^2 + \eta^2 |\xi|^4} \\
I_3 &:= \frac{\eta^2 |\xi|^4}{(|\xi|^2 - \eta^2)^2 + \eta^2 |\xi|^4} \\
I_4 &:= \frac{|\xi|^4}{(|\xi|^2 - \eta^2)^2 + \eta^2 |\xi|^4}
\end{aligned}
\qquad \text{(A.1.1)}
$$

are bounded for all $(\eta, \xi) \in \mathbb{R} \times \mathbb{R}^n$. To this end, fix a period length $\mathcal{T} > 0$. The term $(A.1.1)_3$ is easy to bound by 1, since $(|\xi|^2 - \eta^2)^2 \geq 0$. Due to $|\eta| > \frac{\pi}{\mathcal{T}}$ we have by the same argument

$$
I_4 \leq \frac{1}{\eta^2} < \frac{\mathcal{T}^2}{\pi^2}.
$$

To show that I_1 is bounded, we let $\varepsilon := \frac{\eta^2}{2} > 0$ and consider the two cases

$$
||\xi|^2 - \eta^2| \geq \varepsilon \qquad \text{and} \qquad ||\xi|^2 - \eta^2| < \varepsilon,
$$

where in the second case we further distinguish between $|\xi| > |\eta|$ and $|\xi| \leq |\eta|$. In the case $||\xi|^2 - \eta^2| \geq \varepsilon$ we have that

$$I_1 \leq \frac{1}{(|\xi|^2 - \eta^2)^2} \leq \frac{1}{\varepsilon^2} = 4\frac{1}{\eta^4} < 4\frac{T^4}{\pi^4}, \qquad (A.1.2)$$

as $|\eta| > \frac{\pi}{T}$. In the second case we first consider $|\xi| > |\eta|$ and utilize $\frac{1}{|\xi|^4} < \frac{1}{\eta^4}$ to deduce

$$I_1 \leq \frac{1}{\eta^2 |\xi|^4} < \frac{1}{\eta^6} < \frac{T^6}{\pi^6}. \qquad (A.1.3)$$

The last case is $||\xi|^2 - \eta^2| < \varepsilon$ and $|\xi| \leq |\eta|$. Then we find

$$|\xi|^2 > \eta^2 - \varepsilon = \frac{\eta^2}{2} \Leftrightarrow \frac{1}{|\xi|^4} < 4\frac{1}{\eta^4},$$

and thus

$$I_1 \leq \frac{1}{\eta^2 |\xi|^4} < \frac{4}{\eta^6} < 4\frac{T^6}{\pi^6}. \qquad (A.1.4)$$

Finally, (A.1.2) – (A.1.4) implies (A.1.1)$_1$.

For I_2 we proceed similarly, but choose $\varepsilon := \frac{|\eta|}{2}$ and distinguish between the cases $\left|\frac{|\xi|^2}{\eta} - \eta\right| \geq \varepsilon$ and $\left|\frac{|\xi|^2}{\eta} - \eta\right| < \varepsilon$. Note that

$$I_2 = \frac{1}{\left(\frac{|\xi|^2}{\eta} - \eta\right)^2 + |\xi|^4}.$$

Then, if $\left|\frac{|\xi|^2}{\eta} - \eta\right| \geq \varepsilon$, we obtain

$$I_2 \leq \frac{1}{\left(\frac{|\xi|^2}{\eta} - \eta\right)^2} \leq \frac{1}{\varepsilon^2} < 4\frac{T^2}{\pi^2}. \qquad (A.1.5)$$

As for I_1, we divide the case $\left|\frac{|\xi|^2}{\eta} - \eta\right| < \varepsilon$ into two sub-cases and start by considering $|\xi| > |\eta|$ to obtain

$$I_2 \leq \frac{1}{|\xi|^4} < \frac{1}{\eta^4} < \frac{T^4}{\pi^4}. \qquad (A.1.6)$$

Otherwise, if $|\xi| \leq |\eta|$, we first observe that $\frac{1}{|\xi|^4} < 4\frac{1}{\eta^4}$, and then conclude

$$I_2 \leq \frac{1}{|\xi|^4} < 4\frac{1}{\eta^4} < 4\frac{T^4}{\pi^4},$$

which, in view of (A.1.5) and (A.1.6), yields (A.1.1)$_2$.

A.1.2 Functions in the Context of Fluid-Structure Interaction

Subject of this subsection is to verify the following estimates:

$$I_1 := \frac{1}{(|\zeta|^4 - \theta^2)^2 + \theta^2|\zeta|^4} \leq 4\frac{T^4}{\pi^4},$$

$$I_2 := \theta^2|\zeta|^4 I_1 \leq 1,$$

$$I_3 := (16|\zeta|^8 + 4\theta^2|\zeta|^4)I_1 \leq \frac{16}{\left(1 - \theta^2/|\zeta|^4\right)^2 + \theta^2/|\zeta|^4} + 4 \leq 68, \quad \text{(A.1.7)}$$

$$I_4 := |\zeta|^4 I_1 \leq \frac{T^2}{\pi^2},$$

$$I_5 := (4\theta^4 + \theta^2|\zeta|^4)I_1 \leq \frac{4}{\left(|\zeta|^4/\theta^2 - 1\right)^2 + |\zeta|^4/\theta^2} + 1 \leq 17,$$

where $|\theta| > \frac{\pi}{T}$. We proceed analogously as in the previous subsection. The estimate for I_1 follows by mimicking the exact same steps as for $(A.1.1)_1$ on using $\varepsilon = \frac{\theta^2}{2}$.

As both addends in the denominator of I_1 are nonnegative and $|\theta| > \frac{\pi}{T}$, we conclude that $(A.1.7)_2$ and $(A.1.7)_4$ hold.

To prove $(A.1.7)_3$ and $(A.1.7)_5$ it suffices to show that

$$\frac{1}{\left(1 - \theta^2/|\zeta|^4\right)^2 + \theta^2/|\zeta|^4} \leq 4, \quad \text{(A.1.8)}$$

and

$$\frac{1}{\left(|\zeta|^4/\theta^2 - 1\right)^2 + |\zeta|^4/\theta^2} \leq 4. \quad \text{(A.1.9)}$$

Choosing $\varepsilon = \frac{1}{2}$ and distinguishing the two cases $|1 - \theta^2/|\zeta|^4| \geq \varepsilon$ and $|1 - \theta^2/|\zeta|^4| < \varepsilon$, we obtain (A.1.8) by mimicking the exact same steps as for $(A.1.1)_1$, whereas (A.1.9) follows by considering $||\zeta|^4/\theta^2 - 1| \geq \varepsilon$ and $||\zeta|^4/\theta^2 - 1| < \varepsilon$.

A.2 Fourier Multiplier

This section is dedicated to the investigation of some Fourier multipliers that occur in this doctoral thesis. Since the multiplier appearing herein are fractions, to verify condition (2.3.3) of Marcinkiewicz multiplier theorem, a representation formula for higher order partial derivatives of quotients is essential. This is outlined in the following.

A.2.1 Higher Order Partial Derivatives of Fractions

In order to utilize the multiplier theorem of Marcinkiewicz, it is convenient to find a representation formula for the partial derivatives of the quotient of type

$$\frac{1}{1+f(x)}.$$

The following lemma is used to find such a formula.

Lemma A.2.1. *Let $f \in C^n(\mathbb{R}^n \setminus \{0\})$. For every $\alpha \in \{0,1\}^n$ there exist finitely many constants $c(k)$ such that for all $x \in \mathbb{R}^n \setminus \{0\}$ one has*

$$\partial_x^\alpha \left(\frac{1}{1+f(x)} \right)$$

$$= \frac{-\partial_x^\alpha f}{(1+f(x))^2} + \sum_{k=1}^{|\alpha|-1} \sum_{\substack{\sum_{i=1}^k \beta_i < \alpha, \\ |\beta_i| \geq 1}} c(k) \frac{\left(\partial_x^{\alpha - \sum_{i=1}^k \beta_i} f(x) \right) \prod_{j=1}^k \left(\partial_x^{\beta_j} f(x) \right)}{(1+f(x))^{k+2}}.$$

$$(\text{A.2.1})$$

The constants $c(k)$ are either 0 or $(-1)^{k+1} \prod_{j=0}^k (1+j)$.

Proof. We are going to prove the assertion by induction. Let us first consider the case $|\alpha| = 1$, then

$$\partial_x^\alpha \left(\frac{1}{1+f(x)} \right) = -\frac{\partial_x^\alpha f(x)}{(1+f(x))^2},$$

which proves the initial case since there are no elements in the second sum of (A.2.1). In order to prove the assertion for an arbitrary α with $|\alpha| > 1$ we assume that (A.2.1) holds for any α' with $|\alpha'| = k$. Let $\alpha \in \{0,1\}^n$ with $|\alpha| = k+1$, we then find an $\alpha' \in \{0,1\}^n$ and $s \in \{1, \ldots, n\}$ such that $\alpha = \alpha' + e_s$. We deduce

$$\partial_x^\alpha \left(\frac{1}{1+f} \right) = \partial_x^{e_s} \left(\partial_x^{\alpha'} \left(\frac{1}{1+f} \right) \right)$$

$$= \partial_x^{e_s} \left(\frac{-\partial_x^{\alpha'} f}{(1+f)^2} + \sum_{k=1}^{|\alpha'|-1} \sum_{\substack{\sum_{i=1}^k \beta_i < \alpha', \\ |\beta_i| \geq 1}} c(k) \frac{\left(\partial_x^{\alpha' - \sum_{i=1}^k \beta_i} f \right) \prod_{j=1}^k \left(\partial_x^{\beta_j} f \right)}{(1+f)^{k+2}} \right),$$

and thus

$$\partial_x^\alpha \left(\frac{1}{1+f} \right) = \frac{-\partial_x^\alpha f}{(1+f)^2} + \frac{2\partial_x^{\alpha'} f \partial_x^{e_s} f}{(1+f)^3}$$

$$+ \sum_{k=1}^{|\alpha'|-1} \sum_{\substack{\sum_{i=1}^k \beta_i < \alpha', \\ |\beta_i| \geq 1}} c(k) \partial_x^{e_s} \left(\frac{\left(\partial_x^{\alpha' - \sum_{i=1}^k \beta_i} f \right) \prod_{j=1}^k \left(\partial_x^{\beta_j} f \right)}{(1+f)^{k+2}} \right). \quad \text{(A.2.2)}$$

Computing the derivative in the sum, we obtain

$$\partial_x^{e_s} \left(\frac{\left(\partial_x^{\alpha' - \sum_{i=1}^k \beta_i} f \right) \prod_{j=1}^k \left(\partial_x^{\beta_j} f \right)}{(1+f)^{k+2}} \right)$$

$$= \frac{\left(\partial_x^{\alpha - \sum_{i=1}^k \beta_i} f \right) \prod_{j=1}^k \left(\partial_x^{\beta_j} f \right)}{(1+f)^{k+2}} - (k+2) \frac{\left(\partial_x^{\alpha - e_s - \sum_{i=1}^k \beta_i} f \right) \partial_x^{e_s} f \prod_{j=1}^k \left(\partial_x^{\beta_j} f \right)}{(1+f)^{k+3}}$$

$$+ \frac{\left(\partial_x^{\alpha - e_s - \sum_{i=1}^k \beta_i} f \right) \left(\sum_{j=1}^k \left[\partial_x^{\beta_j + e_s} f \prod_{m=1, m\neq j}^k \partial_x^{\beta_m} f \right] \right)}{(1+f)^{k+2}}.$$

Choosing the constants $\tilde{c}(k)$ suitably, we find by taking the sum in the identity above that

$$\sum_{k=1}^{|\alpha'|-1} \sum_{\substack{\sum_{i=1}^k \beta_i < \alpha', \\ |\beta_i| \geq 1}} c(k) \partial_x^{e_s} \left(\frac{\left(\partial_x^{\alpha' - \sum_{i=1}^k \beta_i} f \right) \prod_{j=1}^k \left(\partial_x^{\beta_j} f \right)}{(1+f)^{k+2}} \right)$$

$$= \sum_{k=1}^{|\alpha|-1} \sum_{\substack{\sum_{i=1}^k \beta_i < \alpha, \\ |\beta_i| \geq 1}} \tilde{c}(k) \frac{\left(\partial_x^{\alpha - \sum_{i=1}^k \beta_i} f \right) \prod_{j=1}^k \left(\partial_x^{\beta_j} f \right)}{(1+f)^{k+2}}.$$

Together with (A.2.2) this implies (A.2.1) and completes the proof. Note that the constants behave in the mentioned way, but as the sum is over all multi-indices β_i, some constants are 0. $\qquad \square$

A.2.2 Fourier Multiplier in the Whole Space

In the following, some Fourier multipliers in $\mathbb{R}^n$ are examined. We shall begin by considering

$$M_{\mathbb{R}}(\eta, \xi) := \frac{\chi(\eta) i^{M_t + M_x} \eta^{M_t} |\eta|^{\frac{\alpha}{2}} \xi^{m_x} (1 + |\xi|^2)^{\frac{\beta}{2}}}{1 + |\eta|^m + |\xi|^{2m}},$$

where $(\eta, \xi) \in \mathbb{R} \times \mathbb{R}^n$. We show that $M_{\mathbb{R}}$ is an L^q-multiplier. Here, $\chi \in C_0^\infty(\mathbb{R}; [0, 1])$ is a cut-off function such that

$$\chi(\eta) = 0 \text{ for } |\eta| \leq \frac{\pi}{\mathcal{T}}, \quad \chi(\eta) = 1 \text{ for } |\eta| \geq \frac{2\pi}{\mathcal{T}},$$

Lemma A.2.2. *Let α, β, m, M_t, M_x and q be as in Theorem 2.6.3. Then $M_{\mathbb{R}}$ is an $L^q\big(\mathbb{R}; L^q(\mathbb{R}^n)\big)$-multiplier.*

Proof. Observe that the denominator does not vanish for any $(\eta, \xi) \in \mathbb{R} \times \mathbb{R}^n$. Therefore, we only have to verify, that $M_{\mathbb{R}}$ stays bounded if (η, ξ) goes to infinity. For this purpose, we write $M_{\mathbb{R}}$ as the fraction

$$M_{\mathbb{R}}(\eta, \xi) := \frac{\rho(\eta) f(\xi)}{\mathcal{N}(\eta, \xi)},$$

where

$$\rho(\eta) := \chi(\eta) i^{M_t + M_x} \eta^{M_t} |\eta|^{\frac{\alpha}{2}} \qquad \text{and} \qquad f(\xi) := \xi^{m_x} (1 + |\xi|^2)^{\frac{\beta}{2}}$$

are polynomials of order $M_t + \frac{\alpha}{2}$ and $M_x + \beta$, respectively, and the denominator is given by

$$\mathcal{N}(\eta, \xi) := 1 + |\eta|^m + |\xi|^{2m}.$$

Let $\gamma \in \{0, 1\}^n$ be a multi-index. Differentiating $M_{\mathbb{R}}$ with respect to ξ and utilizing the product rule, we find that

$$\partial_\xi^\gamma M_{\mathbb{R}} = \rho \sum_{\sigma \in \mathbb{N}_0^n, \sigma \leq \gamma} \binom{\gamma}{\sigma} \partial_\xi^{\gamma - \sigma} f \, \partial_\xi^\sigma (\mathcal{N}^{-1}),$$

where the inequality $\sigma \leq \gamma$ should be understood componentwise and the binomial coefficient is defined as usual as $\frac{\gamma!}{\sigma!(\gamma-\sigma)!}$ with $\gamma! = \gamma_1! \cdots \gamma_n!$. In order to find a representation formula for the derivative of the denominator, we write $\mathcal{N}^{-1}$ as

$$\mathcal{N}^{-1}(\eta, \xi) = \frac{1}{1 + g(\eta, \xi)} \qquad \text{with} \qquad g(\eta, \xi) := |\eta|^m + |\xi|^{2m},$$

and apply formula (A.2.1) from Lemma A.2.1 to get

$$\partial_\xi^\gamma (M_{\mathbb{R}}) = -\rho \sum_{\sigma \in \mathbb{N}_0^n, \sigma \leq \gamma} \binom{\gamma}{\sigma} \frac{\partial_\xi^{\gamma - \sigma} f \partial_\xi^\sigma g}{\mathcal{N}^2}$$

$$+ \rho \sum_{\sigma \in \mathbb{N}_0^n, \sigma \leq \gamma} \sum_{k=1}^{|\sigma|-1} \sum_{\sum_{i=1}^k \zeta_i < \sigma, |\zeta_i| \geq 1} c(k) \binom{\gamma}{\sigma} \frac{\partial_\xi^{\gamma - \sigma} f \big(\partial_\xi^{\sigma - \sum_{i=1}^k \zeta_i} g\big) \prod_{j=1}^k \partial_\xi^{\zeta_i} g}{\mathcal{N}^{k+2}},$$

$$\text{(A.2.3)}$$

where the constant $c(k)$ is either 0 or $(-1)^{k+1}(k+1)!$. Observe that the numerators in (A.2.3) contain derivatives of g only, and since these derivatives are independent of η, we have that the numerators only depend on ξ. Moreover, to verify condition (2.3.3) of Marcinkiewicz multiplier theorem, we can study any term occurring in the sums above, *i.e.*,

$$\frac{\rho \partial_\xi^{\gamma-\sigma} f \partial_\xi^\sigma g}{\mathcal{N}^2} \qquad \text{and} \qquad \frac{\rho \partial_\xi^{\gamma-\sigma} f \big(\partial_\xi^{\sigma-\sum_{i=1}^k \zeta_i} g\big) \prod_{j=1}^k \partial_\xi^{\zeta_i} g}{\mathcal{N}^{k+2}},$$

separately. Employing Leibniz' formula again, we obtain for the partial derivatives appearing in the two fractions above that

$$\partial_\xi^\gamma f = \sum_{\tilde\gamma \in \mathbb{N}_0^n, \tilde\gamma \leq \gamma} c_0 \prod_{j=1}^n \big(m_{x_j}^{\tilde\gamma_j} \xi_j^{\tilde\gamma_j(m_{x_j}-1)} \xi_j^{(1-\tilde\gamma_j)m_{x_j}} \big) \xi^{\gamma-\tilde\gamma} (1+|\xi|^2)^{\frac{\beta}{2}-|\gamma-\tilde\gamma|},$$

$$\partial_\xi^\gamma g = \partial_{\xi_1}^{\gamma_1} \cdots \partial_{\xi_n}^{\gamma_n} g = \begin{cases} c_1 |\xi|^{2(m-|\gamma|)} \xi^\gamma & \text{if } |\gamma| \geq 1, \\ g & \text{if } |\gamma| = 0, \end{cases}$$

whit $c_0, c_1 > 0$ independent of η and ξ. Observe, $m_{x_j} - 1 \geq 0$. Otherwise we have that $\partial_\xi^\gamma f = 0$, hence $M_x - |\gamma| \geq 0$. With this observation and the representation formulas for $\partial_\xi^{\gamma-\sigma} f$ and $\partial_\xi^\sigma g$ at hand we are able to estimate the partial derivative in (A.2.3), beginning with the first sum. As preparation, note that $\partial_\xi^{\gamma-\sigma} f$ and $\partial_\xi^\sigma g$ satisfy

$$|\partial_\xi^{\gamma-\sigma} f| \leq c_2 \sum_{\tilde\gamma \in \mathbb{N}_0^n, \tilde\gamma \leq \gamma-\sigma} |\xi|^{M_x - 2|\tilde\gamma| + |\gamma-\sigma|} (1+|\xi|^2)^{\frac{\beta}{2}-|\gamma-\sigma-\tilde\gamma|}$$

$$\leq c_3 \sum_{\tilde\gamma \in \mathbb{N}_0^n, \tilde\gamma \leq \gamma-\sigma} (1+|\xi|^2)^{\frac{M_x+\beta}{2}-\frac{|\gamma|-|\sigma|}{2}} \leq c_4 (1+|\xi|^2)^{\frac{M_x+\beta}{2}-\frac{|\gamma|-|\sigma|}{2}},$$

$$|\partial_\xi^\sigma g| \leq c_5 |\xi|^{2m-|\sigma|} \leq c_5 (1+|\xi|^2)^{m-\frac{|\sigma|}{2}},$$

hence, utilizing Young's inequality with $p = \frac{4m}{2M_t+\alpha}$ and $p' = \frac{4m}{4m-2M_t-\alpha} = \frac{4m}{2m+M_x+\beta}$, we deduce

$$\left| \xi^\gamma \rho \sum_{\sigma \in \mathbb{N}_0^n, \sigma \leq \gamma} \binom{\gamma}{\sigma} \frac{\partial_\xi^{\gamma-\sigma} f \partial_\xi^\sigma g}{\mathcal{N}^2} \right| \leq \sum_{\sigma \in \mathbb{N}_0^n, \sigma \leq \gamma} \binom{\gamma}{\sigma} |\rho| \frac{|\xi^{\gamma-\sigma} \partial_\xi^{\gamma-\sigma} f||\xi^\sigma \partial_\xi^\sigma g|}{|\mathcal{N}|^2}$$

$$\leq c_6 \frac{|\eta|^{\frac{2M_t+\alpha}{2}} (1+|\xi|^2)^{\frac{2m+M_x+\beta}{2}}}{(1+|\eta|^m+|\xi|^{2m})^2} \leq c_7 \frac{|\eta|^{2m} + (1+|\xi|^2)^{2m}}{(1+|\eta|^m+|\xi|^{2m})^2} \leq C.$$

$$\text{(A.2.4)}$$

Similarly, one may verify that for the second summand in (A.2.3) the estimate

$$
\left| \xi^\gamma \frac{\rho \partial_\xi^{\gamma-\sigma} f \left(\partial_\xi^{\sigma-\sum_{i=1}^k \zeta_i} g \right) \prod_{j=1}^k \partial_\xi^{\zeta_i} g}{\mathcal{N}^{k+2}} \right|
$$

$$
\leq c_8 \frac{|\rho| |\xi^{\gamma-\sigma} \partial_\xi^{\gamma-\sigma} f| |\xi^{\sigma-\sum_{i=1}^k \zeta_i} \partial_\xi^{\sigma-\sum_{i=1}^k \zeta_i} g| |\xi^{\sum_{i=1}^k \zeta_i} \prod_{j=1}^k \partial_\xi^{\zeta_i} g|}{|\mathcal{N}|^{k+2}}
$$

$$
\leq c_9 \frac{|\rho| (1+|\xi|^2)^{\frac{M_x+\beta}{2}} |\xi|^{|\sigma|} (1+|\xi|^2)^{m-\frac{|\sigma|-\sum_{i=1}^k |\zeta_i|}{2}} (1+|\xi|^2)^{km-\frac{\sum_{j=1}^k |\zeta_j|}{2}}}{|\mathcal{N}|^{k+2}}
$$

$$
\leq c_{10} \frac{|\eta|^{\frac{2M_t+\alpha}{2}} (1+|\xi|^2)^{\frac{2(k+1)m+M_x+\beta}{2}}}{|\mathcal{N}|^{k+2}}
$$

$$
\leq c_{11} \frac{|\eta|^{m(k+2)} + (1+|\xi|^2)^{m(k+2)}}{|1+|\eta|^m + |\xi|^{2m}|^{k+2}} \leq C.
$$

$$
(\mathrm{A}.2.5)
$$

Note that the fourth inequality follows from an application of Young's inequality with $p = \frac{2m(k+2)}{2M_t+\alpha}$ and $p' = \frac{2m(k+2)}{2m(k+2)-2M_t-\alpha}$. Since all the sums in (A.2.3) are finite, we obtain from (A.2.4) and (A.2.5) that

$$
|\xi^\gamma \partial_\xi^\gamma \mathrm{M}_\mathbb{R}| \leq c_{12} < \infty.
$$

To verify the condition of Marcinkiewicz multiplier theorem it remains to differentiate $\mathrm{M}_\mathbb{R}$ with respect to η and check that it decay properly. For this purpose we recall the representation formula (A.2.3) and that the numerators therein only contain derivatives of g that are independent of η. Hence, we differentiate the terms with respect to η and deduce

$$
\partial_\eta \partial_\xi^\gamma \mathrm{M}_\mathbb{R} = \mathrm{M}_{\mathbb{R},1} + \mathrm{M}_{\mathbb{R},2}
$$

with

$$
\mathrm{M}_{\mathbb{R},1} := \partial_\eta \rho \Bigg(- \sum_{\sigma \in \mathbb{N}_0^n, \sigma \leq \gamma} \binom{\gamma}{\sigma} \frac{\partial_\xi^{\gamma-\sigma} f \partial_\xi^\sigma g}{\mathcal{N}^2}
$$

$$
+ \sum_{\sigma \in \mathbb{N}_0^n, \sigma \leq \gamma} \sum_{k=1}^{|\sigma|-1} \sum_{\sum_{i=1}^k \zeta_i < \sigma, |\zeta_i| \geq 1} c(k) \binom{\gamma}{\sigma} \frac{\partial_\xi^{\gamma-\sigma} f \left(\partial_\xi^{\sigma-\sum_{i=1}^k \zeta_i} g \right) \prod_{j=1}^k \partial_\xi^{\zeta_i} g}{\mathcal{N}^{k+2}} \Bigg),
$$

and

$$M_{\mathbb{R},2} := \rho \frac{\partial_\eta \mathcal{N}}{\mathcal{N}} \left(-2 \sum_{\sigma \in \mathbb{N}_0^n, \sigma \leq \gamma} \binom{\gamma}{\sigma} \frac{\partial_\xi^{\gamma-\sigma} f \partial_\xi^\sigma (1+g)}{\mathcal{N}^2} \right.$$

$$+ \sum_{\sigma \in \mathbb{N}_0^n, \sigma \leq \gamma} \sum_{k=1}^{|\sigma|-1} \sum_{\sum_{i=1}^k \zeta_i < \sigma, |\zeta_i| \geq 1} \tilde{c}(k) \binom{\gamma}{\sigma} \frac{\partial_\xi^{\gamma-\sigma} f \left(\partial_\xi^{\sigma-\sum_{i=1}^k \zeta_i} g\right) \prod_{j=1}^k \partial_\xi^{\zeta_i} g}{\mathcal{N}^{k+2}} \right),$$

Here, the constant $\tilde{c}$ is either 0 or $c(k)(k+2)$, and the derivatives $\partial_\eta \rho$ and $\partial_\eta \mathcal{N}$ are given by

$$\partial_\eta \rho = \chi'(\eta) i^{M_t+M_x} \eta^{M_t} |\eta|^{\frac{\alpha}{2}} + M_t \chi(\eta) i^{M_t+M_x} \eta^{M_t-1} |\eta|^{\frac{\alpha}{2}}$$
$$+ \frac{\alpha}{2} \chi(\eta) i^{M_t+M_x} \eta^{M_t+1} |\eta|^{\frac{\alpha}{2}-2},$$

$$\partial_\eta \mathcal{N} = m\eta |\eta|^{m-2}.$$

With this, it is straightforward to verify

$$|\eta \partial_\eta \rho| \leq c_{13}(\mathcal{T}) |\eta|^{\frac{2M_t+\alpha}{2}} \qquad \text{and} \qquad \left| \eta \frac{\partial_\eta \mathcal{N}}{\mathcal{N}} \right| \leq m \frac{|\eta|^m}{|\mathcal{N}|} \leq m.$$

Hence, we obtain in the same way as in (A.2.4) and (A.2.5) that

$$|\eta \xi^\gamma \partial_\eta \partial_\xi^\gamma M_{\mathbb{R}}| \leq c_{14} < \infty.$$

Finally, Marcinkiewicz multiplier theorem (Theorem 2.3.2) implies that $M_{\mathbb{R}}$ is an $L^q(\mathbb{R} \times \mathbb{R}^n)$-multiplier. $\qquad\square$

A.2.3 Fourier Multiplier on the Torus Group

Here, we outline some of the calculations used in Chapter 4 to carry out the L^q estimates for $(v_{\text{tp}}, \mathfrak{p}_{\text{tp}}, \eta_{\text{tp}})$. Precisely, we will show that $m_{\mathbb{T}}, M_{\mathbb{T}} : \frac{2\pi}{\mathcal{T}}\mathbb{Z} \times \left(\frac{2\pi}{L}\mathbb{Z}\right)^2 \to \mathbb{C}$ given by

$$m_{\mathbb{T}}(k, \xi) := \rho_{\mathbb{Z}}(k) \, \mathfrak{m}_{\mathbb{T}}^{-1}(k, \xi), \tag{A.2.6}$$

$$M_{\mathbb{T}}(k, \xi) := \frac{\rho_{\mathbb{Z}}(k) \, |\xi| \mathfrak{m}_{\mathbb{T}}(k, \xi)}{|\xi| \mathfrak{m}_{\mathbb{T}}(k, \xi) - k^2 + ik|\xi| \left(|\xi| + \sqrt{|\xi|^2 + ik}\right)} \tag{A.2.7}$$

are $L^q(\mathbb{T} \times \mathbb{T}_0^2)$-multipliers, where $\mathfrak{m}_{\mathbb{T}}$ and $\rho_{\mathbb{Z}}$ are given by

$$\mathfrak{m}_{\mathbb{T}}(k, \xi) := |\xi|^4 - k^2 + ik|\xi|^2,$$
$$\rho_{\mathbb{Z}}(k) := 1 - \delta_{\frac{2\pi}{\mathcal{T}}\mathbb{Z}}(k).$$

Recall that $\delta_{\frac{2\pi}{\mathcal{T}}\mathbb{Z}}(k)$ denotes the Dirac measure on $\frac{2\pi}{\mathcal{T}}\mathbb{Z}$ defined in (2.3.1). This will be done by an application of the transference principle (Theorem 2.3.1) and the multiplier theorem of Marcinkiewicz (Theorem 2.3.2).

Lemma A.2.3. *Let $q \in (1, \infty)$ and $(\alpha, \beta) \in \mathbb{N}_0 \times \mathbb{N}_0^2$, with $\alpha \leq 2$ and $|\beta| \leq 4$. For any $(k, \xi) \in \frac{2\pi}{\mathcal{T}}\mathbb{Z} \times \left(\frac{2\pi}{L}\mathbb{Z}\right)^2$ we have that $(ik)^\alpha m_{\mathbb{T}}$ and $(i\xi)^\beta m_{\mathbb{T}}$ are $\mathrm{L}^q(\mathbb{T} \times \mathbb{T}_0^2)$-multipliers and the estimate*

$$\left\| \mathscr{F}_{\mathbb{T} \times \mathbb{T}_0^2}^{-1} \left[(ik)^\alpha m_{\mathbb{T}} \mathscr{F}_{\mathbb{T} \times \mathbb{T}_0^2}[f] \right] \right\|_{\mathrm{L}_\perp^q(\mathbb{T} \times \mathbb{T}_0^2)}$$
$$+ \left\| \mathscr{F}_{\mathbb{T} \times \mathbb{T}_0^2}^{-1} \left[(i\xi)^\beta m_{\mathbb{T}} \mathscr{F}_{\mathbb{T} \times \mathbb{T}_0^2}[f] \right] \right\|_{\mathrm{L}_\perp^q(\mathbb{T} \times \mathbb{T}_0^2)} \leq C_{54} \|f\|_{\mathrm{L}_\perp^q(\mathbb{T} \times \mathbb{T}_0^2)}, \tag{A.2.8}$$

holds for any $f \in \mathrm{L}_\perp^q(\mathbb{T} \times \mathbb{T}_0^2)$, with $C_{54} = C_{54}(q, \mathcal{T}) > 0$.

Proof. Following an analogous approach as in the proof of Lemma 3.2.5, we let $\chi \in C_0^\infty(\mathbb{R}; [0, 1])$ be a cut-off function with

$$\chi(\theta) = 1 \quad \text{for } |\theta| \leq \frac{\pi}{\mathcal{T}} \qquad \text{and} \qquad \chi(\theta) = 0 \quad \text{for } |\theta| \geq \frac{2\pi}{\mathcal{T}},$$

and put

$$m_{\mathbb{R}} \colon \mathbb{R} \times \mathbb{R}^2 \to \mathbb{C}, \quad m_{\mathbb{R}}(\theta, \zeta) := \rho_{\mathbb{R}}(\theta) \mathfrak{m}_{\mathbb{R}}^{-1}(\theta, \zeta) \tag{A.2.9}$$

with

$$\mathfrak{m}_{\mathbb{R}}(\theta, \zeta) := |\zeta|^4 - \theta^2 + i\theta|\zeta|^2,$$
$$\rho_{\mathbb{R}}(\theta) := 1 - \chi(\theta).$$

In order to employ Theorem 2.3.1 we put

$$\Phi \colon \frac{2\pi}{\mathcal{T}}\mathbb{Z} \times \left(\frac{2\pi}{L}\mathbb{Z}\right)^2 \to \mathbb{R} \times \mathbb{R}^2, \qquad \Phi(\theta, \zeta) := (\theta, \zeta). \tag{A.2.10}$$

Clearly, Φ is a (continuous) homomorphism of topological groups. Moreover, $m_{\mathbb{T}} = m_{\mathbb{R}} \circ \Phi$. Consequently, it suffices to show that $m_{\mathbb{R}}$ is an $\mathrm{L}^q(\mathbb{R} \times \mathbb{R}^2)$-multiplier to conclude the assertion for $m_{\mathbb{T}}$. To verify the condition (2.3.3) of Marcinkiewicz multiplier theorem, we set $I_1 - I_5$ as in (A.1.7), that is

$$\begin{aligned}
I_1 &= \frac{1}{(|\zeta|^4 - \theta^2)^2 + \theta^2 |\zeta|^4}, & I_4 &= |\zeta|^4 I_1, \\
I_2 &= \theta^2 |\zeta|^4 I_1, & I_5 &= 4\theta^4 I_1 + I_2, \\
I_3 &= (16|\zeta|^8 + 4\theta^2 |\zeta|^4) I_1,
\end{aligned} \tag{A.2.11}$$

and use the boundedness of these terms to deduce

$$|m_{\mathbb{R}}|^2 = |\rho_{\mathbb{R}}(\theta)|^2 I_1 \leq c_0(\mathcal{T}), \tag{A.2.12}$$

for all $(\theta, \zeta) \in \mathbb{R} \times \mathbb{R}^2$. Differentiating $m_{\mathbb{R}}$ with respect to θ we see

$$\partial_\theta m_{\mathbb{R}} = -\chi' \mathrm{m}_{\mathbb{R}}^{-1} - \rho_{\mathbb{R}} \mathrm{m}_{\mathbb{R}}^{-2}(-2\theta - i|\zeta|^2)$$
$$= -\chi' \mathrm{m}_{\mathbb{R}}^{-1} - m_{\mathbb{R}} \mathrm{m}_{\mathbb{R}}^{-1}(-2\theta - i|\zeta|^2)$$

and deduce from this representation formula that

$$|\theta \partial_\theta m_{\mathbb{R}}|^2 \leq |\chi'|^2 I_1 + |m_{\mathbb{R}}|^2 (I_4 + I_5) \leq c_1(\mathcal{T}). \tag{A.2.13}$$

Similarly, the derivatives with respect to ζ_1 and ζ_2 as well as the mixed derivatives can be determined and estimated in the following way:

$$|\zeta_j \partial_{\zeta_j} m_{\mathbb{R}}|^2 = \left| -\zeta_j \frac{\rho_{\mathbb{R}}(4\zeta_j |\zeta|^2 - 2i\theta\zeta_j)}{(|\zeta|^4 - \theta^2)^2 + \theta^2 |\zeta|^4} \right|^2 \leq |m_{\mathbb{R}}|^2 I_3,$$

$$|\zeta_1 \zeta_2 \partial_{\zeta_1} \partial_{\zeta_2} m_{\mathbb{R}}|^2 \leq 4|m_{\mathbb{R}}|^2 (16I_4 + I_3^2),$$

$$|\theta\zeta_j \partial_\theta \partial_{\zeta_j} m_{\mathbb{R}}|^2 \leq |\chi'(\theta)|^2 \theta^2 I_1 I_3 + 4|m_{\mathbb{R}}|^2 (I_2 + I_5 I_3), \tag{A.2.14}$$

$$|\theta\zeta_1 \zeta_2 \partial_\theta \partial_{\zeta_1} \partial_{\zeta_2} m_{\mathbb{R}}|^2 \leq 256\big(|\chi'(\theta)|^2 \theta^2 (I_4^2 + I_1 I_3^2)$$
$$+ |m_{\mathbb{R}}|^2 I_3(I_2 + I_5 I_3) + |\rho_{\mathbb{R}}(\theta)|^2 I_5 I_4^2\big).$$

Owing to (A.1.7), and using the construction of χ, we deduce that the inequalities above are uniform, which is condition (2.3.3) of the multiplier theorem of Marcinkiewicz. Hence, Theorem 2.3.2 yields that $m_{\mathbb{R}}$ is an $L^q(\mathbb{R} \times \mathbb{R}^2)$-multiplier, and therefore $m_{\mathbb{T}}$ is an $L^q(\mathbb{T} \times \mathbb{T}_0^2)$-multiplier according to the transference principle (Theorem 2.3.1). Moreover, we obtain

$$\|\mathscr{F}_{\mathbb{T} \times \mathbb{T}_0^2}^{-1}[m_{\mathbb{T}} \mathscr{F}_{\mathbb{T} \times \mathbb{T}_0^2}[f]]\|_q \leq c_2 \|f\|_q.$$

In order to repeat the previous argumentation for $(i\theta)^\alpha m_{\mathbb{R}}$, observe first that

$$I_6 := \theta^2 I_1 \leq 4 \frac{\mathcal{T}^2}{\pi^2} \tag{A.2.15}$$

holds, with $(\theta, \zeta) \in \mathbb{R} \times \mathbb{R}^2$ such that $|\theta| > \frac{\pi}{\mathcal{T}}$. In view of (A.2.14), we deduce from (A.1.7) in combination with (A.2.15) that

$$|\theta^\gamma \xi^\mu \partial_\theta^\gamma \partial_\xi^\mu ((i\theta)^\alpha m_{\mathbb{R}})| \leq c_3 \big(\gamma |\theta^{\alpha+\gamma-1} \xi^\mu \partial_\xi^\mu m_{\mathbb{R}}| + |\theta^{\alpha+\gamma} \xi^\mu \partial_\theta^\gamma \partial_\xi^\mu m_{\mathbb{R}}|\big)$$
$$\leq c_4 \big(I_6(16I_4 + I_3) + |\chi'(\theta)| \theta^{\frac{\alpha+2}{2}} (I_4 + I_1 I_3^2)$$
$$+ I_6 I_3(I_2 + I_5 I_3) + |\rho_{\mathbb{R}}(\theta)|^2 I_2 I_4 I_5\big)$$

with $\gamma \in \{0,1\}$ and $\mu \in \{0,1\}^2$. Hence, Theorem 2.3.2 and Theorem 2.3.1 yield that $(ik)^\alpha m_{\mathbb{T}}$ is an $L^q(\mathbb{T} \times \mathbb{T}_0^2)$-multiplier. Estimate (A.2.15) can be established by mimicking the exact same steps as for (A.1.1)$_1$ with $\varepsilon = \frac{\theta}{2}$.

In order to verify that $(i\zeta)^\beta m_{\mathbb{R}}$ is an $L^q(\mathbb{R} \times \mathbb{R}^2)$-multiplier, we proceed similarly as for $(i\theta)^\alpha m_{\mathbb{R}}$ and employ the estimates (A.1.7) and (A.2.14) to obtain

$$|\theta^\gamma \zeta^\mu \partial_\theta^\gamma \partial_\zeta^\mu ((i\zeta)^\beta m_{\mathbb{R}})| \le C < \infty. \tag{A.2.16}$$

Observe that we have used

$$I_7 := |\zeta|^2 I_1 = \sqrt{I_1 I_4} \le 4\frac{\mathcal{T}^3}{\pi^3},$$

$$I_8 := |\zeta|^5 I_1 \le \frac{3}{4}(I_4 + I_{11}) \le \frac{3}{4}(\frac{\mathcal{T}^2}{\pi^2} + 4),$$

$$I_9 := |\zeta|^6 I_1 \le \frac{3}{4}(I_4 + I_{11}) \le \frac{3}{4}(\frac{\mathcal{T}^2}{\pi^2} + 4),$$

$$I_{10} := |\zeta|^7 I_1 \le \frac{3}{4}(I_4 + I_{11}) \le \frac{3}{4}(\frac{\mathcal{T}^2}{\pi^2} + 4),$$

$$I_{11} := |\zeta|^8 I_1 = \frac{1}{16}(I_3 - 4I_2) \le 4$$

to get (A.2.16). Hence, the transference principle yields that also $(i\xi)^\beta m_{\mathbb{T}}$ is an $L^q(\mathbb{T} \times \mathbb{T}_0^2)$-multiplier, since Φ is a (continuous) homomorphism. Moreover, Theorem 2.3.2 implies that the L^q estimate (A.2.8) is valid. $\square$

Lemma A.2.4. *Let $q \in (1,\infty)$. For any $(k,\xi) \in \frac{2\pi}{\mathcal{T}}\mathbb{Z} \times (\frac{2\pi}{L}\mathbb{Z})^2$ we have that $M_{\mathbb{T}}$ is an $L^q(\mathbb{T} \times \mathbb{T}_0^2)$-multiplier and the estimate*

$$\|\mathscr{F}_{\mathbb{T} \times \mathbb{T}_0^2}^{-1}[M_{\mathbb{T}} \mathscr{F}_{\mathbb{T} \times \mathbb{T}_0^2}[f]]\|_{L_\perp^q(\mathbb{T} \times \mathbb{T}_0^2)} \le C_{55}\|f\|_{L_\perp^q(\mathbb{T} \times \mathbb{T}_0^2)} \tag{A.2.17}$$

holds for any $f \in L_\perp^q(\mathbb{T} \times \mathbb{T}_0^2)$, with $C_{55} = C_{55}(q,\mathcal{T}) > 0$.

Proof. We prove the assertion of this lemma in the same way as for $m_{\mathbb{T}}$. For this purpose we put

$$M_{\mathbb{R}} : \mathbb{R} \times \mathbb{R}^2 \to \mathbb{C},$$

$$M_{\mathbb{R}}(\theta,\zeta) := \frac{\rho_{\mathbb{R}}(\theta)|\zeta|m_{\mathbb{R}}(\theta,\zeta)}{|\zeta|m_{\mathbb{R}}(\theta,\zeta) - \theta^2 + i\theta|\zeta|\left(|\zeta| + \sqrt{|\zeta|^2 + i\theta}\right)},$$

with χ, $\rho_{\mathbb{R}}$ and $\mathfrak{m}_{\mathbb{R}}$ as in the proof of the previous lemma, and show that $M_{\mathbb{R}}$ is a $L^q(\mathbb{R} \times \mathbb{R}^2)$-multiplier, which together with the transference principle yields the assertion. Moreover, we define

$$\mathcal{N}_1 := |\zeta|(|\zeta|^4 - \theta^2) - \theta^2 - \theta|\zeta|\operatorname{Im}\sqrt{|\zeta|^2 + i\theta},$$

$$\mathcal{N}_2 := \theta|\zeta|\left(|\zeta|^2 + |\zeta| + \operatorname{Re}\sqrt{|\zeta|^2 + i\theta}\right),$$

$$\mathcal{N} := \mathcal{N}_1 + i\mathcal{N}_2, \tag{A.2.18}$$

$$\partial_{s_1,\cdots,s_n}\widetilde{\mathcal{N}} := \frac{\partial_{s_1}\cdots\partial_{s_n}\mathcal{N}}{\mathcal{N}},$$

$$\partial_{s_1,\cdots,s_n}\widetilde{m} := \frac{\partial_{s_1}\cdots\partial_{s_n}(|\zeta|\mathfrak{m}_{\mathbb{R}})}{\mathcal{N}}$$

with $s_j \in \{\theta, \zeta_1, \zeta_2\}$, $j = 1, \ldots, n$, and observe that $M_{\mathbb{R}}$ can be written as

$$M_{\mathbb{R}} = \frac{\rho_{\mathbb{R}}|\zeta|\mathfrak{m}_{\mathbb{R}}}{\mathcal{N}}. \tag{A.2.19}$$

Furthermore, the derivatives of $M_{\mathbb{R}}$ with respect to θ and ζ are given by

$$\partial_\theta M_{\mathbb{R}} = \partial_\theta \widetilde{m} - M_{\mathbb{R}}\partial_\theta\widetilde{\mathcal{N}},$$

$$\partial_{\zeta_j} M_{\mathbb{R}} = \partial_{\zeta_j}\widetilde{m} - M_{\mathbb{R}}\partial_{\zeta_j}\widetilde{\mathcal{N}}, \tag{A.2.20}$$

$$\partial_{\zeta_1}\partial_{\zeta_2} M_{\mathbb{R}} = \partial_{\zeta_1,\zeta_2}\widetilde{m} - \partial_{\zeta_1} M_{\mathbb{R}}\partial_{\zeta_2}\widetilde{\mathcal{N}} - \partial_{\zeta_2} M_{\mathbb{R}}\partial_{\zeta_1}\widetilde{\mathcal{N}} - M_{\mathbb{R}}\partial_{\zeta_1,\zeta_2}\widetilde{\mathcal{N}},$$

whereas the mixed derivatives are given by

$$\partial_\theta\partial_{\zeta_j} M_{\mathbb{R}} = \partial_{\theta,\zeta_j}\widetilde{m} - \partial_\theta M_{\mathbb{R}}\partial_{\zeta_j}\widetilde{\mathcal{N}} - \partial_{\zeta_j} M_{\mathbb{R}}\partial_\theta\widetilde{\mathcal{N}} - M_{\mathbb{R}}\partial_{\theta,\zeta_j}\widetilde{\mathcal{N}},$$

$$\begin{aligned}\partial_\theta\partial_{\zeta_1}\partial_{\zeta_2} M_{\mathbb{R}} = {} &\partial_{\theta,\zeta_1,\zeta_2}\widetilde{m} - \partial_{\zeta_1}\partial_{\zeta_2} M_{\mathbb{R}}\partial_\theta\widetilde{\mathcal{N}} - \partial_\theta\partial_{\zeta_1} M_{\mathbb{R}}\partial_{\zeta_2}\widetilde{\mathcal{N}} \\ &- \partial_\theta\partial_{\zeta_2} M_{\mathbb{R}}\partial_{\zeta_1}\widetilde{\mathcal{N}} - \partial_{\zeta_1} M_{\mathbb{R}}\partial_{\theta,\zeta_2}\widetilde{\mathcal{N}} - \partial_{\zeta_2} M_{\mathbb{R}}\partial_{\theta,\zeta_1}\widetilde{\mathcal{N}} \\ &- \partial_\theta M_{\mathbb{R}}\partial_{\zeta_1,\zeta_2}\widetilde{\mathcal{N}} - M_{\mathbb{R}}\partial_{\theta,\zeta_1,\zeta_2}\widetilde{\mathcal{N}}.\end{aligned} \tag{A.2.21}$$

To find appropriate estimates for these terms, we first observe that the real and imaginary part of $\sqrt{|\zeta|^2 + i\theta}$ occurring in $\mathcal{N}_1$ and $\mathcal{N}_2$ are given by

$$\mathfrak{Re} := \operatorname{Re}\left[\sqrt{|\zeta|^2 + i\theta}\right] = \frac{\sqrt{\sqrt{|\zeta|^4 + \theta^2} + |\zeta|^2}}{\sqrt{2}},$$

$$\mathfrak{Im} := \operatorname{Im}\left[\sqrt{|\zeta|^2 + i\theta}\right] = \operatorname{sgn}(\theta)\frac{\sqrt{\sqrt{|\zeta|^4 + \theta^2} - |\zeta|^2}}{\sqrt{2}}.$$

Before we show that $M_{\mathbb{R}}$ satisfies condition (2.3.3) of Marcinkiewicz's multiplier theorem, we prove the boundedness of

$$J_1 := \frac{|\zeta|^{10}}{\mathcal{N}_1^2 + \mathcal{N}_2^2}, \qquad J_2 := \frac{\theta^2 |\zeta|^6}{\mathcal{N}_1^2 + \mathcal{N}_2^2}, \qquad J_3 := \frac{\theta^4}{\mathcal{N}_1^2 + \mathcal{N}_2^2},$$
$$J_4 := \frac{\theta^4 |\zeta|^2}{\mathcal{N}_1^2 + \mathcal{N}_2^2}, \qquad J_5 := \frac{\theta^2 |\zeta|^4}{\mathcal{N}_1^2 + \mathcal{N}_2^2}, \qquad J_6 := \frac{|\zeta|^2 (|\zeta|^4 - \theta^2)^2}{\mathcal{N}_1^2 + \mathcal{N}_2^2}, \qquad (\text{A.2.22})$$
$$J_7 := \frac{\theta^2 |\zeta|^2 (\mathfrak{Re})^2}{\mathcal{N}_1^2 + \mathcal{N}_2^2}, \qquad J_8 := \frac{\theta^2 |\zeta|^2 (\mathfrak{Im})^2}{\mathcal{N}_1^2 + \mathcal{N}_2^2}$$

for all $(\theta, \zeta) \in \mathbb{R} \times \mathbb{R}^2$ with $|\theta| > \frac{\pi}{\mathcal{T}}$. Observe that since $|\zeta| \geq 0$ and $\sqrt[4]{\theta^2} > \sqrt{\frac{\pi}{\mathcal{T}}} > 0$, the only zeros of $\mathcal{N}_2$ are $(\theta, 0)$ with arbitrary $\theta \in \mathbb{R}$. But in this case we have that

$$|\mathcal{N}_1(\theta, 0)| = \theta^2 > \frac{\pi^2}{\mathcal{T}^2} > 0.$$

Hence, the denominator of $J_1 - J_8$ does not vanish for any $(\theta, \zeta) \in \mathbb{R} \times \mathbb{R}^2$ with $|\theta| > \frac{\pi}{\mathcal{T}}$, and therefore they are continuous on that set; in fact $J_1 - J_8$ are smooth, and bounded as long as the numerator stays bounded. Hence, for J_1 it only remains to exclude the case $|\zeta| \to \infty$. But in this limiting case we conclude

$$\lim_{|\zeta| \to \infty} J_1 = \lim_{|\zeta| \to \infty} \left[\left(1 - \frac{\theta^2}{|\zeta|^4} - \frac{\theta^2}{|\zeta|^5} - \frac{\theta}{|\zeta|^4} \mathfrak{Im} \right)^2 \right.$$
$$\left. + \theta^2 \left(\frac{1}{|\zeta|^2} + \frac{1}{|\zeta|^3} + \frac{1}{|\zeta|^4} \mathfrak{Re} \right)^2 \right]^{-1} = 1,$$

for any fixed $\theta \in \mathbb{R}$, which implies the boundedness of J_1. Here we have used that

$$\lim_{|\zeta| \to \infty} \frac{\mathfrak{Re}}{|\zeta|^4} = 0 = \lim_{|\zeta| \to \infty} \frac{\mathfrak{Im}}{|\zeta|^4}.$$

For J_2 and J_5, observe that for $|\zeta| = 0$ and arbitrary $\theta \in \mathbb{R}$ (with $|\theta| > \frac{\pi}{\mathcal{T}}$), J_2 and J_5 vanish, hence we determine for these two terms

$$J_2 \leq \frac{1}{(1 + \frac{1}{|\zeta|} + \frac{1}{|\zeta|^2} \mathfrak{Re})^2} \leq 1 < \infty,$$
$$J_5 \leq \frac{1}{\frac{1}{|\zeta|^2} \left(|\zeta|^2 + |\zeta| + \mathfrak{Re} \right)^2} \leq 1 < \infty. \qquad (\text{A.2.23})$$

In the case of J_3 and J_4 a more complicated, but somehow straightforward calculation yields

$$J_3 \leq 1 < \infty, \qquad \text{and} \qquad J_4 \leq \frac{4}{3} < \infty. \qquad (\text{A.2.24})$$

To be more precise, calculating $\mathcal{N}_1^2 + \mathcal{N}_2^2$ and observe that even there appears negative addend in the denominator, it is possible to write the denominator as

$$1 + \sum_{i=1}^{m} a_i > 1,$$

with $m \in \mathbb{N}$ finite and $a_i \in \mathbb{R}_+$ nonnegative. Hence, we find (A.2.24). Due to the boundedness of J_1 we deduce from (A.2.23) and (A.2.24) that

$$J_6 = J_1 - 2J_2 + J_4 \le J_1 + 2 + \frac{4}{3} \le c_0(\mathcal{T}) < \infty, \tag{A.2.25}$$

and for the remaining terms we find

$$J_7 \le \frac{1}{\left(\frac{|\zeta|^2}{\mathfrak{Re}} + \frac{|\zeta|}{\mathfrak{Re}} + 1\right)^2} \le 1, \quad \text{and} \quad J_8 \le \frac{\mathfrak{Im}^2}{\mathfrak{Re}^2} \frac{1}{\left(\frac{|\zeta|^2}{\mathfrak{Re}} + \frac{|\zeta|}{\mathfrak{Re}} + 1\right)^2} \le 1, \tag{A.2.26}$$

for all $(\theta, \zeta) \in \mathbb{R} \times \mathbb{R}^2$ with $|\theta| > \frac{\pi}{\mathcal{T}}$. Obviously, $J_1 - J_8$ can be used to rewrite and estimate $M_\mathbb{R}$ and all its partial derivatives collected in (A.2.19) – (A.2.21). For this purpose, we first turn to (A.2.18) and show that the partial derivatives $\partial_{s_1,\cdots,s_n}\widetilde{\mathcal{N}}$ and $\partial_{s_1,\cdots,s_n}\widetilde{m}$ satisfy the boundedness condition of Marcinkiewicz's multiplier theorem by exploiting (A.2.23) – (A.2.26). Observe that, since we investigate $M_\mathbb{R}$ in $\mathbb{R} \times \mathbb{R}^2$, it suffices to consider partial derivatives up to order three. For the first order derivatives of $\widetilde{m}$ we deduce

$$|\theta\partial_\theta\widetilde{m}|^2 = J_2 + 4J_4 \quad \text{and} \quad |\zeta_j\partial_{\zeta_j}\widetilde{m}|^2 \le |M_\mathbb{R}|^2 + 16J_1 + 4J_2, \tag{A.2.27}$$

whereas for those of $\widetilde{\mathcal{N}}$, it holds that

$$|\theta\partial_\theta\widetilde{\mathcal{N}}|^2 \le c_1(J_2 + J_3 + J_4 + J_5 + J_7 + J_8),$$
$$|\zeta_j\partial_{\zeta_j}\widetilde{\mathcal{N}}|^2 \le c_2(|M_\mathbb{R}|^2 + J_1 + J_2 + J_5 + J_7 + J_8). \tag{A.2.28}$$

In the case $n = 2$, $\partial_{\zeta_1,\zeta_2}\widetilde{m}$ and $\partial_{\zeta_1,\zeta_2}\widetilde{\mathcal{N}}$ obey

$$|\zeta_1\zeta_2\partial_{\zeta_1,\zeta_2}\widetilde{m}|^2 \le c_3(|M_\mathbb{R}|^2 + J_1 + J_2),$$
$$|\zeta_1\zeta_2\partial_{\zeta_1,\zeta_2}\widetilde{\mathcal{N}}|^2 \le c_4(|M_\mathbb{R}|^2 + J_1 + J_2 + J_5 + J_7 + J_8), \tag{A.2.29}$$

179

and the second order mixed derivatives can be estimated as

$$
\begin{aligned}
|\theta\zeta_j\partial_{\theta,\zeta_j}\widetilde{m}|^2 &\le c_5(J_2 + J_4 + J_5), \\
|\theta\zeta_j\partial_{\theta,\zeta_j}\widetilde{\mathcal{N}}|^2 &\le c_6(J_1 + J_2 + J_4 + J_5 + J_7 + J_8).
\end{aligned}
\tag{A.2.30}
$$

Finally, the highest-order partial derivatives of $\widetilde{m}$ and $\widetilde{\mathcal{N}}$, that is, $\partial_{\theta,\zeta_1,\zeta_2}\widetilde{m}$ and $\partial_{\theta,\zeta_1,\zeta_2}\widetilde{\mathcal{N}}$, are bounded due to

$$
\begin{aligned}
|\theta\zeta_1\zeta_2\partial_{\theta,\zeta_1,\zeta_2}\widetilde{m}|^2 &\le c_7(J_2 + J_4), \\
|\theta\zeta_1\zeta_2\partial_{\theta,\zeta_1,\zeta_2}\widetilde{\mathcal{N}}|^2 &\le c_8(J_1 + J_2 + J_4 + J_5 + J_7 + J_8).
\end{aligned}
\tag{A.2.31}
$$

With (A.2.27) – (A.2.31) at hand we now show that $M_{\mathbb{R}}$ satisfies condition (2.3.3) of Marcinkiewicz multiplier theorem (Theorem 2.3.2), but first observe that due to (A.2.23) and (A.2.25), we determine

$$
|M_{\mathbb{R}}|^2 = |\rho_{\mathbb{R}}|^2(J_2 + J_6) \le |\rho_{\mathbb{R}}|^2(1 + c_0) < \infty.
$$

For the first order derivatives we employ (A.2.27) and (A.2.28), with view to (A.2.23) – (A.2.26), to find that

$$
\begin{aligned}
|\theta\partial_\theta M_{\mathbb{R}}|^2 &\le |\theta\partial_\theta\widetilde{m}|^2 + |M_{\mathbb{R}}|^2|\theta\partial_\theta\widetilde{\mathcal{N}}|^2 \le c_9 < \infty, \\
|\zeta_j\partial_{\zeta_j}M_{\mathbb{R}}|^2 &\le |\zeta_j\partial_{\zeta_j}\widetilde{m}|^2 + |M_{\mathbb{R}}|^2|\zeta_j\partial_{\zeta_j}\widetilde{\mathcal{N}}|^2 \le c_{10} < \infty,
\end{aligned}
$$

and utilizing (A.2.28) – (A.2.30), this further implies that

$$
\begin{aligned}
|\zeta_1\zeta_2\partial_{\zeta_1}\partial_{\zeta_2}M_{\mathbb{R}}|^2 &\le |\zeta_1\zeta_2\partial_{\zeta_1,\zeta_2}\widetilde{m}|^2 + |\zeta_1\partial_{\zeta_1}M_{\mathbb{R}}|^2|\zeta_2\partial_{\zeta_2}\widetilde{\mathcal{N}}|^2 \\
&\quad + |\zeta_2\partial_{\zeta_2}M_{\mathbb{R}}|^2|\zeta_1\partial_{\zeta_1}\widetilde{\mathcal{N}}|^2 + |M_{\mathbb{R}}|^2|\zeta_1\zeta_2\partial_{\zeta_1,\zeta_2}\widetilde{\mathcal{N}}|^2 \le c_{11} < \infty, \\
|\theta\zeta_j\partial_\theta\partial_{\zeta_j}M_{\mathbb{R}}|^2 &\le |\theta\zeta_j\partial_{\theta,\zeta_j}\widetilde{m}|^2 + |\theta\partial_\theta M_{\mathbb{R}}|^2|\zeta_j\partial_{\zeta_j}\widetilde{\mathcal{N}}|^2 \\
&\quad + |\zeta_j\partial_{\zeta_j}M_{\mathbb{R}}|^2|\theta\partial_\theta\widetilde{\mathcal{N}}|^2 + |M_{\mathbb{R}}|^2|\theta\zeta_j\partial_{\theta,\zeta_j}\widetilde{\mathcal{N}}|^2 \le c_{12} < \infty
\end{aligned}
$$

holds. Note that the constants occurring above depend at most on the period length $\mathcal{T}$. Collecting all these information, we finally deduce

$$
\begin{aligned}
|\theta\zeta_1\zeta_2\partial_\theta\partial_{\zeta_1}\partial_{\zeta_2}M_{\mathbb{R}}|^2 &\le |\theta\zeta_1\zeta_2\partial_{\theta,\zeta_1,\zeta_2}\widetilde{m}|^2 + |\zeta_1\zeta_2\partial_{\zeta_1}\partial_{\zeta_2}M_{\mathbb{R}}|^2|\theta\partial_\theta\widetilde{\mathcal{N}}|^2 \\
&\quad + |\theta\zeta_1\partial_\theta\partial_{\zeta_1}M_{\mathbb{R}}|^2|\zeta_2\partial_{\zeta_2}\widetilde{\mathcal{N}}|^2 + |\theta\zeta_2\partial_\theta\partial_{\zeta_2}M_{\mathbb{R}}|^2|\zeta_1\partial_{\zeta_1}\widetilde{\mathcal{N}}|^2 \\
&\quad + |\zeta_1\partial_{\zeta_1}M_{\mathbb{R}}|^2|\theta\zeta_2\partial_{\theta,\zeta_2}\widetilde{\mathcal{N}}|^2 + |\zeta_2\partial_{\zeta_2}M_{\mathbb{R}}|^2|\theta\zeta_1\partial_{\theta,\zeta_1}\widetilde{\mathcal{N}}|^2 \\
&\quad + |\theta\partial_\theta M_{\mathbb{R}}|^2|\zeta_1\zeta_2\partial_{\zeta_1,\zeta_2}\widetilde{\mathcal{N}}|^2 + |M_{\mathbb{R}}|^2|\theta\zeta_1\zeta_2\partial_{\theta,\zeta_1,\zeta_2}\widetilde{\mathcal{N}}|^2 \le c_{13} < \infty.
\end{aligned}
$$

Since the $M_{\mathbb{R}}$ and all its partial derivatives are uniformly bounded, which is (2.3.3), Theorem 2.3.2 yields that $M_{\mathbb{R}}$ is an $L^q(\mathbb{R} \times \mathbb{R}^2)$-multiplier that satisfies

$$\left\| \mathscr{F}^{-1}_{\mathbb{R} \times \mathbb{R}^2}\left[M_{\mathbb{R}} \mathscr{F}_{\mathbb{R} \times \mathbb{R}^2}[f] \right] \right\|_{L^q_\perp(\mathbb{R} \times \mathbb{R}^2)} \leq c_{14} \| f \|_{L^q_\perp(\mathbb{R} \times \mathbb{R}^2)},$$

for all $f \in L^q_\perp(\mathbb{R} \times \mathbb{R}^2)$. Consequently the assertion follows by employing the transference principle (Theorem 2.3.1). $\qquad \square$

Bibliography

[1] T. Abe and Y. Shibata. On a resolvent estimate of the Stokes equation on an infinite layer. II. $\lambda = 0$ case. *J. Math. Fluid Mech.*, 5(3):245–274, 2003. 146, 147

[2] R. A. Adams and J. J. F. Fournier. *Sobolev spaces*, volume 140 of *Pure and Applied Mathematics (Amsterdam)*. Elsevier/Academic Press, Amsterdam, second edition, 2003. 61, 84, 152

[3] S. Agmon, A. Douglis, and L. Nirenberg. Estimates near the boundary for solutions of elliptic partial differential equations satisfying general boundary conditions. I. *Comm. Pure Appl. Math.*, 12:623–727, 1959. 48

[4] J. Bergh and J. Löfström. *Interpolation spaces. An introduction.* Springer-Verlag, Berlin-New York, 1976. Grundlehren der Mathematischen Wissenschaften, No. 223. 22, 23

[5] D. Bernoulli. Réflexions et éclaircissements sur les nouvelles vibrations des cordes, exposdées dans les mémoires de l'académie, de 1747 et 1748. page 147ff, 1755. 3

[6] R. Beyer. Parameter of nonlinearity in fluids. *J. Acoust. Soc. Am.*, 32:719–721, 1960. 2, 39

[7] D. Blackstock. Approximate equations governing finite-amplitude sound in thermoviscous fluids. GD/E report GD-1463-52, General Dynamics Coporation, 1963. 1, 4, 39, 43, 44

[8] F. Bruhat. Distributions sur un groupe localement compact et applications à l'étude des représentations des groupes $\wp$-adiques. *Bull. Soc. Math. France*, 89:43–75, 1961. 13

[9] R. Brunnhuber. Well-posedness and exponential decay of solutions for the Blackstock-Crighton-Kuznetsov equation. *J. Math. Anal. Appl.*, 433(2):1037–1054, 2016. 4, 44, 74, 75

[10] R. Brunnhuber and B. Kaltenbacher. Well-posedness and asymptotic behavior of solutions for the Blackstock-Crighton-Westervelt equation. *Discrete Contin. Dyn. Syst.*, 34(11):4515–4535, 2014. 4, 75

[11] R. Brunnhuber and S. Meyer. Optimal regularity and exponential stability for the Blackstock-Crighton equation in L_p-spaces with Dirichlet and Neumann boundary conditions. *J. Evol. Equ.*, 16(4):945–981, 2016. 4, 75

[12] J. M. Burgers. A mathematical model illustrating the theory of turbulence. In *Advances in Applied Mechanics*, pages 171–199. Academic Press, Inc., New York, N. Y., 1948. edited by Richard von Mises and Theodore von Kármán,. 4

[13] A. Celik and M. Kyed. Nonlinear wave equation with damping: Periodic forcing and non-resonant solutions to the Kuznetsov equation. *ZAMM Z. Angew. Math. Mech.*, 98(3):412–430, 2018. 41

[14] A. Celik and M. Kyed. Nonlinear acoustics: Blackstock-Crighton equations with a periodic forcing term. *J. Math. Fluid Mech.*, 21(3):Art. 45, 12, 2019. 41

[15] A. Celik and M. Kyed. Time-periodic Stokes equations with inhomogeneous Dirichlet boundary conditions in the half-space. *Math. Meth. Appl. Sci.*, pages 1–17, 2019. 102

[16] A. Chambolle, B. Desjardins, M. J. Esteban, and C. Grandmont. Existence of weak solutions for the unsteady interaction of a viscous fluid with an elastic plate. *J. Math. Fluid Mech.*, 7(3):368–404, 2005. 5, 6, 87, 89

[17] I. Chueshov and I. Ryzhkova. A global attractor for a fluid-plate interaction model. *Commun. Pure Appl. Anal.*, 12(4):1635–1656, 2013. 89

[18] D. G. Crighton and J. F. Scott. Asymptotic solutions of model equations in nonlinear acoustics. *Philos. Trans. Roy. Soc. London Ser. A*, 292(1389):101–134, 1979. 1, 4, 44

[19] H. B. da Veiga. On the existence of strong solutions to a coupled fluid-structure evolution problem. *J. Math. Fluid Mech.*, 6:21–52, 2004. 5, 6

[20] H. B. da Veiga. On the existence of strong solutions to a coupled fluid-structure evolution problem. *J. Math. Fluid Mech.*, 6(1):21–52, 2004. 87, 89, 100, 124

[21] J. L. R. d'Alembert. Recherches sur le courbe que forme une corde tendue raise en vibration. page 214ff, 1747. 3

[22] K. de Leeuw. On L_p multipliers. *Ann. of Math. (2)*, 81:364–379, 1965. 9, 15, 46

[23] L. Debnath. *Sir James Lighthill and modern fluid mechanics.* Imperial College Press, London, 2008. 4

[24] R. Denk and J. Saal. L_p Theory for a Fluid-Structure Interaction Model. 09 2019. 6, 87, 100

[25] B. Desjardins, M. Esteban, C. Grandmont, and P. Tallec. Weak solutions for a fluid-elastic structure interaction model. *Rev. Mat. Complut.*, 14, 06 2000. 5, 6, 87

[26] B. Desjardins and M. J. Esteban. Existence of weak solutions for the motion of rigid bodies in a viscous fluid. *Arch. Ration. Mech. Anal.*, 146(1):59–71, 1999. 5, 87

[27] R. E. Edwards and G. I. Gaudry. *Littlewood-Paley and multiplier theory.* Springer-Verlag, Berlin-New York, 1977. Ergebnisse der Mathematik und ihrer Grenzgebiete, Band 90. 9, 15, 46

[28] T. W. Eiter and M. Kyed. *Time-periodic linearized Navier-Stokes equations: an approach based on Fourier multipliers.*, pages 77–137. Cham: Birkhäuser/Springer, 2017. 13

[29] L. Euler. De la propagation du son. *Mém. Acad. Sci.*, 15:185 – 209, 1766. 3

[30] E. C. Everbach. *Parameters of Nonlinearity of Acoustic Media*, chapter 20, pages 219–226. John Wiley & Sons, Ltd, 2007. 39

[31] R. Farwig and H. Sohr. Generalized resolvent estimates for the Stokes system in bounded and unbounded domains. *J. Math. Soc. Japan*, 46(4):607–643, 1994. 125

[32] R. D. Fay. Plane sound waves of finite amplitude. *J. Acoust. Soc. Am.*, 3:222–241, 1931. 4

[33] F. Flori and P. Orenga. On a nonlinear fluid-structure interaction problem defined on a domain depending on time. *Nonlinear Analysis: Theory, Methods & Applications*, 38(5):549 – 569, 1999. 87

[34] H. Fujita and N. Sauer. On existence of weak solutions of the navier-stokes equations in regions with moving boundaries. *Journal of the Faculty of Science. Section I A*, 17, 01 1970. 5

[35] G. Galdi and H. Sohr. Existence and uniqueness of time-periodic physically reasonable Navier-Stokes flow past a body. *Arch. Ration. Mech. Anal.*, 172(3):363–406, 2004. 5

[36] G. P. Galdi. *An introduction to the mathematical theory of the Navier-Stokes equations*. Springer Monographs in Mathematics. Springer, New York, second edition, 2011. Steady-state problems. 38, 101, 103, 104, 115, 125, 135, 138, 141, 150, 154

[37] G. P. Galdi and M. Kyed. Time-periodic flow of a viscous liquid past a body. In *Partial differential equations in fluid mechanics*, volume 452 of *London Math. Soc. Lecture Note Ser.*, pages 20–49. Cambridge Univ. Press, Cambridge, 2018. 5, 32, 135, 136, 140

[38] G. P. Galdi, M. Mohebbi, R. Zakerzadeh, and P. Zunino. Hyperbolic-parabolic coupling and the occurrence of resonance in partially dissipative systems. In *Fluid-structure interaction and biomedical applications*, Adv. Math. Fluid Mech., pages 197–256. Birkhäuser/Springer, Basel, 2014. 2, 89, 100

[39] G. P. Galdi and R. Rannacher, editors. *Fundamental trends in fluid-structure interaction*, volume 1 of *Contemporary Challenges in Mathematical Fluid Dynamics and Its Applications*. World Scientific Publishing Co. Pte. Ltd., Hackensack, NJ, 2010. 5

[40] G. P. Galdi and A. L. Silvestre. Existence of time-periodic solutions to the Navier-Stokes equations around a moving body. *Pacific J. Math.*, 223(2):251–267, 2006. 5

[41] G. P. Galdi and A. L. Silvestre. On the motion of a rigid body in a Navier-Stokes liquid under the action of a time-periodic force. *Indiana Univ. Math. J.*, 58(6):2805–2842, 2009. 5

[42] M. Geißert, H. Heck, and M. Hieber. *On the Equation div u = g and Bogovskii's Operator in Sobolev Spaces of Negative Order*, pages 113–121. Birkhäuser Basel, Basel, 2006. 38

[43] G. Geymonat. Sui problemi ai limiti per i sistemi di equazioni lineari ellittici. *Atti Accad. Naz. Lincei, VIII. Ser., Rend., Cl. Sci. Fis. Mat. Nat.*, 37:35–39, 1964. 81

[44] L. Grafakos. *Classical Fourier analysis*, volume 249 of *Graduate Texts in Mathematics*. Springer, New York, second edition, 2008. 15, 28, 29, 47, 148

[45] L. Grafakos. *Modern Fourier Analysis*. Graduate Texts in Mathematics. Springer New York, 2014. 31

[46] C. Grandmont. Existence of weak solutions for the unsteady interaction of a viscous fluid with an elastic plate. *SIAM J. Math. Anal.*, 40(2):716–737, 2008. 89

[47] Grandmont, C. and Maday, Y. Existence for an unsteady fluid-structure interaction problem. *ESAIM: M2AN*, 34(3):609–636, 2000. 87

[48] P. Grisvard. Interpolation non commutative. *Atti Accad. Naz. Lincei Rend. Cl. Sci. Fis. Mat. Natur. (8)*, 52:11–15, 1972. 35

[49] E. Hopf. Über die Anfangswertaufgabe für die hydrodynamischen Grundgleichungen. *Math. Nachr.*, 4:213–231, 1951. 5

[50] M. S. Howe. *Acoustics of fluid-structure interactions*. Cambridge Monographs on Mechanics. Cambridge University Press, Cambridge, 1998. 93

[51] B. Kaltenbacher. Mathematics of nonlinear acoustics. *Evol. Equ. Control Theory*, 4(4):447–491, 2015. 4

[52] B. Kaltenbacher and I. Lasiecka. Well-posedness of the Westervelt and the Kuznetsov equation with nonhomogeneous Neumann boundary conditions. *Discrete Contin. Dyn. Syst.*, 2011:763–773, 2011. 4, 70

[53] B. Kaltenbacher and I. Lasiecka. An analysis of nonhomogeneous Kuznetsov's equation: Local and global well-posedness; exponential decay. *Math. Nachr.*, 285(2-3):295–321, 2012. 4, 70

[54] R. Kamakoti and W. Shyy. Fluid-structure interaction for aeroelastic applications. *Progress in Aerospace Sciences*, 40(8):535 – 558, 2004. 2, 5, 87

[55] S. Kaniel and M. Shinbrot. A reproductive property of the Navier-Stokes equations. *Arch. Rational Mech. Anal.*, 24:363–369, 1967. 5

[56] P. Kokocki. Effect of resonance on the existence of periodic solutions for strongly damped wave equation. *Nonlinear Anal.*, 125:167–200, 2015. 45, 70

[57] V. Kuznetsov. Equations of nonlinear acoustics. *Sov. Phys. Acoust.*, 16:467–470, 1971. 2, 4, 39, 44

[58] M. Kyed. Time-Periodic Solutions to the Navier-Stokes Equations. *Habilitationsschrift, Technische Universität Darmstadt*, 2012. 9

[59] M. Kyed. The existence and regularity of time-periodic solutions to the three-dimensional Navier-Stokes equations in the whole space. *Nonlinearity*, 27(12):2909–2935, 2014. 5

[60] M. Kyed. Maximal regularity of the time-periodic linearized Navier-Stokes system. *J. Math. Fluid Mech.*, 16(3):523–538, 2014. 5, 101

[61] M. Kyed and J. Sauer. A method for obtaining time-periodic L^p estimates. *J. Differential Equations*, 262(1):633–652, 2017. 75, 78, 79, 80, 112, 113, 114, 118

[62] M. Kyed and J. Sauer. On time-periodic solutions to parabolic boundary value problems. *Math. Ann.*, 374(1):37–65, 2019. 9, 112

[63] J. L. Lagrange. Sec. 42 in nouvelles recerces sur la nature et la propagation du son. *Miscellanea Taurinensis II*, pages 11 – 172, 1761. 3

[64] D. Lengeler and M. Růžička. Weak solutions for an incompressible Newtonian fluid interacting with a Koiter type shell. *Arch. Ration. Mech. Anal.*, 211(1):205–255, 2014. 93

[65] J. Leray. *Étude de diverses équations intégrales non linéaires et de quelques problèmes que pose l'hydrodynamique.* 1933. 5

[66] J. Leray. Sur le mouvement d'un liquide visqueux emplissant l'espace. *Acta Math.*, 63(1):193–248, 1934. 5

[67] M. J. Lighthill. Viscosity effects in sound waves of finite amplitude. In *Surveys in mechanics*, pages 250–351 (2 plates). Cambridge, at the University Press, 1956. 4, 44

[68] R. B. Lindsay. The story of acoustics. *J. Acoust. Soc. Am.*, 39:629–644, 1966. 3, 4

[69] Y. Maekawa and J. Sauer. Maximal regularity of the time-periodic Stokes operator on unbounded and bounded domains. *J. Math. Soc. Japan*, 69(4):1403–1429, 2017. 101

[70] S. Meyer and M. Wilke. Global well-posedness and exponential stability for Kuznetsov's equation in L_p-spaces. *Evol. Equ. Control Theory*, 2(2):365–378, 2013. 4, 70

[71] M. Mitrea, S. Monniaux, and M. Wright. The Stokes operator with Neumann boundary conditions in Lipschitz domains. *Journal of Mathematical Sciences*, 176:409–457, 07 2011. 19

[72] T. Miyakawa and Y. Teramoto. Existence and periodicity of weak solutions of the Navier-Stokes equations in a time dependent domain. *Hiroshima Math. J.*, 12(3):513–528, 1982. 5

[73] B. O. Enflo and C. Hedberg. Theory of nonlinear acoustics in fluids. *Theory of Nonlinear Acoustics in Fluids: Fluid Mechanics and Its Applications, Volume 67. ISBN 978-1-4020-0572-5. Kluwer Academic Publishers, 2004*, 67, 2004. 3, 42, 43, 44

[74] S. D. Poisson. Mémoire sur la théorie du son. *J. l'école polytech.*, Paris, 7:319 – 392, 1808. 3

[75] G. Prodi. Qualche risultato riguardo alle equazioni di Navier-Stokes nel caso bidimensionale. *Rend. Sem. Mat. Univ. Padova*, 30:1–15, 1960. 5

[76] G. Prouse. Soluzioni periodiche dell'equazione di Navier-Stokes. *Atti Accad. Naz. Lincei Rend. Cl. Sci. Fis. Mat. Nat. (8)*, 35:443–447, 1963. 5

[77] J. Prüss and G. Simonett. *Moving interfaces and quasilinear parabolic evolution equations*, volume 105 of *Monographs in Mathematics*. Birkhäuser/Springer, [Cham], 2016. 103

[78] A. Quarteroni, M. Tuveri, and A. Veneziani. Computational vascular fluid dynamics: Problems, models and methods. *Computing and Visualization in Science*, 2:163–197, 03 2000. 2, 5, 87, 93

[79] P. J. Rabier. A complement to the Fredholm theory of elliptic systems on bounded domains. *Bound. Value Probl.*, 2009:9, 2009. 81

[80] T. Rossing, editor. *Springer Handbook of Acoustics*. Springer New York, 2007. 4

[81] J. O. Sather. *The Initial-Boundary Value Problem for the Navier-Stokes Equations in Regions with Moving Boundaries*. ProQuest LLC, Ann Arbor, MI, 1963. Thesis (Ph.D.)–University of Minnesota. 5

[82] J. Serrin. A note on the existence of periodic solutions of the Navier-Stokes equations. *Arch. Rational Mech. Anal.*, 3:120–122, 1959. 5

[83] C. G. Simader. The weak dirichlet and neumann problem for the laplacian in lq for bounded and exterior domains. applications. In M. Krbec, A. Kufner, B. Opic, and J. Rákosník, editors, *Nonlinear Analysis, Function Spaces and Applications Vol. 4: Proceedings of the Spring School held in Roudnice nad Labem 1990*, pages 180–223. Vieweg+Teubner Verlag, Wiesbaden, 1990. 136, 137

[84] E. M. Stein. *Singular integrals and differentiability properties of functions*. Princeton Mathematical Series, No. 30. Princeton University Press, Princeton, N.J., 1970. 15

[85] G. G. Stokes. On the Steady Motion of Incompressible Fluids. *Transactions of the Cambridge Philosophical Society*, 7:439, Jan 1848. 5

[86] A. Tani. Mathematical analysis in nonlinear acoustics. *AIP Conference Proceedings*, 1907(1):020003, 2017. 75

[87] H. Triebel. *Interpolation theory, function spaces, differential operators*. VEB Deutscher Verlag der Wissenschaften, Berlin, 1978. 34, 35, 47, 110

[88] H. Triebel. *Theory of function spaces*. Basel: Birkhäuser, reprint of the 1983 original edition, 2010. 16, 107, 109

[89] J. Wloka. *Partial differential equations*. Cambridge University Press, Cambridge, 1987. Translated from the German by C. B. Thomas and M. J. Thomas. 48, 81, 82

[90] V. I. Yudovič. Periodic motions of a viscous incompressible fluid. *Soviet Math. Dokl.*, 1:168–172, 1960. 5

Curriculum Vitæ

15/03/88   Born in Nusaybin, Turkey

08/04 - 07/08   **Secondary school,** *Pelizaeus-Gymnasium Paderborn,* Paderborn, Germany, Abitur (3.3 / satisfying)

10/09 - 03/13   **Bachelor's studies,** *Universität Paderborn,* Paderborn, Germany, Bachelor of Science Mathematics (2.2 / good) Bachelor's thesis: *Der Satz von Artin über induzierte Darstellungen*

04/13 - 03/15   **Master's studies,** *Universität Paderborn,* Paderborn, Germany, Master of Science Mathematics (1.6 / good) Master's thesis: *Der Liouville-Satz für eine degenerierte parabolische partielle Differentialgleichung*

04/15 - 02/20   **Doctoral studies,** *Technische Universität Darmstadt,* Darmstadt, Germany, research assistant in the working group *Partielle Differentialgleichungen*

06/02/20   **Submission of the doctoral thesis (Dissertation),** *Non-resonant Solutions in Hyperbolic-Parabolic Systems with Periodic Forcing* at *Technische Universität Darmstadt,* Darmstadt, Germany

26/02/20   **Defense of the doctoral thesis,** overall assessment *magna cum laude*